Introducing Proteomics

Introducing Proteomics

From concepts to sample separation, mass spectrometry and data analysis

Josip Lovrić

Faculty of Life Sciences, University of Manchester

WILEY-BLACKWELL

A John Wiley & Sons, Ltd., Publication

This edition first published 2011, © 2011 by John Wiley & Sons, Ltd.

Wiley-Blackwell is an imprint of John Wiley & Sons, formed by the merger of Wiley's global Scientific, Technical and Medical business with Blackwell Publishing.

Registered office: John Wiley & Sons Ltd, The Atrium, Southern Gate, Chichester, West Sussex, PO19 8SQ, UK

Editorial Offices:
9600 Garsington Road, Oxford, OX4 2DQ, UK
111 River Street, Hoboken, NJ 07030-5774, USA
The Atrium, Southern Gate, Chichester, West Sussex, PO19 8SQ, UK

For details of our global editorial offices, for customer services and for information about how to apply for permission to reuse the copyright material in this book please see our website at www.wiley.com/wiley-blackwell

Library of Congress Cataloguing-in-Publication Data
Lovrić, Josip.
 Introducing proteomics : from concepts to sample separation, mass spectrometry, and data analysis / Josip Lovrić.
      p. ; cm.
 Includes bibliographical references and index.
 ISBN 978-0-470-03523-8 (cloth) – ISBN 978-0-470-03524-5 (pbk.)
 1. Proteomics. I. Title.
 [DNLM: 1. Proteomics. QU 58.5]
 QP551.L875 2011
 572'.6 – dc22
                                                                                    2010029400

A catalogue record for this book is available from the British Library.

This book is published in the following electronic formats: ePDF ISBN 978-0-470-67021-7.

Typeset in 9/12pt Times-Roman by Laserwords Private Limited, Chennai, India

First Impression 2011

To my family, I hope they'll have me back . . . . . .

A narućito naknađujem ovo knjigu za moje roditelje.

Contents

Preface

The term proteomics was coined in the mid-1990s by the Australian (then post-doctoral) researcher Marc Wilkins. The term arose in response to the spirit of the day; researchers working in genetics developed genome-wide approaches and were very successful at the time. Researchers working on proteins rather than genes also felt that the time was right for a more holistic approach – rather than working on a single protein at a time, many (if not all) proteins in a single biological system should be analysed in one experiment. While surely the will was there and some good foundational work was done at the time, it still took about another five years before technologies were developed far enough, so that proteomics became a concept that could deliver some (but still not all) of the answers researchers hoped to be able to get by using it.

Historically proteomics was driven mainly by researches coming from the field of 2D gel electrophoresis. These 'bluefingers' joined forces with experts in mass spectrometry and bioinformatics. It was the combination of these fields together with the genomic revolution that created the first proteomic approaches. These were inevitably studies using 2D gel electrophoresis in combination with mass spectrometry of 'isolated' spots, often using MALDI-ToF mass spectrometry. In the beginning the development of the ionization technologies of MALDI, ESI and nano ESI were critical steps to allow the mass spectrometric analysis of biological material with reasonable sensitivity.

Together with advances in all fields concerned, it was major developments in gel free, hyphenated peptide separation technologies that allowed proteomics to prosper in more recent times. Recent developments in gel based proteomics were confined mainly to more convenient sample handling and more pre-fabricated devices and most important computer based image analysis and new protein dyes, allowing for less variable results in a shorter time with less manual input. 2D HPLC in combination with tandem mass spectrometry is a hallmark of the development of hyphenated technologies. Modern proteomics is driven by the development of ever improving software to deal with the huge amount of data generated, allowing better and more efficient data mining; new mass spectrometers allowing new imaging approaches or qualitatively better approaches through improvements in versatility, accuracy, resolution or sensitivity. Also developments in labelling reagents and affinity matrices allow more intelligent approaches, more tailored to specific questions, such as quantitative analyses and analyses of phosphorylations and other posttranslational modifications. Nano-separation methods become more routine and combination of multi-dimensional separation approaches become feasible, allowing 'deeper' views of the proteome. And if all these developments were not enough, there is a plethora of more specialized developments, like the molecular scanner (Binz *et al.*, 2004), MALDI imaging mass spectrometry for tissues, organs and whole organisms such as the mouse or rat (Caldwell and Caprioli, 2005) or Laser Capture Microdissection (Jain, 2002) which enables proteomics analysis from just a dozen of cells (Nettikadan *et al.*, 2006). In a book like this it is impossible to do justice to all these developments, and they will be mentioned as we go along, especially in Chapter 5 on strategies in proteomics. Sadly, some fields such as 3D structural analyses have had to be omitted.

Next to complete 'work floors', the mass spectrometers and separation devices (e.g. nano HPLC, free flow electrophoresis equipment) that come with the territory represent the biggest capital investment for laboratories getting involved in proteomics, ranging from some US$ 160 000 to more than a million dollars per item. In the early days of proteomics, many developments were driven by scientists rather than industries. Since 2000, proteomics has become big business, with the potential for companies to sell hundreds of mass spectrometers instead of a dozen a year to the scientific community.

Away from 'classical' approaches there have been huge developments in very diverse fields such as protein fluorescent staining, chemical peptide modification, ultra-accurate mass spectrometers, microscope assisted sample collection, improved sample treatment, isobaric peptide tagging and of course bioinformatics, to name just a few, that have opened up a whole range of new possibilities to tackle biological problems by proteomic approaches. It is this diverse group of fields that contribute towards making proteomics such a vibrant and interesting field, on the one hand, but also a field that may seem difficult to get started in, on the other hand.

This is where we aim to place this book: to give an introduction to the complete field of proteomics without delving too deeply into every single area within it, because for most of these areas there is excellent specialist literature available.

In this respect the book aims primarily to give a basic understanding of the most important technologies. At the same time it intends to allow the reader to develop an understanding of the possibilities, but also the limitations, of each of the technologies or their combinations. All this is presented with the aim of helping the reader to develop proteomic approaches that are suited to the needs of their specific research challenge.

WHO WILL BENEFIT FROM THE BOOK

This book is aimed at diverse groups of potential users. In the academic world it is written easy enough to be useful and aimed at undergraduate students to give an introduction to the field of proteomics; so many biochemical/physical principals are explained at that level.

On the other hand, this book will also be useful for postgraduate students and more senior researchers in academia and industry. While it brings an overview and an explanation of principles to postgraduate students who may be about to start to work on a proteomic project, it will also explain the possibilities and limitations of a potential proteomics approach for a principal investigator and give them an idea of the sort of financial and intellectual commitments necessary.

It will be a useful tool for experienced researchers in the field of proteomics to 'catch up' on areas that were outside their focus for a while or have developed only recently. It may also help scientists to understand the needs of a certain approach and help them with their planning; be it for starting collaborating with someone in the field of proteomics, or to help such a collaboration to be successful or for writing a new grant in this field.

While this book does not contain recipes or manuals for instruments, it will be of great benefit in helping people to get trained practically in the field, since it explains all the major principles and puts them in a wider perspective.

I hope it will also help researchers from (apparently) distant areas of research to develop new approaches and identify fields in which further research into technologies might be necessary and possible to help proteomics to become and remain one of the sharpest tools in the box of biological and medico-pharmaceutical research.

REFERENCES

Binz, P.A., Mueller, M., Hoogland, C. *et al.* (2004) The molecular scanner: concept and developments. *Curr Opin Biotech*, **15**, 17–23.

Caldwell, R.L. and Caprioli, M.R. (2005) Tissue profiling by mass spectrometry. *Mol Cell Proteomics*, **4** (4), 394–401.

Jain, K.K. (2002) Application of laser capture microdissection to proteomics. *Methods Enzymol*, **356**, 157–167.

Nettikadan, S., Radke, K., Johnson, J. *et al.* (2006) Detection and quantification of protein biomarkers from fewer than 10 cells. *Mol Cell Proteomics*, **5** (5), 895–901.

Acknowledgements

My sincere thanks go to all those individuals who helped in the making of this book, be it by advice, by letting me have their data or by helping in the production process of this book. Special thanks are due to Dr Alistair McConnell, Dr Chris Storey, Dr David Knight, Fiona Woods, Gill Whitley, Dr Guido Sauer, Haseen Khan and her amazing project team, Izzy Canning, Nicky McGirr, Dr Paul Sims and Dr Songbi Chen.

1

Introduction

1.1 WHAT ARE THE TASKS IN PROTEOMICS?

1.1.1 The proteome

In genomics, one of the main aims is to establish the composition of the genome (i.e. the location and sequence of all genes in a species), including information about commonly seen polymorphisms and mutations. Often this information is compared between different species and local populations. In functional genomics, scientists mainly aim to analyze the expression of genes, and proteomic is even regarded by some as part of functional genomics. In proteomics we aim to analyze the whole proteome in a single experiment or in a set of experiments. We will shortly look at what is meant by the word analysis. Performing any kind of proteomic analysis is quite an ambitious task, since in its most comprehensive definition the proteome consists of all proteins expressed by a certain species. The number of these proteins is related to the number of genes in an organism, but this relation is not direct and there is much more to the proteome than that. This comprehensive definition of the proteome would also account for the fact that not a single individual of a species will express all possible proteins of that species, since the proteins might exist in many different isoforms, with variations and mutations, differentiating individuals. An intriguing example are antibodies, more specifically their antigen binding regions, which exist in millions of different sequences, each created during the lifetime of individuals, without their sequence being predictable by a gene. Antibodies are also a good example of the substantial part played by external influences, which define the proteome; for example, the antibody-mixture present in our bodies is strictly dependent on which antigens we have encountered during our lives. But of

course a whole host of more obvious external factors influence our proteome, but not the genome (Figure 1.1).

Furthermore, the proteome also contains all possible proteins expressed at all developmental stages of a given species; obvious examples are different proteins in the life cycle of a malaria parasite, or the succession of oxygen binding species during human development, from fetal haemoglobin to adult haemoglobin (Figure 1.2).

On top of all these considerations, there are possible modifications to the expression of a protein that are not encoded by the sequence of its gene alone; for example, proteins are translated from messenger RNAs, and these mRNAs can be spliced to form different final mRNAs. Splicing is widespread and regulated during the development of every single individual, for example during the maturation of specific cell types. Changes in differential splicing can cause and affect various diseases, such as cancer or Alzheimer's (Figure 1.3).

As if all this was not enough variability within the proteome, most proteins show some form of posttranslational modification (PTM). These modifications can be signs of ageing of the protein (e.g. deamidation or oxidation of old cellular proteins; Hipkiss, 2006) or they can be added in an enzymatically regulated fashion after the proteins are translated, and are fundamental to its function. For example, many secreted proteins in multicellular organisms are glycosylated. In the case of human hormones such as erythropoietin this allows them to be functional for longer periods of time (Sinclair and Elliott, 2004). In other cases proteins are modified only temporarily and reversibly, for example by phosphorylation or methylation. This constitutes a very important mechanism of functional regulation, for example during signal transduction, as we will see in more detail later. In summary, there are a host of

Introducing Proteomics: from Concepts to Sample Separation, Mass Spectrometry and Data Analysis, Josip Lovrić
© 2011 John Wiley & Sons Ltd

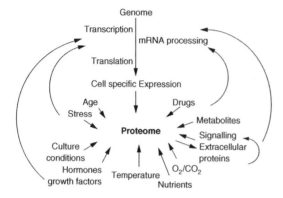

Figure 1.1   Influences on the proteome. The proteome is in a constant state of flux. External factors constantly influence the proteome either directly or via the genome.

relevant modifications to proteins that cannot be predicted by the sequence of their genes. These modifications are summarized in Figure 1.4.

Moreover, it is important to remember that the proteome is not strictly defined by the genome. While most possible protein sequences might be predicted by the genome (except antibodies, for example), their expression pattern, PTMs and protein localization are not strictly predictable from the genome. All these factors define a proteome and each protein in it. The genome is the basic foundations for the 'phenotype' of every protein, but intrinsic regulations and external influences also have a strong influence (Figure 1.5).

1.1.2 A working definition of the proteome

For all the above mentioned reasons most researchers use a more practical definition of the word 'proteome'; they use it for the proteins expressed in a given organism, tissue/organ (or most likely cell in culture), under a certain, defined condition. These 'proteomes' are then compared with another condition, for example two strains of a microorganism, or cells in culture derived from a healthy or diseased individual. This so-called differential proteomics approach has more than a description of the proteome in mind; its aim is to find out which proteins are involved in specific functions. This is of course hampered by the number of proteins present (some changes may occur as mere coincidences) and by the many parameters that influence the functionality of proteins, expression, modification, localization and interactions. While differential proteomics seems a prudent way to go, we have to keep in

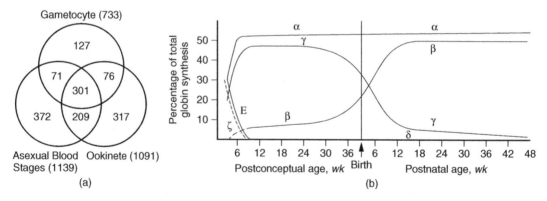

Figure 1.2   The composition of the proteome changes during ontology. (a) Plasmodium, the agent causing malaria, has a complex life cycle. Its asexual blood stage cycle lasts about 24 hours, then the sexual stages (gametocytes) develop within 30 hours and develop into the ookinetes after fertilization. A comprehensive proteomic study of these and other stages of the life cycle detected more than 5 000 proteins. The Venn diagram shows the number of total proteins identified in each specific stage in parentheses. The numbers in the Venn diagram represent the number of proteins involved in sexual development exclusive to one of the three stages shown in the picture. Over a third of the proteins in each state were found exclusively in one stage only, about 30–50% were common to all stages and about 10–20% were found in more than one of the three stages. (b) Humans express different globin species during their ontogenesis. These globin proteins come from different genes and bind the haeme group to form haemoglobins with specific characteristics essential for different stages of development. The figure shows how the relative production of different globin species changes in early human development. (a) Hall *et al.* (2007). © 2005 American Association for the Advancement of Science. (b) Modified from Wood (1976) and reproduced with permission. © 1976 Oxford University Press.

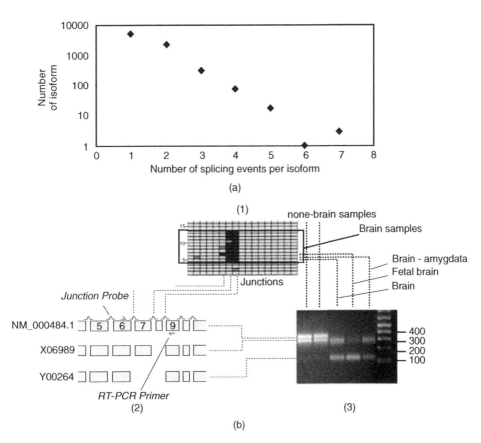

Figure 1.3 The importance of splicing. (a) The known frequency of splicing events for human proteins (Wang *et al.*, 2005). Splicing events were extracted form the SWISS-PROT database, one of the best-annotated databases for proteins. It can be assumed that there are a huge number of non-annotated splicing events. The number of proteins showing a certain number of splicing isoforms is shown. In the case of one splicing event per isoform, no alternative splicing isoform is annotated. (b) The mRNA for human β-amyloid precursor protein is spliced in brain tissues as compared to non-brain tissues. Alternative splicing of amyloid precursor protein may play a role in the development of human Alzheimer's disease. Screens for alternative splicing were performed on mRNAs microarrays (1) using splice event specific probes spanning two exons (2) and then confirmed by specific PCR reactions (3), using primers whose product length is influenced by splicing events. (a) Wang *et al.* (2005). © 2005 National Academy of Sciences, USA. (b) From Johnson *et al.*, *Science*, 2003; 302:2141–44. Reprinted with permission from the American Association for the Advancement of Science.

mind that the methods chosen for proteomic analyses will also determine the results; for example, if we use a gel-based approach, membrane proteins are almost completely excluded from the analyses. Furthermore, most analyses have a certain cut off level for the low abundant proteins. This means that proteins below (say) 10 000 copies expressed per cell are not easily measurable, because the approaches are usually not sensitive enough.

Even within this limited definition of proteomics we still face substantial tasks, as the proteome is defined not only by the physical state of the proteins in it (expression and modifications) but also by their subcellular location and their membership in protein–protein complexes of ever changing compositions. For instance, it makes a big functional difference to its activity if a transcription factor is inside or outside the nucleus and a proteomic study that fails to analyze the transcription factor's sub-cellular location will miss major changes in the activity of this transcription factor (Figure 1.7). A kinase that needs to be in a multiprotein complex to be active will be inactive when it is only bound to parts of that complex, an important difference that will be missed if we analyze only the

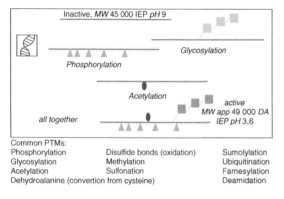

Common PTMs:

Phosphorylation	Disulfide bonds (oxidation)	Sumolylation
Glycosylation	Methylation	Ubiquitination
Acetylation	Sulfonation	Farnesylation
Dehydroalanine (convertion from cysteine)		Deamidation

Figure 1.4   Proteins are regulated by posttranslational modifications. Genes and splicing define the primary sequence of proteins. The primary sequence contains motives that allow different PTMs. Which of them are actually found on a protein at any given time in a specific tissue cannot be predicted. Often a combination of PTMs is necessary for active proteins. PTMs can change the 3D structure of proteins. They also change parameters such as apparent molecular weight and isoelectric point in gel-based protein separations.

presence of a protein but not the interaction partners. The same holds true for kinases that switch complexes and thereby regulate their target specificity (Kolch, 2005).

1.1.3 The tasks in proteomics

Most proteomic studies aim to correlate certain functions with the expression or modification of specific proteins; only few aim to describe complete proteomes or compare them between different species. For a functional correlation we need to analyze the most important protein features of functional relevance. We have already mentioned the analysis of proteins in proteomic studies – just what does this mean? Proteomic analyses can be summarized in terms of specific goals:

1. detection and quantification of protein level;
2. detection and quantification of protein modifications;
3. detection and quantification of sub-cellular protein localization;
4. detection and quantification of protein interactions.

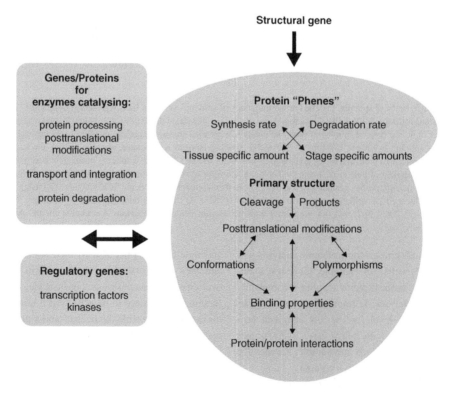

Figure 1.5   Proteins have a 'phenotype'. Similar to whole organisms, proteins can be regarded as having observable traits that are derived by genetic factors as well as influences from the surroundings they experience during their 'life'.

Historically, protein expression has been the first parameter analyzed by proteomics. While this involves a certain form of quantification (present/not present means usually at least a three- to tenfold difference in expression level), it is much harder to quantify proteins on a proteomic scale and many of the latest technological developments focus on this aspect (see Chapters 2–5). Since the abundance of proteins can vary from presumably a single protein to over a million proteins per cell, the quantifications have to cover a dynamic range of over 6 orders of magnitude in cells and up to 10 orders of magnitude in plasma (Patterson and Aebersold, 2003).

PTMs are very important for the function of proteins, and proteomics is the only approach to analyze them on a global scale. Nevertheless, the current approaches (e.g. phosphoproteomics) are by no means able to analyze all possible PTMs, and this remains a hot topic in the development of new technologies.

Before the onset of life cell imaging technology, fractionation of cells was the only method to analyze the subcellular localization of proteins. While being relatively crude and error-prone due to long manipulation times, fractionation studies are very successful in defining protein function. This holds true especially when not only organelles but also functional structures such as ribosomes (Takahashi *et al.*, 2003) or mitotic spindles can be intelligently isolated (Sauer *et al.*, 2005).

The detection of protein interactions is surely the most challenging of proteomic targets, but also a very rewarding one. In single studies the goal is often to identify all interacting partners of a single protein (see Figure 1.8), and several studies taken together can be used to identify, for instance, all interactions within a single signalling module (Bader *et al.*, 2003). Interactions on a truly proteomic scale have been analyzed only in some exceptional studies (Ho *et al.*, 2002; Krogan *et al.*, 2006) and the results are by no means complete, given the temporal and fragile nature of protein–protein interactions, the different results reached with different methods and their complexity.

Non-covalent and hence the most difficult to analyze are localization and interactions of proteins – although none of the above tasks is easily reached, considering the shear number of proteins involved, the minute amounts of sample usually available and the temporal resolution that might be required. Proteomic parameters can change from seconds or minutes (e.g. in signalling) to hours, days and even longer time periods (e.g. in degenerative diseases).

1.2 CHALLENGES IN PROTEOMICS

1.2.1 Each protein is an individual

Nucleotides are made up of four different bases each, and the structure of DNA is usually very uniform. Even if RNA forms more complex structures, we have many different buffers in which we can solubilise all known nucleotides. No such thing exists in proteomics. There is no buffer (and there probably never will be) that can solubilize all proteins of a cell or organism (Figure 1.6). Proteins are made out of 20 amino acids, which allows even a peptide that is 18 amino acids long to acquire more different sequences than there are stars in the galaxy or a hundred times more different sequences than there are grains of sand on our planet!

The average length of proteins is about 450 amino acids. The complexity that can be reached by such a

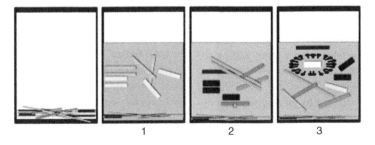

Figure 1.6 Protein solubilization. Complex mixtures of proteins (e.g. cellular lysates) can be solubilized in a variety of buffers (e.g. different ionic strength, pH). Some proteins will dissolve in one or the other buffer, but not in both, while some or most protein interactions are preserved (1/2). Adding detergents allows most proteins to be dissolved, but protein interactions are disrupted (3). Strong detergents even interfere with further manipulation or analysis of the proteins.

protein is beyond the imagination. More to the point, while almost every sequence of DNA will have fairly similar biochemical properties to any other sequence of similar length, with proteins the situation is totally different. Some proteins will bind to materials used for their extraction and so get lost in analyses, others will appear predominant in a typical mass spectrometry (MS) analysis because they contain optimal amounts and distributions of arginine and lysine. If proteins are very hydrophobic, they will not even get dissolved without the help of detergents. Some proteins show aberrant behaviour with dye; either they are stained easily or very badly. This behaviour makes absolute quantifications and even relative comparisons of protein abundances very difficult. Proteins can display highly dynamic characteristics; their abundances can change dramatically within minutes, by either rapid new synthesis or degradation. Some proteins are more susceptible to degradation by either specific ubiquitin dependent or independent proteolysis than others. These processes in turn can be triggered during cellular processes such as differentiation or apoptosis (active cell death). There are more than 360 known chemical modifications of proteins (see the 'Delta Mass' listing on the Association of Biomolecular Resource Facilities website, http://www.abrf.org). These include natural PTMs such as phosphorylation, glycosylation and acetylation, as well as artefacts such as oxidation or deamidation that might occur naturally inside cells but also as artefacts during protein preparation. There are of course also totally artificial modifications occurring exclusively during protein isolation, such as the addition of acrylic acid.

1.2.2 The numbers game

This variety explains how relatively complex organisms can manage to rely on a relative small amount of genes. The least complex forms of life are found among the viruses; in a typical example, a dozen genes will encode about 40 proteins by means of alternative RNA processing and controlled proteolysis. On top of this, these proteins are alternatively processed (e.g. by glycosylation) to regulate their function in different phases of the viral life cycle. In these relatively simple life forms the proteome is much more complex than the genome would suggest, and the more complex the life form, the more this gap widens. Bacteria have about 3 000–4 500 genes. In a typical example (if there are any 'typical' examples of these

fascinating organisms!) like *Escherichia coli* there are 4 290 protein encoding genes plus about 90 only producing RNA. Splicing of mRNA is rare; PTMs are present in a variety of forms, but do occur rarely. In yeast (*Saccharomyces cerevisiae*) we detect about 6 000 genes and these are moderately modified. Splicing is a regular event, and so are differential glycosylation, phosphorylation, methylation and a host of other PTMs, resulting in a much higher number of protein isoforms than the pure addition of nuclear and mitochondrial genes would suggest. In multicellular organisms such as insects (e.g. the fruit fly, *Drosophila melanogaster*) or worms (e.g. the roundworm, *Caenorhabditis elegans*) we encounter about 13 400 and 19 000 genes, respectively. All known popular mechanisms to enlarge the number of proteins from one gene are observed. Finally, let us have a look at the highest evolved life forms, as we wish to see ourselves. Only a couple of years ago, before the completion of the human genome project phase 1, it was widely accepted that we might have about 100 000 genes. The human genome project still does not know the exact answer, but we assume between 20 000 and 40 000 genes for our species, and most scientist agree on a figure of about 25 000. We are left wondering how we manage to be so much more complex than worms with just slightly increased numbers of genes. The answer lies within the increasing complexity on the way from the genome to the proteome (see Table 1.1).

Assuming we have about 30 000 genes, a single individual will have about 200 000 differentially spliced forms of mRNA and roughly the same number of proteins, as identified by identical sequence, over the course of his or her development. Adding all found or presumed common polymorphisms (e.g. different alleles or single-nucleotide polymorphisms) we encounter on the DNA level, we might well speak of twice the number of 400 000 proteins. If we include the PTMs, numbers increase further. It seems a conservative estimate that on average about five posttranslationally modified isoforms exist per protein, leading to about 2 million different proteins that one might consider analysing in a comprehensive proteomic experiment! There are, of course, no methods at hand to do any such experiment at present!

Obviously, not all possible proteins encoded for by the genome will be expressed at all times in a given practical sample. It is safe to assume that a mammalian cell line expresses some 10 000–15 000 genes at any

Table 1.1 Numbers in proteomics. From a fixed (and in humans still only estimated) number of genes, a larger number of mRNA splice variants is generated. The number of proteins is larger than the number of mRNAs due to N-terminal processing, removal of signal peptides and proteolysis. Each protein can carry various PTMs. The most popular analysis method in proteomics performs analyses on the level of tryptic peptides (MS and MS/MS), as peptides are more informative with the instruments/strategies available. Peptides can be chemically modified by PTMs or by one or more of several hundred known chemical modifications. All figures are estimates.

Number of human genes (tentatively)	3×10^4
Number of mRNAs	$1\text{--}2 \times 10^5$
Number of proteins	$1\text{--}2 \times 10^5$
Number of protein isoforms with differential PTM	2×10^6
Number of all detectable tryptic peptide (no PTM)	$>1 \times 10^6$
Number of all detectable tryptic peptides with natural PTM	1×10^7
Number of all different tryptic peptides including PTMs and artificial chemical modifications	$>3 \times 10^7$

given time, or slightly less than half the proteome of the species. Tissues consist of several cell types (plus blood cells, arteries, lymph nodes, etc.) and have a larger complexity. Thus we could encounter the products of perhaps 15 000–20 000 genes in a given tissue sample, or about half of the proteome.

Another problem in numbers arises from the dynamic range in which proteins are encountered. Proteins can be expressed from the rare one protein per cell up to several million proteins per cell (Futcher *et al.*, 1999), whereas there are usually only one or two genes per cell. And of course the Nobel prize winning invention of the polymerase chain reaction allows the amplification of one single molecule of DNA or RNA to any amount needed for repetitive analyses; there is no such thing for proteins. Researchers face the challenge of analysing a small number of proteins (one per cell?) in the presence of very abundant ones (10 million copies per cell; Ghaemmaghami *et al.*, 2003), and it is obviously difficult to quantify any measurements with results ranging over seven orders of magnitude! The most sensitive way to analyze unknown proteins is the use of mass spectrometers, which is another reason why they are so popular in proteomics. Most proteomic approaches can measure peptides down to the low femtomole level, more advances and complex approaches might reach attomole levels, and well characterized proteins can be detected down to the zeptomole level.

1.2.3 Where do proteins hang out?

Apart from other parameters, the location of each protein is most important for its function. Good examples are transcription factors, which might be in an inactive conformation in the cytoplasm and have to translocate to the nucleus to get activated (Kawamori, 2006). So to define a proteome functionally we need to know exactly where proteins are ... very exactly indeed. A protein being inside or outside an organelle makes a difference of about 20 nm in position, for example! The spatial distribution is also regulated within short time scales; as a typical example we can think about growth factor receptors accumulating within minutes of stimulation in degrading vesicles (e.g. epidermal growth factor: Aguilar and Wendland, 2005). These different locations cannot all be addressed equally well; it is, for instance, difficult to compare protein distribution in cells with different polarity (e.g. apical and distal in epithelial cells). Proteins might be located not only outside or inside an organelle (e.g. the nucleus – Figure 1.7), but also inside its membrane(s) or in other sub-cellular structures (e.g. ribosomes, or skeletal components). Most organelles and many sub-cellular structures can be isolated to quite high purity to analyze the proteins contained in/on them. However, the higher the purity, the longer and more complicated the isolation procedure (usually involving differential centrifugation), and the more time there is for the samples to acquire artefactual changes, as the example from work in our laboratory shows: we label cells radioactively to investigate phosphorylations and a two-hour cellular fractionation procedure allows about 90% of the label to be removed (by phosphatases) when compared to a direct lysis of whole cells in high concentration urea sample buffer. Other possible artefacts include proteolysis or deglycosylation. Together, they can result in proteins dissociating from their 'correct' position. Even without

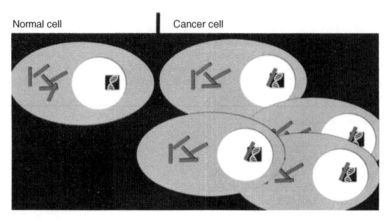

Figure 1.7 Importance of localization in proteomics. The cell in the left panel contains the same amount of red proteins as the cancer cell on the right. However, some of the proteins are in the nucleus, where they can activate transcription and cause cancer. If sub-cellular localization were not analyzed, a quantitative proteomic approach would miss this important difference.

this it is often difficult to judge if a protein is specifically associated with an organelle, or if it just 'sticks' non-specifically to the organelle, as a result of the cell lysis and often mediated by artificial associations with other, perhaps denatured, proteins or nucleic acids. On the other hand, proteins, which are in the living cell (let us call this *in vivo* for our purposes) associated with certain organelles, might get lost during the isolation process. These proteins can be small proteins that 'leak' out through artefactual damage in the membranes or pores in the organelles; for

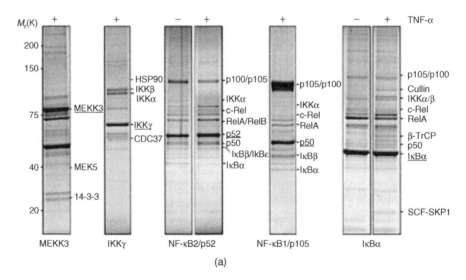

(a)

Figure 1.8 Analyzing protein interaction on proteomic levels. To analyze the complex interaction in the human TNF-α/NF-κB signal transduction pathway selected components were tagged and affinity-purified using a tandem affinity tag approach (see Chapter 5). The affinity tagged proteins (underlined) as well as co-purifying (i.e. physically interacting) proteins were resolved by sodium dodecyl sulfate polyacrylamide gel electrophoresis (SDS PAGE), and unknown protein bands were cut out from gels and identified by liquid chromatography (LC) coupled MS/MS analyses. To cover as many as possible of the interactions some components were 'knocked down' from the human cells used for the experiments by RNAi. Parts of the results of hundred such experiments are combined in a database and presented graphically (b). Presentations follow internationally agreed rules for easier interpretation. Even with this amount of work, not all the physical interaction of the proteins involved has actually been analyzed. Reproduced from Bouwmeester *et al.* (2004) courtesy of Nature Publishing Group. © 2004 Nature Publishing Group.

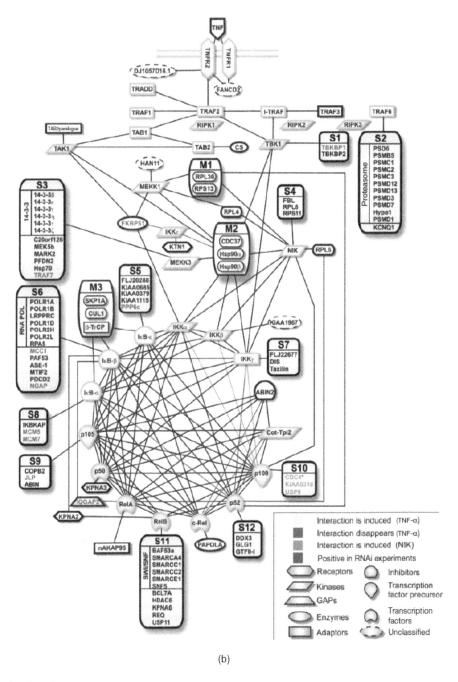

(b)

Figure 1.8 (*continued*)

example, proteins below 45 kDa can diffuse freely in and out of the nucleus, in addition to any specific mechanism for importing, exporting or retaining them. Another species of proteins that can get lost are weakly associated proteins on the outside of organelles (as opposed to transmembrane proteins or internal proteins); they are held in place by delicate protein–protein interactions, which will be discussed in the next chapter.

Proteomic studies on sub-cellular structures have been very successful in mapping their composition and function and they have been hugely helped by the onset of gel-free proteomic methods such as free flow electrophoresis and especially multidimensional protein identification technology, known as MudPIT (also called shotgun proteomics; see Chapter 3).

1.2.4 Proteins always hang out with their mates

No protein can exert its function alone – there always has to be an interaction with another protein. Structural proteins are often found in huge complexes, and even if they only contain one protein their structure and composition are an important functional feature. As an example just think of tubulin in microtubules – it can be found in long microtubules, short fragments and also in combination with other proteins, often regulating its association/ dissociation parameters. Enzymes are often activated and/or kept in place by their association with other proteins. They often even have to be assembled in close association with other proteins (chaperones) in order to fold into a functional form, and that is subject to intricate regulations. For example, a specific class of so-called heat shock proteins (proteins that generally stabilize correct protein folding) has to be associated with some fragile kinases in order to keep them active (e.g. Raf/HSP90: Kolch, 2000) and the regulation of this association is signalling and cell cycle dependent (Lovric *et al.*, 1994). It can be so specific that blocking the function of the heat shock protein kills the cells by inactivation the kinases. Other typical interactions are enzymes and their substrates; often even the substrate preference or specificity is regulated by protein interactions (e.g. Jun binding by extracellular signal-regulated kinase). A very good example of this is the KSR1 protein within the MAP kinase module: using the same kinases with different adaptor proteins, different substrates get phosphorylated (Casar *et al.*, 2009). It is impossible to analyze all these interaction on a proteomic scale, but several proteomic studies have added impressively to our understanding

of either the interactive partners and functions of single proteins (Figure 1.8) or whole protein complexes (e.g. ribosome or transcription complexes). However, results from interaction studies are very complex, and it can be difficult to understand their significance. Depending on the methods used, it might be difficult to understand whether, for example a protein shows a weak but specific interaction or a strong but unspecific interaction (e.g. one that does not occur in living cells) and one has to be careful comparing and combining data from different studies, because they might have been derived using different technologies.

1.3 PROTEOMICS IN RELATION TO OTHER -omics AND SYSTEM BIOLOGY

At the moment there are an ever growing number of new -omics coming into being, next to the classical genomics (Figure 1.9). The main ones are transcriptomics, phosphoproteomics, glycomics and metabolomics.

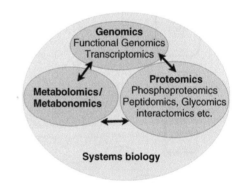

Figure 1.9 The new biology: -omics and systems. Each of the -omics tries to analyze its own sphere of components in a quantitative and qualitative manner (e.g. metabonomics), trying to understand regulatory processes. Related -omics are pharmacogenomics (the study of how genetics affects drug responses) and physiomics (physiological dynamics/functions of whole organisms). Studies in each of the -omics seem troublesome enough, but since the members of all three major -omics are interconnected and influence each other, system biology tries to reach an understanding of the quantitative and qualitative properties of a whole organism or system. An important part of systems biology is the study of how organisms respond to changes (internal or external perturbations) on every level. Mathematical models are often derived to test or expand understanding. Based on findings from the -omics, systems biology depends on rigorous quantitative information (e.g. rate constants of all enzymes, involving signalling kinases, under physiological conditions) to feed its models.

Clearly genomics is a pre-requisite for proteomics. Mass spectrometry is the analytical tool of choice in proteomics, because it is fast, cheap and accurate. However, no one really identifies a protein or a modification by MS, as is always stated; most of the time the mass spectrometer produces data that are highly likely to match the data derived by computer from genomic data. On the other hand, genomic databases can be corrected by data derived from proteomic studies (from mass spectrometers). Proteomic data can discover faults in the genomic database and deliver proof that an inferred gene (and the gene product!) really exists. Going down the information hierarchy, transcriptomics analyses the transcription of DNA into (mainly) mRNA. Transcriptomics derives most of its interest from the assumption that changes in transcript levels are reflected at the functional level, that is, at the level of proteins. Many studies have shown that this is on average not strictly true, as shown in Figure 1.10.

Usually, if an mRNA equilibrium changes, this will be reflected in some sort of change at the protein level; it has to be controlled, however, because of controls on the level of mRNA stability, splicing and translational control. Of course, just because there is more of a protein, that does not necessarily mean it is more active, so transcriptomic studies should really be backed up by proteomic evidence. Combining both technologies, it is also possible in many cases to back up proteomic data and to find the mechanism that led to the changes in protein levels, for example. There are also other reasons why combining proteomics and transcriptomics is beneficial; it is virtually impossible to measure all proteins in proteomics studies as usually the less abundant ones are missed or poorly characterized. Since transcriptomics can be very sensitive, but miss out on several regulation levels, combining technologies has the advantage of increasing coverage of the analyses.

Phosphoproteomics and glycomics are special fields in proteomics; they deserve their names (like other more specialized -omics) since it is impossible with standard proteomic technologies to achieve any reasonable coverage for phosphorylation or glycosylation of proteins. If we estimate that in a typical proteomic approach using a cultured cell line we can analyze about 30–50% of all the different protein species (covering perhaps more than 95% of the total amount of proteins), it is a reasonable estimate that we would be able to analyze maybe around a dozen or so phosphorylated proteins or peptides. Using the best current approaches we still would not be able to

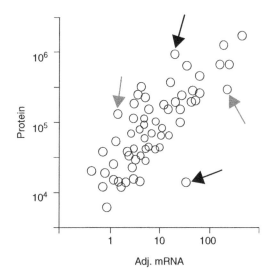

Figure 1.10 Correlation between mRNA and protein levels. Amounts of mRNA and proteins per yeast cell are compared for about 80 genes. Only relatively abundant proteins can be used for this measurement, as reliable data for absolute protein amount are more difficult to obtain for low abundance proteins. On average there are 4 000 protein molecules present per mRNA molecule. The correlation coefficient is 0.76. Although this is a good trend, the variation between mRNA and protein amount is on average 10-fold. The grey (black) arrows show that for identical amounts of protein (mRNA) the mRNA level (protein level) can vary about 100-fold (Futcher *et al.*, 1999). Thus it is not possible to reliably predict the amount of protein based on mRNA analyses. Similar relationships between mRNA and protein level can be observed for the mRNA/protein relationship for one gene, when compared in between different tissues in higher organisms. It is prudent to assume that variations are even larger for rare mRNAs or proteins. Adapted from 'A Sampling of the Yeast Proteome' B. Futcher, G.I. Latter, P. Monardo, C.S. McLaughlin, and J.I. Garrels, *Mol Cell Biol*, 0270-7306/99/$04.0010 Nov. 1999, p. 7357–7368, Vol. **19**, No. 11 Copyright © 1999, American Society for Microbiology.

detect more than about 2 000 phosphorylated peptides or proteins, and we would still not be able to analyze more than perhaps 200 in a quantitative way (i.e. which residues at which ratio are phosphorylated at any given time). If we start with an estimated 10 000 different proteins expressed in a certain cell type in a typical experiment, a look at Table 1.2 shows that we would expect some 50 000 different phospho-isoforms of these proteins; in other words, our coverage in detection of phospho-isoforms is 4% and far lower in quantitative analysis of phospho-proteins. Surely the analysis of PTMs is a field in which still a lot of further development is needed!

Metabolomics is very different from the -omics discussed so far. It is nearly impossible to link metabolites to single genes directly; they do not encode for metabolites, many different genes are involved in the regulation of each single metabolite, and many metabolites are derived from external sources, like other organisms. Metabolomics has been used very successfully to monitor diseases in newborns and to describe the state of microorganisms. If you look at it from a clinical perspective, screening of metabolites is a very efficient way to screen for dysfunctional genes and proteins. On average, more than 100 genes and their products influence one metabolite. In a typical study about 500 metabolites are controlled – barring redundancies, enough for a potential 50 000 proteins to be controlled! Given the complexity of metabolomics, each combination of metabolite concentrations can be derived from different scenarios on the level of regulation, so it is difficult to find out exactly which dysfunctional enzymes might be responsible for a given metabolic pathology.

This is a good time to have a look at the relatively new field of systems biology. One way to describe systems biology is to say that it is the research field that collects all information available on a system (say, a cell or organism) in order to figure out how the whole system (involving every signalling pathway, every executive pathway, every metabolite) works and is controlled. Since no regulatory circuit is entirely separated from the rest (in fact most seem intensively interconnected) we cannot look at a single pathway; we have to have a look at the whole system, hence the term 'systems biology'.

An important aspect of systems biology is the aim to simulate a complete system in the computer (*in silico*). For this an enormous amount of data needs to be known; all the enzymes and proteins involved, all concentrations of all metabolites and regulators, all ratios of synthesis, breakdown and half-life for all components, all binding constants and distributions, to name the most important. If a system can be modelled, we can try to unbalance it. If the system reacts like the *in silico* approximation, we might just have a correct understanding of the system. For some systems impressive results have been achieved, from complete imitations of bacterial metabolism to explanations of how signalling pathways in higher organisms regulate differentiation and growth (von Kriegsheim *et al.*, 2009). Only if we can understand cells and organisms in this way will we be able to understand

and cure cancer or metabolic diseases or viral infections. Therefore, in a way, proteomics should be delivering a lot of data to systems biology so that we can understand functional relationships on a truly systemic scale (Figure 1.9).

1.4 SOME GENERAL APPLICATIONS OF PROTEOMICS

Before the term 'proteomics' was coined some of its typical technologies were already in use in isolation – for example, the comparison of different maize specimens for their identification and control of variability. To distinguish different variants it is enough to generate a good separation of some marker proteins; using two-dimensional gel electrophoresis, one can usually chose from about 600–2 000 protein 'spots' (Figure 1.11). For this kind of analysis it is not even necessary to know why the proteins migrated in different 'spots'. The spots can arise from different proteins being expressed, or from slight sequence variations of the same (homologue) proteins or from different PTMs on proteins with the same sequence.

Proteomics can also be used for the comparison of species to analyze evolutionary relationships. Humans and chimpanzees are said to be 98.7% identical at the genomic level; when you look at a chimpanzee you would certainly feel (or hope) that the differences are somewhat larger than 1.3%. Genomic studies are very powerful for establishing evolutionary relationships between different strains, species or even higher evolutionary units such as kingdoms. However, at the genomic level the evolution of regulatory differences such as splicing or gene regulation is not very good. Using proteomics, or even organ specific proteomics, this level of evolution can be analyzed. The proteomic study of brain proteins from humans and chimpanzees showed that about 40% of the brain proteins showed either quantitative or qualitative differences (Figure 1.12). This result is a lot more in keeping with our expectations when comparing humans and chimpanzees.

The previous examples showed us the main application of proteomics, the so-called differential proteomics approach. In differential proteomics one is not interested so much in analysing every protein encountered; rather, two sets of proteins are compared, arising from similar but distinct samples. Differential proteomics involves the screening or quantitative/qualitative analysis of as many proteins as possible. However, only a part of these proteins will later be analyzed in any depth, for example to

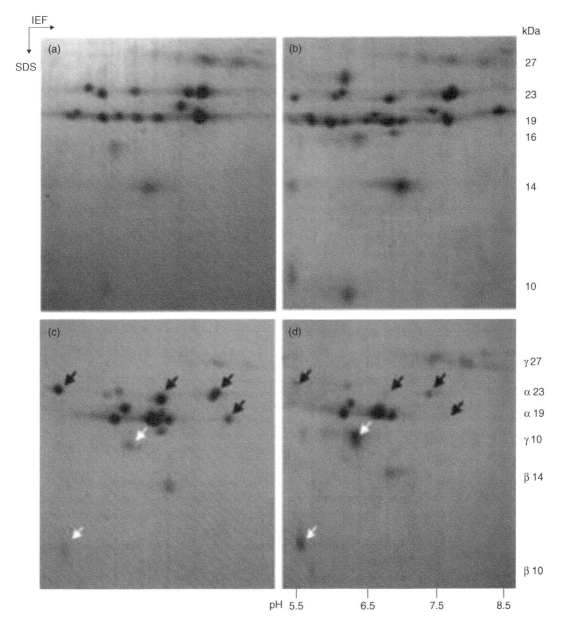

Figure 1.11 Proteomics for the analysis of genetic variability in maize. Several genetic traits influence the quality of maize corns, affecting the group of zein proteins. Zeins are the main proteins in mature seeds; their sequences are not known. A 2D gel electrophoresis of zein proteins isolated from maize powder. Proteins are separated by their isoelectric point horizontally and by their apparent molecular weight vertically. The differences in migration pattern could be based on entirely different amino acid sequences, different modifications, or (most likely) a mixture of both. The zein proteins are affected in the o2 maize line (panel d), with lower quality corns. The arrows in panel d show the zein proteins. Panels a, b and c show other inbred maize strains. The o2 mutant shows increased levels in some zein proteins (white arrows) and diminishing amounts in others (black arrows). By comparing similarities from these gels the variability of 45 inbred maize lines was analyzed in this study, to help and breed the best quality maize lines. Reproduced with permission from Consoli & Damerval (2001). Copyright © 2001 Wiley-VCH Verlag GmbH & Co. KGaA.

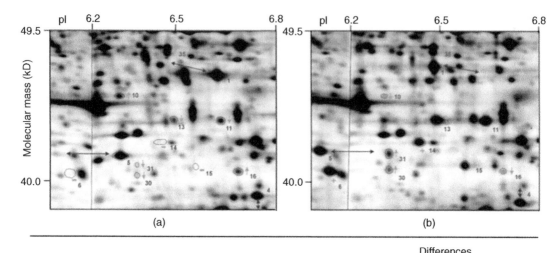

Comparison	Analyzed spots	Differences	
		Qualitative	Quantitative
Human-chimpanzee	538	41 (7.6%)	169 (31.4%)
M. musculus-M. spretus	8767	668 (7.6%)	656 (7.5%)

Figure 1.12 Differential proteomics for evolutionary studies. Brain proteins from humans (a) and chimpanzees (b) were separated by 2D SDS PAGE (see legend of Figure 1.11) and gel images were analyzed for qualitative changes (i.e. presence or absence of 'spots', indicated by + or −, or shift in position, indicated by double-headed arrows) or quantitative changes (i.e. more or less of the spot is present, indicated by up or down arrows). Results of repeat experiments were analyzed and are shown in the table below the images. Note that qualitative changes between humans and chimpanzees were as low as changes between two strains of mouse, while quantitative changes were about 4.5 times higher. For this type of analysis the genomes do not need to be known − it is enough to analyze spots on gel images. In (a) and (b) only 200 spots each out of 8 500 spots visible on a large scale gel are shown. From Enard *et al.*, *Science*, 2002; 296: 340–3. Reprinted with permission from AAAS.

identify the gene, analyze PTMs, establish the purity of seeds or distinguish pathological from harmless bacteria (Figure 1.13) − in other words, to identify a biological marker for a pathogen.

However, most differential proteomics studies are designed not only to detect differential proteins but also to identify them or their differential modifications by matching mass spectrometric data to predictions from databases. Typical applications for differential proteomics are the comparison of body fluids or cells or tissues from healthy and diseased states. The diseases range from hydrocephalus to cardiovascular disorders, genetic disorders, dementia (e.g. Alzheimer's) and diverse cancers. When it comes to analysing cancers, differential proteomics is also an important tool for cancer classification. The tissue of origin, the grade of de-differentiation and the level of spread throughout the body are used for classical cancer classification. This is

sometimes helped by limited genetic analysis (e.g. test for chromosomal abnormalities, like losses or translocations of big regions of the chromosomes) or the expression of certain antigens (e.g. specific proteins or glycosylation) known to be tumour or tumour-stage specific. Some cancers have been analyzed by genomic or transcriptomic analysis, and this has delivered a better understanding of their development inside the body. The same holds true for proteomic classifications; instead of having a dozen parameters as in the case of standard classifications, proteomics analyzes thousands of potential tumour markers. This also allows the occurrences of certain changes to be grouped (clustered), when changes in individual proteins are not being very helpful. These new and better classifications are important for choosing different potential treatments and predicting their outcome.

Related to differential proteomics is the field of biomarker discovery. The biggest surge in proteomic

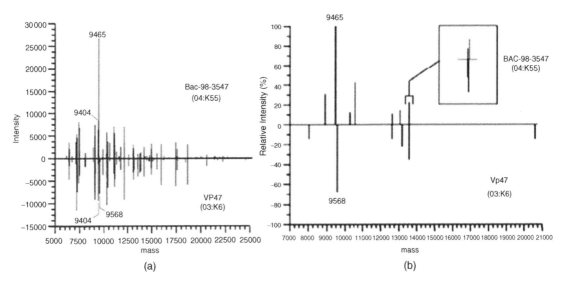

Figure 1.13 Proteomics for the rapid identification of pathogenic bacterial strains. The *Vibrio parahaemolyticus* strain O3:K6 is a dangerous contaminant in sea food, while Q4:K55 is a commonly found harmless strain of the bacterium. To distinguish them, an online LC MS approach was set up. The proteins were separated by LC and the masses of all proteins eluting from the LC were measured by MS. The intensities for both strains were joined for analysis in one graph (a) with O3:K6 derived masses plotted in the negative direction and subtracted from the Q4:K55 derived signals. Thus the graph in (a) shows the differential signal of both strains. A detailed view (b) reveals that the differences in mass are sometimes very low, only 1 Da, reflecting mutations at the protein level (PTMs are rare in bacteria). Additional LC MS and LC MS/MS analyses revealed which proteins are differential and the nature of the mutations. These differences can now be used for very fast and specific detection of this dangerous pathogen in food. Reprinted with permission from Williams *et al.*, *Journal of Clinical Microbiology*, 2004; 42:1657–1665. © 2004 American Society for Microbiology.

applications is the analysis of potential biomarkers using proteomic methods. At the time of writing, thousands of studies have been undertaken to detect biomarkers for diverse conditions, ranging from various cancers, to resistance to chemotherapy, heart conditions, kidney function and the function of the immune system (Figure 1.14). It is very significant that while most studies are very promising, few evaluated biomarkers have yet emerged. On the one hand, this seems to reflect the lack of maturity of proteomic methods, the lack of standardization and thus the problem of world-wide collaboration in studies with statistically significant sample sizes. On the other hand, this lack of immediate results reflects the complexity of the task; we do not know enough about the interactions in complex organisms to understand the huge amount of variability we encounter. It is the aim of this book to contribute to an appreciation of this complexity from the proteomic side of the analysis; the biological samples bring their own complexity as well.

Other fields to which differential proteomics is applied with great success include the study of signalling events and the elucidation of other cellular processes such as DNA replication, transcriptional control, translation, differentiation and the cell cycle. One important feature of proteomics in this setting is that it can analyze the composition of sub-cellular structures with high spatial and temporal resolution. By correlating changes in the composition of structures during biological processes, it is possible to obtain detailed knowledge of the functions of the proteins involved.

Proteomics is regularly used to analyze the reaction of organisms and cells to a changed environment, for example growth under different culture conditions and different food sources or for the analysis of stress response. The stresses analyzed can be very different in nature (e.g. temperature, nutrients, oxygen, osmotic stress, toxins), some of which are very interesting (e.g. during transplantations or more generally for the survival of operations).

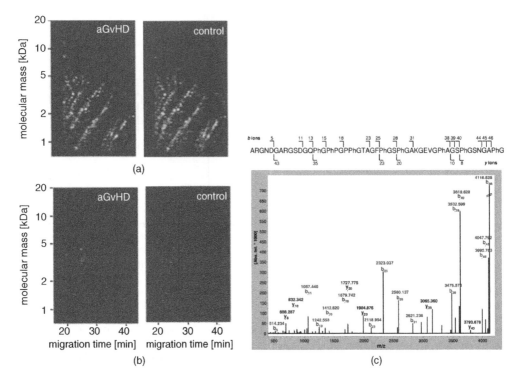

Figure 1.14 Proteomics in biomarker discovery. After treatment for some haematopoietic diseases like cancers, patients undergo stem cell transplantation. The transplanted stem cells can initiate a fatal immune reaction (acute graft versus host disease) against the new host. Biomarkers for the early stages of this disease were found by comparing peptides of serum samples from healthy and diseased individuals. Peptides were separated by capillary electrophoresis coupled with MS, for peptide analysis/identification. In (a) all peptides detected in the samples are shown as white dots, indicating their elution time and mass. In (b) only the differential peptides are shown, after extensive data analysis, and in (c) the identification of one of the diagnostic peptides from its fragmentation pattern in tandem MS is shown. Peptide fragments and their breakpoint inside the deduced peptide sequence are indicated. Reprinted with permission from Weissinger *et al.*, *Blood*, 2007; 109: 5511–19. © 2007 American Society of Hematology.

Using similar approaches, differential proteomics has also been put to good use in pharmacological studies, so a new term, 'pharmacoproteomics', has been coined. The main challenges here are to identify the modes of action of drugs, identify new drug targets and evaluate possible toxicities, side effects and resistances. One disease that has been tackled by different proteomic studies is diabetes. Diabetes affects some 200 million people worldwide. It is caused in 90% of cases by decreased pancreatic insulin production or resistance to insulin in the target tissues (e.g. muscle, adipose tissue and liver), where insulin normally induces increased uptake of blood glucose, leading to hyperglycaemia. Different reference maps of 2D gels have been published (e.g. from insulin producing and target tissues), with the aim of helping to understand the effects of anti-diabetic drugs and their side effects.

Differential proteomics is also very helpful in comparing different strains of microorganisms; it delivers more levels of complexity on top of genetics at which homologies and differences can be analyzed. These new complexity levels include the actual expression of similar or identical genes and their pattern of PTM. This is very helpful when it comes to deciding just how close strains of microorganisms are to one another and where the differences derive from. It has been shown that surface enhanced LASER desorption ionization (SELDI), a rapid MALDI MS based technology using an array of different absorbing surfaces for sample preparation, is a fast tool for discriminating different bacterial strains (Barzaghi *et al.*, 2006). Different strains of bacteria can also be analyzed using proteomics, for example to find markers that correlate with different pathologies, as

Table 1.2 Common applications of proteomics. This list of applications or references is by no means comprehensive, nor is the systematic mandatory. For example, the study of parasites by proteomics is listed under fundamental biological processes, but could equally well be listed under biomarker discovery. The references also do not necessarily cover all aspects of the particular applications – rather, they are examples.

Fundamental biological processes	
Which genes are expressed into proteins?	Zougman *et al.* (2008) and de Godoy *et al.* (2008)
Relation between genome, transcriptome and proteome	Kislinger *et al.* (2006) and Ambrosio *et al.* (2009)
Study of model organism	Washburn *et al.* (2001)
Study of certain compartments/organs	Anderson *et al.* (2004)
Study of parasites	Nett *et al.* (2009)
Molecular mechanism of cellular processes	
Physiological adaptations	Hecker *et al.* (2008)
Correlation of composition and function of organelles	Batrakou *et al.* (2009)
Study of signal transduction events	Lovrić *et al.* (1998) and Casey *et al.* (2010)
Protein structure and function analysis	
Study of the associations of proteins	Paul *et al.* (2009), De Bodt *et al.* (2009) and Ho *et al.* (2002)
Analysis of posttranslational modifications	Shu *et al.* (2004) and Choudhary *et al.* (2009)
Analyzing the effects of protein KO/suppression	LaCourse *et al.* (2008) and Chen *et al.* (2009)
Product analysis	
Detection of food contaminations	Mamone *et al.* (2009)
Analysis of seeds	Guo *et al.* (2008)
Optimization of products	Lücker *et al.* (2009) and Wang *et al.* (2002)
Comparison of strains and species	
Evolutionary studies	Arnesen *et al.* (2009), Roth *et al.* (2009), Dworzanski *et al.* (2006) and Pe'er *et al.* (2004)
Breeding	Davoli and Braglia (2008)
Rapid detection of bacteria	Barzaghi *et al.* (2006)
Biomarker discovery	
Diagnostic markers for cancers	Sodek *et al.* (2008) and Lau *et al.* (2010)
Biomarkers for a variety of diseases, for example cardiovascular or infections	Kussmann *et al.* (2006), de la Cuesta *et al.* (2009) and Mini *et al.* (2006)
Biomarkers for the function of organs, for example kidneys	Cummins *et al.* (2010)
Markers for drug response	Okano *et al.* (2007)
System analysis	
Drug development/toxicity	Sung *et al.* (2006) and Gao *et al.* (2009)
Development of drug targets	Rix and Superti-Furga (2009)
Personalized medicine	Marko-Varga *et al.* (2007)

exemplified in proteomic studies of *Helicobacter pylori*, which causes ulcers (Mini *et al.*, 2006).

In a similar approach differential proteomics can also be used in evolutionary studies, to compare different species and deduce their development and relationships (Dworzanski *et al.*, 2006) or even analyze more comprehensively how proteomes evolved in different phyla (Pe'er *et al.*, 2004) to improve our understanding of long-term evolution.

Proteomics can also be used in some very straightforward commercial activities, for example for the improvement of bio-processing (Wang *et al.*, 2003) and hence the rapid optimization of the production and processing of biomaterials by microorganisms.

Examples for all these applications, together with a rough classification, are given in Table 1.2.

1.5 STRUCTURE OF THE BOOK

The book begins with an overview of the more 'classical' approach to proteomics, that is, the isolation of the sub-proteome of interest, separation of all the proteins involved, visualization and analysis by mass spectrometry and database searches. Alternatively, after isolation of the sub-proteome of interest, the proteins can be digested into peptides and these are separated by hyphenated technologies and visualized/analyzed by mass spectrometry followed by database searches. These 'basics' are covered in Chapters 2–4. Since proteomics has to be as varied as the proteins and questions we are dealing with, some practical examples will be discussed in Chapter 5. Note that the isolation of the sub-proteome for 'deeper' analysis of the proteome is only covered briefly at the beginning of Chapter 2. For a book of this nature it is impossible to cover all the special and often functional approaches to sample preparation. These will usually be the expertise of the researcher wanting to use proteomics. Some hints will be given on sample preparation, in order to avoid the destruction of any chance for a meaningful proteomic analysis in this first, immensely important step, even before the analysis begins.

REFERENCES

Aguilar, R.C. and Wendland, B. (2005) Endocytosis of membrane receptors: two pathways are better than one. *Proc Natl Acad Soc*, **102** (8), 2679–2680.

Ambrosio, D.L., Lee, J.H., Panigrahi, A.K., Nguyen, T.N., Cicarelli, R.M.B. and Gunzl, A. (2009) Spliceosomal proteomics in Trypanosoma brucei reveal new RNA splicing factors. *Eukaryotic Cell*, **8** (7), 990–1000.

Anderson, N.L., Polanski, M., Pieper, R. *et al.* (2004) The human plasma proteome. *Mol Cell Proteomics*, **3** (4), 311–326.

Arnesen, T. *et al.* (2009) Proteomics analyses reveal the evolutionary conservation and divergence of N-terminal acetyltransferases from yeast and humans. *Proc Natl Acad Soc*, **106** (20), 8157–8162.

Bader, G.D., Heilbut, A., Andrews, B. *et al.* (2003) Functional genomics and proteomics: charting a multidimensional map of the yeast cell. *Trends Cell Biol*, **13** (7), 344–356.

Barzaghi, D., Isbister, J.D., Lauer, K.P. and Born, T.L. (2006) Use of surface-enhanced LASER desorption/ionization – time of flight to explore bacterial proteomes. *Proteomics*, **4**, 2624–2628.

Batrakou, D.G., Kerr, A.R.W. and Schirmer, E.C. (2009) Comparative proteomics analyses of the nuclear envelope and pore complex suggests a wide range of heretofore unexpected functions. *J Proteomics*, **72**, 56–70.

Bouwmeester, T. *et al.* (2004) A physical and functional map of the human TNF-alpha/NFkappaB signal transduction pathway. *Nat Cell Biol*, **6** (2), 97–105.

Casar, B., Arozarena, I., Sanz-Moreno, V. *et al.* (2009) RAS subcellular localization defines ERK1/2 substrate specificity through distinct utilization of scaffold proteins. *Mol Cell Biol*, **29** (5), 1338–1353.

Casey, T., Solomon, P.S., Bringans, S. *et al.* (2010) Quantitative proteomic analysis of G-protein signalling in Stagonora nodorum using isobaric tags for relative and absolute quantification. *Proteomics*, **10**, 38–47.

Chen, S., Martin, C., Maya-Mandoza, A. *et al.* (2009) Reduced expression of lamin A/C results in modified cell signalling and metabolism coupled with changes in expression of structural proteins. *J Proteomics Res*, **8** (11), 5196–5211.

Choudhary, C., Kumar, C., Gnad, C. *et al.* (2009) Lysine acetylation targets protein complexes and co-regulates major cellular functions. *Science*, **325**, 834–840.

Consoli, L. and Damerval, C. (2001) Quantification of individual zein isoforms resolved by 2D electrophoresis: genetic variability in 45 maize inbred lines. *Electrophoresis*, **22**, 2983–2989.

de la Cuesta, F., Alvarez-Llamas, G., Gil-Dones, F. *et al.* (2009) Tissue proteomics in atherosclerosis: elucidating the molecular mechanisms of cardiovascular diseases. *Expert Rev Proteomics*, **6** (4), 395–409.

Cummins, T.D., Barati, M.T., Coventry, S.C. *et al.* (2010) Quantitative mass spectrometry of diabetic kidney tubules identifies GRAP as a novel regulator of TGF-beta signalling. *Biochim Biophys Acta*, **1804** (4), 653–661.

Davoli, R. and Braglia, S. (2008) Molecular approaches in pig breeding to improve meat quality. *Brief Funct Genomic Proteomics*, **6** (4), 313–321.

De Bodt, S., Proost, S., Vandepoele, K. *et al.* (2009) Predicting protein-protein interactions in Arabidopsis thaliana through integration of orthology, gene ontology and co-expression. *BMC Genomics*, **10**, 288.

Dworzanski, J.P., Deshpande, S.V., Chen, R. *et al.* (2006) Mass spectrometry-based proteomics combined with bioinformatic tools for bacterial classification. *J Proteomics Res*, **5** (1), 76–87.

Enard, W., Khaitovich, P., Klose, J. *et al.* (2004) Intra and interspecific variation in primate gene expression patterns. *Science*, **296**, 340–343.

Futcher, B., Latter, G.I., Monardo, P. *et al.* (1999) A sampling of the yeast proteome. *Mol Cell Biol*, **19** (11), 7357–7368.

Gao, Y., Holland, R.D. and Yu, L.R. (2009) Quantitative proteomics for drug toxicity. *Brief Funct Genomic Proteomics*, **8** (2), 158–166.

Ghaemmaghami, S., Huh, W.K., Bower, K. *et al.* (2003) Global analysis of protein expression in yeast. *Nature*, **425**, 737–741.

de Godoy, L.M., Olsen, J.V., Cox, J. *et al.* (2008) Comprehensive mass spectrometry-based proteome quantification of haploid versus diploid yeast. *Nature*, **455**, 1251–1254.

Guo, B., Liang, X., Chung, S.Y. *et al.* (2008) Proteomic analysis of peanut seed storage proteins and genetic variation in a potential peanut allergen. *Protein Pept Lett*, **15** (16), 567–577.

Hall, N. *et al.* (2007) A comprehensive survey of the plasmodium life cycle by genomic, transcriptomic, and proteomic analyses. *Science*, **307**, 82–86.

Hecker, M., Antelman, H., Buttner, K. and Bernhardt, J. (2008) Gel based proteomics of gram positive bacteria: a powerful tool to address physiological questions. *Proteomics*, **8**, 4958–4975.

Hipkiss, R.A. (2006) Accumulation of altered proteins and ageing: causes and effects. *Exp Geront*, **41**, 464–473.

Ho, Y. *et al.* (2002) Systematic identification of protein complexes in Saccharomyces cerevisiae by mass spectrometry. *Nature*, **415**, 180–183.

Johnson, J.M., Castle, J., Garrett-Engele, P. *et al.* (2003) Genome-wide survey of human alternative pre-mRNA splicing, with exon junction microarrays. *Science*, **302**, 2141–2144

Kawamori, D., Kaneto, H., Nakatani, Y. *et al.* (2006) The forkhead transcription factor Foxo1 bridges the JNK pathway and the transcription factor PDX-1 through its intracellular translocation. *J Biol Chem*, **281**, 1091–1098.

Kislinger, T. *et al.* (2006) Global survey of organ and organelle protein expression in mouse: combined proteomic and transcriptomic profiling. *Cell*, **125**, 173–186.

Kolch, W. (2000) Meaningful relationships: the regulation of the Ras/Raf/MEK/ERK pathway by protein interactions. *Biochem J*, **351**, 289–305.

Kolch, W. (2005) Coordinating ERK/MAPK signaling through scaffolds and inhibitors. *Nat Rev Mol Cell Biol*, **6**, 827–839.

von Kriegsheim, A., Baiocchi, D., Birtwistel, M. *et al.* (2009) Cell fate decisions are specified by the dynamic ERK interactome. *Nat Cell Biol*, **11** (12), 1458–1466.

Krogan, N.J. *et al.* (2006) Global landscape of protein complexes in the yeast Saccharomyces cerevisiae. *Nature*, **440**, 637–643.

Kussmann, M., Raymond, F. and Affolter, M. (2006) OMICS-driven biomarker discovery in nutrition and health. *J Biotech*, **124** (4), 758–787.

LaCourse, E.J., Perally, S., Hernandez-Viadel, M. *et al.* (2008) A proteomic approach to quantify protein levels following RNA interference: case study with glutathione transferase superfamily from the model metazoan Caenorhabditis elegans. *J Proteomics Res*, **7**, 3314–3318.

Lau, T.Y.K., Power, K., Dijon, S. *et al.* (2010) Prioritization of candidate protein biomarkers from in vitro model system of breast tumor progression towards clinical verification. *J Proteomics Res*, **9** (3), 1450–1459.

Lovrić, J., Bischof, O., Moelling, K. *et al.* (1994) Cell cycle-dependent association of Gag-Mil and hsp90. *JFEBS Letters*, **343** (35), 15–21.

Lovrić, J., Dammeier, S., Kieser, A. *et al.* (1998) Activated Raf induces the hyperphosphorylation of stathmin and the reorganization of the microtubule network. *J Biol Chem*, **273** (35), 22848–22855.

Lücker, J., Laszczak, M., Smith, D. and Lund, S.T. (2009) Generation of a predicted protein database from EST data and application to iTRAQ analyses in grape (*Vitis vinifera* cv. Cabernet Sauvignon) berries at ripening initiation. *BMC Genomics*, **10**, 50.

Mamone, G., Picariello, G., Caira, S. *et al.* (2009) Analysis of food proteins and peptides by mass spectrometry-based techniques. *J Chromatogr A*, **1216**, 7130–7142.

Marko-Varga, G. *et al.* (2007) Personalized medicine and proteomics: lessons from non-small cell lung cancer. *J Proteomics Res*, **6**, 2925–2935.

Mini, R., Bernadini, G., Salzano, A.M. *et al.* (2006) Comparative and immunoproteomics of Helicobacter pylori related to different gastric pathologies. *J Chromatogr B Analyt Technol Biomed Life Sci*, **833** (1), 63–79.

Nett, I.R.E., Martin, D.M.A., Miranda-Saavedra, D. *et al.* (2009) The phosphoproteome of Trypanosoma brucei, causative agent of African sleeping sickness. *Mol Cell Proteomics*, **8** (7), 1527–1538.

Okano, T., Kondo, T., Fuji, K. *et al.* (2007) Proteomic signature corresponding to the response to gefitinib (Iressa, ZD1839), an epidermal growth factor receptor tyrosine kinase inhibitor in lung adenocarcinoma. *Clin Cancer Res*, **13** (3), 799–805.

Patterson, S.D. and Aebersold, R.H. (2003) Proteomics, the first decade and beyond. *Nat Genet*, **33** (Suppl.), 311–324.

Paul, A.L., Liu, L., McClung, S. *et al.* (2009) Comparative interactomics; analysis of Arabidopsis 14-3-3 complexes reveals highly conserves 14-3-3 interactions between humans and plants. *J Proteomics Res*, **8**, 1913–1924.

Pe'er, I., Felder, C.E., Man, O. *et al.* (2004) Proteomic signatures: amino acid and oligopeptide compositions differentiate among phyla. *Proteins Struct Funct Bioinfo*, **54**, 20–40.

Rix, U. and Superti-Furga, G. (2009) Target profiling of small molecules by chemical proteomics. *Nat Chem Biol*, **5** (9), 616–623.

Roth, S., Fromm, B., Gade, G. and Predel, R. (2009) A proteomic approach for studying insect phylogeny: CAPA peptides of ancient insect taxa. *BMC Evol Biol*, **9**, 50.

Sauer, G., Korner, R., Hanisch, A. *et al.* (2005) Proteome analysis of the human mitotic spindle. *Mol Cell Proteomics*, **4** (1), 35–44.

Shu, H., Chen, S., Bi, Q. *et al.* (2004) Identification of phosphoproteins and their phosphorylation sites in the WEHI-231 B Lymphoma cell line. *Mol Cell Proteomics*, **3** (3), 279–286.

Sinclair, A.M. and Elliott, S. (2004) Glycoengineering: the effect of glycosylation on the properties of therapeutic proteins. *J Pharm Sci*, **94** (8), 1626–1636.

Sodek, K.L., Evangelou, A.I., Ignachenko, A. *et al.* (2008) Identification of pathways associated with invasive behaviour by ovarian cancer cells using multidimensional protein identification technology (MudPIT). *Mol Biosyst*, **4**, 762–773.

Sung, F.L., Pang, R.T., Ma, B.B. *et al.* (2006) Pharmacoproteomic study of cetuximab in nasopharyngal carcinoma. *J Proteomics Res*, **5** (12), 3260–3267.

Takahashi, N., Yanagida, M., Fujiyama, S. *et al.* (2003) Proteomic snapshot analyses of preribosomal ribonucleoprotein complexes formed at various stages of ribosome biogenesis in yeast and mammalian cells. *Mass Spectrom Rev*, **22** (5), 287–317.

Wang, P., Yan, B., Guo, J.T. *et al.* (2005) Structural genomics analysis of alternative splicing and application to isoform structure modeling. *Proc Natl Acad Soc*, **102** (52), 18920–18925.

Wang, W., Sun, J., Hartlep, M. *et al.* (2003) Combined use of proteomic analysis and enzyme activity assays for metabolic pathway analysis of glycerol fermentation by Klebsiella pneumoniae. *Biotech Bioengin*, **83** (5), 525–536.

Washburn, M.P., Wolters, D. and Yates, J.R. III (2001) Large-scale analysis of the yeast proteome by multidimensional protein identification technology. *Nat Biotech*, **19**, 242–247.

Weissinger, E.M., Schiffer, E., Hertenstein, B. *et al.* (2007) Proteomic patterns predict acute graft-versus-host disease after allogeneic hematopoietic stem cell transplantation. *Blood*, **109**, 5511–5519.

Williams, T.L., Musser, S.M., Nordstrom, J.L. *et al.* (2004) Identification of a protein biomarker unique to the pandemic O3:K6 clone of Vibrio parahaemolyticus. *J Clin Microbiol*, **42** (4), 1657–1665.

Wood, W.G. (1976) Haemoglobin synthesis during human fetal development. *Br Med Bull*, **32** (3), 282–287.

Zougman, A., Ziolkowski, P., Mann, M. and Wisniewski, J.R. (2008) Evidence for insertional RNA editing in humans. *Curr Biol*, **18** (22), 1760–1765.

2

Separation and Detection Technologies

2.1 INTRODUCTION TO EXPERIMENTAL STRATEGIES IN PROTEOMICS

2.1.1 Overview of the main experimental strategies

In order to provide an overview, we introduce two typical proteomic approaches. These approaches are a mixture of many technologies, which will follow us in various guises throughout the book.

1. Proteins are isolated from cells using a high concentration urea buffer, then separated by high resolution 2D gel electrophoresis (see Figure 2.1). Proteins are visualized in the gels by one of various methods (coomassie stain, silver stain, fluorescence stain) and gel images are analyzed for spots of interest (SPOs). SPOs are usually differential in some way (differential proteomics), for example absent from harmless bacteria but present in a pathogenic strain and compromise usually less than 5% of the total number of spots (or less than 100 of 2 000 protein 'spots' on a typical gel). Gel pieces containing the SPOs are cut out from the gels and proteins are digested with trypsin in the gel. The resulting peptides are eluted and extracted from the gel and analyzed using one or more mass spectrometers (see later chapters). This MS data (list of peptide masses) is compared with that from trypsin digests performed 'in silico' (i.e. simulated on the computer) using the peptide sequences of hundreds of thousands of proteins derived from RNA/DNA and protein databases. Often the MS analyses involve the analysis of characteristic fragments of a peptide (tandem MS or MS/MS or MS^2). If the actual experimental data are reasonably similar to one of the results of any of the

thousands of 'in silico' experiments for one or more of the database entries, we say we have identified a protein, maybe even with its posttranslational modifications (PTMs). We will refer to this approach as a gel-based protein identification approach.

2. Proteins are again isolated from cells using high concentration urea buffer but this time subjected straight away to proteolysis using trypsin (see Figure 2.1). This generates an immense amount of peptides, which are separated into 7–12 individual fractions according to their charge in a slightly acidic environment using a high pressure/performance liquid chromatography (HPLC) column, e.g. a strong cation exchange (SCX) column. Each fraction is subjected to a reversed phase separation, according to its hydrophobicity in buffers of changing polarity. The second separation results in a continuum of peptides eluting from an HPLC column (temporally separated, the more hydrophobic the later the elution from the column) directly in one of several possible mass spectrometers. The MS data are collected and this time it is essential to create MS/MS date. Similar to approach 1, the experimental data on different peptides (thousands per single experiment) are compared to the 'in silico' generated data mimicking the experiment and using the same huge RNA/DNA and protein databases as in approach 1. Again a reasonable similarity between experimental and simulated data means we have identified a protein with MS/MS data from more than one of its peptides, again maybe even with its PTMs. This experiment will give a good idea of all the protein present in a sample. If we repeat the experiment several times with two different samples we can compare the lists of proteins that

Introducing Proteomics: from Concepts to Sample Separation, Mass Spectrometry and Data Analysis, Josip Lovrić
© 2011 John Wiley & Sons Ltd

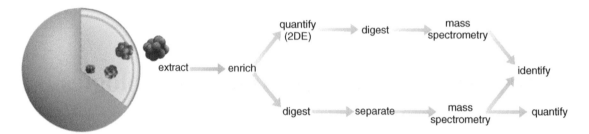

Figure 2.1 Gel based and shotgun proteomic approaches. Complex samples can be analyzed by gel-based approaches (top) or MudPIT approaches (bottom). Quantifications are based on image analysis of single spots or MS data of peptide. Protein identifications are always based on peptide analysis. Peptides are either derived from isolated spots, or from the complete protein mixture. Reproduced from Patterson & Aebersold, *Nat Gen*, 2003; Suppl. 33: 311–23; courtesy of Nature Publishing Group. © 2003 Nature Publishing Group.

are identified in each sample and again find proteins that are specific for one of the samples. We will call this peptide based approach multidimensional protein identification technology (MudPIT) or the shotgun (all proteins shot to pieces) proteomics approach.

Both approaches are described as bottom-up approaches, as we generate peptides to identify or characterize the proteins of interest. In top-down approaches the complete proteins are measured by mass spectrometer to generate data to identify or characterize them. If one omits the tryptic digest (and uses different MS analyses) approach number 2 could be transformed into a top-down approach.

All these basic approaches can be divided into a separation stage (of proteins in 1 and peptides in 2), visualization (of proteins in 1 and peptides in 2) and analysis by MS (of peptides derived from a single protein in 1 and from multiple proteins in 2, but of proteins in the top-down approach, modified from 2). In approach 1 a relatively simple MS analysis may give good results or the results may be confirmed and extended by MS/MS results. For approach 2, MS results nearly always have to be supplemented by MS/MS results. In any case the mass spectrometric analyses are always followed by analyses of the data using computers and databases. Finally, these elements can be combined with other methods and changed to develop specific strategies for answering specific proteomic questions.

This division into separation (quantification), visualization, mass spectrometry (quantification), data analysis and how the different elements fit together is reflected in the structure of this book and the sequence of its chapters.

2.1.2 Some remarks on sample preparation

Before we can even start a proteomic analysis we might already have destroyed all chances of success by choosing the wrong sample preparation approach. Sample preparation has to meet four main criteria:

1. the sample should contain as many of the proteins originally present in the sample as possible;
2. all additional components (other protein, salts and buffer, etc.) should be removed as far as possible;
3. the PTMs of the proteins in the sample should be preserved;
4. the sample composition should be compatible with all subsequent analysis procedures.

Of course there are also other considerations, such as sample stability or concentration, but they are usually not the most critical considerations.

Concerning point 1, no proteomic analysis can be complete if we do not get all the proteins into the analysis pipeline in the first place. This is one of the biggest challenges that we face during sample preparation for proteomics. Often it might be easy (in mammalian cells) to solubilize most if not all proteins and the solubilization buffer is also the lysis and sample buffer. However, if your proteins are 'better protected', as in yeast or bacteria and plant cells with thick cell walls (like meat in a tin!), you might have to spend some time and effort opening the tin and getting to the proteins without getting into trouble with points 2–4. As mentioned in the introduction, one of the main challenges in proteomics is something that appears rather trivial; keeping all the proteins of interest in solution. We currently have no complete answer to this problem, although there is a lot of literature available regarding this issue. Basically, we do not know how to do it perfectly, but we do know how to do it incorrectly, and sample preparation is the single most important step in proteomics. If this step is not optimized it will become the limiting element of the analysis, no matter how sophisticated all other parts of the analysis are, and therefore we will have this short chapter on sample preparation even before we discuss different proteomics technologies.

Concerning point 3, another very important aspect of preparing samples for proteomics is the modification of proteins and the amino acids that form them. This problem is quite important. If, for example, you wish to analyze the phosphorylation of proteins it is of course of the utmost importance to conserve the protein phosphorylation pattern throughout protein isolation. This is by no means a trivial problem that applies in the same way to other modifications such as glycosylations, ubiquitinylations, methylations or deamidations and the formation of sulfur bridges. The last modification leads us to another important phenomenon: proteins can be chemically modified during the preparation process. One of the most common modifications is oxidation, which affects mainly methionine and cysteine residues, but also lysine, histidine, tryptophan and proline. In other common forms of protein analysis this might not affect the results and might be totally irrelevant (e.g. denaturing sodium dodecyl sulfate polyacrylamide gel electrophoresis (SDS-PAGE), Western blot or immunofluorescence). However, if we do not control this artefact in proteomics we could end up missing all peptides containing either of the amino acids, because their masses will increase by distinct values from 2 to 48 Da. Since all proteins might be equally affected, we might miss out on up to about 30% of the peptides, no matter how sophisticated later steps of the analysis are; it is all lost in the initial steps! Imagine another scenario: we wish to analyze cellular proteins, some of which are prone to hydrolysis during preparation. How can we separate them from proteins that get targeted by proteases in a biological process like apoptosis? Since in a typical real life proteomic experiment we analyze about 2 000 proteins, if only 2% of the proteins get artificially hydrolyzed we are dealing with 40 proteins that have to be tested and analyzed in further, often very complicated experiments.

In the 'classical' gel-based protein identification approach problems 1 and 3 can often be addressed at the same time by using high concentrations of chaotropic agents such as urea and thiourea in combination with 2D-compatible detergents like 3-[(3-cholamidopropyl)dimethylammonio]-2-hydroxy-1-propanesulfonate (CHAPS)

Table 2.1 Problems in sample preparation 1.

Problem	Method	Sample buffer	Solution
Protein protection from proteases	2D electrophoresis	High concentration urea/thiourea with CHAPS	Sample buffer denatures all possible proteases, early addition to sample during sample preparation, e.g. ideally straight onto living cells
	2D HPLC	Low concentration urea buffer with SDS	Fast and complete proteolysis by added specific protease (e.g. trypsin) degrades all possible proteases
Retaining the original PTM status	2D electrophoresis	High concentration urea/thiourea with CHAPS	Sample buffer denatures all possible enzymes, early addition to sample during sample preparation, e.g. ideally straight onto living cells
Not all proteins are solubilized	2D electrophoresis	High concentration urea/thiourea with CHAPS	High conc. chaotropic agents total end concentration 8 M, add 0.1% SDS, lysis in liquid nitrogen
	2D HPLC		Digest with trypsin need only be partial to help solubilization, acetonitrile (up to 10%)
Minor components not preserved in sample mix	2D electrophoresis	High concentration urea/thiourea with CHAPS	Pre-fractionation, sub-cellular compartment isolation

as soon as possible in the preparation protocol; most proteins are solubilized and denatured at the same time, so there is no chance for phosphatases, proteases or other enzymes to artefactually change our protein composition.

An account of problems during sample preparation and possible solutions (by no means comprehensive) is given in Table 2.1.

Turning to point 4, we need to exclude any non-protein contaminations from the natural environment of the proteins and avoid the introduction of new contaminations (e.g. by fulfilling criteria 1, 2 and 3), which would jeopardize further proteomic analysis. The requirements for protein purity vary for different technologies and proteomic approaches. The purity can be defined on two levels, that is regarding the non-protein contaminations of our preparation and regarding how many different proteins are present in a protein preparation. The latter problem will be dealt with in more detail in the relevant chapters describing specific proteomic technologies, and the former will be dealt with now.

From a proteomic point of view proteins are always contaminated by non-protein impurities. These may be biological contaminations, such as salts from body fluids (e.g. eyes are extremely salty, up to 1.5 M, and make it difficult to analyze samples) or RNA or more often DNA or other biological components like sugars and acids or tough components of extracellular basal laminas or the glycocalyx. Other contaminations are often put in by the experimenter with the best intentions – for example, detergents to keep proteins in solution or sugars and

salts from density gradient isolations of organelles. Other substances may be salts to keep proteins in a native state or buffer substances to keep the pH stable, again to keep the proteins in solution and as close to a native state as possible. There may be many good reasons for each of these substances to be used and for most applications these substances cause no problems, but they just might in proteomic applications. Therefore, it is worth dwelling a little on the most common snags with contaminants.

Typically there are three major stages where artefacts from prior protein isolation/handling cause problems: either with protein hydrolysis by proteases, protein/peptide separation or with protein/peptide analysis in mass spectrometers. Once inside a proteomic analysis there are certain steps that are less compatible with others, but this is usually accounted for by the design of the analysis, and tried and tested before by other researchers if you follow established procedures. You will at this stage only get into trouble if you try and combine different methods as the 'pick and mix' might not be very compatible.

There are principally two different ways to feed proteins in proteomic separation methods, either as full-length proteins or by hydrolysis into shorter peptides (usually 4–30 amino acids). These methods are dealt with in more detail later, and Table 2.2 tries to give some idea of the main problems in sample preparation for proteomics. However, it is by no means exhaustive!

Table 2.2 Problems in sample preparation 2.

Method	Problem	Solution
2D electrophoresis	Too much salt in sample	Limit amount of sample
		Reduce salt during preparation, wash sample prior to addition of sample buffer (e.g. wash cells in low salt buffer)
		Acetone precipitation of sample prior to solubilization in final buffer
2D electrophoresis	Too much DNA in sample	Shearing of DNA by passing through syringe or crushing in shredders
		Ultracentrifugation prior to electrophoresis
2D electrophoresis	Too much RNA in sample	Incubation with RNAse prior to solubilization in final buffer
		Acetone precipitation of sample prior to solubilization in final buffer
All methods	Oxidation of proteins	DTT/iodoacetamide treatment of sample in immobilized pH gradient gel or solution
All methods	Chemical modifications of proteins prior to MS	Use reagents (gels, stains) of ultra high purity, minimize handling steps, avoid temperatures above 37°C in urea, account for putative chemical modifications in database search strategy

So what general conclusions can we draw from all these modifications/contaminations for the design of a proteomic experiment?

- Plan your experiments with proteomic methods in mind, adapt buffers and procedures.
- Keep it simple, avoiding any steps and buffer components that are not absolutely necessary.
- Talk to your co-workers: how was the sample prepared, what is going to happen to it next?
- Make sure you know the composition of each buffer, and delete components you do not really need.
- Design your experiments so that everything happens quickly: the less time the sample has to degrade, the better (e.g. shorten time from cell lysis to chaotropic buffer addition).
- Work towards convenient storage points to stop an experiment, for example once in chaotropic buffer your sample can be stored for a long time.
- Take precautions to deal with oxidation (degas solutions and liquids!), degradation and artificial removal of amino acid modifications that might affect your results (e.g. phosphorylations).
- Make sure reagents are of good quality and within their shelf life, control for decomposition wherever possible (e.g. acrylamide solutions; measure pH, which should be about pH 6.8, and use only for non-proteomic applications if it is pH 4–5).

2.1.3 Some remarks on protein digestion

In bottom-up proteomic approaches, which comprise over 90% of all proteomic analyses, proteins are digested into shorted peptides before the analysis by mass spectrometer. This can be done after the separation (e.g. gel-based approaches) or before the separation (e.g. MudPIT-like approaches). In the overwhelming majority of cases this digestion into shorter peptides is performed using trypsin, as trypsin has a number of advantages:

1. It is cheap and easy to isolate.
2. It has a high specificity which is easy to predict.
3. It is a robust enzyme, works over a range of concentrations and conditions.
4. It works very well on denatured or native proteins.
5. It generates peptides of a good length for mass spectrometric analyses.
6. All tryptic peptides contain at least one Arg or Lys, with their positive charge retaining amino groups.

Table 2.3 shows some other commonly used proteolytic procedures in proteomics with their specificities and, as we can see, trypsin hydrolyses ('cuts') the peptide bond carboxyterminal of Arg or Lys. With all these advantages there are some steric influences of neighbouring amino acids on the preference of tryptic digestions, as we will see in Section 4.2.1.

Trypsin is not an exopeptidase, it is a pure endopeptidase. This means that sequences with Arg or Lys in consecutive positions of the primary amino acid sequence will not be cut completely: either one or the other position is cut, after which it becomes a tryptic site at the end of a peptide, which cannot be cut.

If all amino acids were present with the same frequency and trypsin cuts after 2 out of 20 amino acids, we would expect tryptic peptides to be about 10 amino acids long on average. In practice there is a wide variety of length of tryptic peptides that are predicted, and on average the ones that are detected are slightly shorter than in the theoretical prediction. Most tryptic peptides used in proteomics are

Table 2.3 Proteolysis in proteomics.

Enzyme/process	Specificity	Remarks
Trypsin	C-terminal of Arg, Lys	Endopeptidase only
Chymotrypsin	C-terminal of Trp, Tyr, Phe, Leu, Met	Cleavage after some amino acids only partial
Lysyl endopeptidase = Lys C	C-terminal of Lys only	Creates longer peptides than trypsin
Arg-C	C-terminal of Arg	–
Glu-C	C-terminal of Glu	Enzymes from different sources are affected by amino acid C-terminal from cleavage site
CNBr	C-terminal of Met	Not if followed by Ser/Thr

between 4 and 40 amino acids long, with most of them being 8 to 23 amino acids in length. Peptides outside this size range are not detected very efficiently by the mass spectrometric methods applied. The smaller peptides are less informative in database searches as they are often encountered in more than one protein.

Trypsin works best in a slightly basic environment and under mildly denaturing conditions, which deactivate most other enzymes but make all tryptic sites in the substrates more accessible. Trypsin can be stored in an inactive form and then activated before use by a pH, temperature and buffer shift, ensuring that not too much self-digestion occurs.

In gel-based approaches or for the analysis of purified proteins, single proteins or mixtures of two to three proteins are digested and the resulting peptides are directly loaded onto a mass spectrometer (Figure 2.1). In principle the digestion can be performed either inside the separating gel or in solution. The resulting peptides can also be loaded onto a liquid chromatography (LC) column, separated and, transfered directly (online) into a mass spectrometer. In MudPIT/shotgun approaches a complex mixture of proteins is digested in solution and than loaded onto an LC system online coupled to a mass spectrometer (Figure 2.1).

While most tryptic digestions are performed at 37 °C for 3–16 hours, trypsin works best at higher temperatures, but at these temperature urea adducts can be formed. As digestions are performed for prolonged times in low volumes (10–20 µl for gel bands/spot, 50 µl for MudPIT) in large vessels it is important to stop condensations at lids.

With all these different approaches, trypsin generates peptides in a highly reproducible manner. The digestion of single proteins from gels and of complex mixtures (containing several thousand different proteins) is most common in proteomics. In these reactions the amount of protein per digestion varies from about 0.1 ng (faint or invisible spot in a 2D gel) to about 50 µg for complex mixtures in solution. In complex mixtures the dynamic range of the different proteins encountered exceeds 6 orders of magnitude for cellular proteins and even 10 orders of magnitude for plasma samples. 'Star' activity – that is, proteolytic activity that is not from trypsin – is very rare under all these conditions. Several adaptations to the digestion procedure ensure complete digestion as well as compatibility of the sample with mass spectrometry or later separations. Most reactions are buffered using ammonium bicarbonate; the pH is about right for trypsin and this buffer is volatile and can

be easily removed by evaporation. This is particularly important for gel-based digests, as these can be loaded onto a matrix assisted LASER desorption/ionization (MALDI) target plate without much sample preparation. Buffers like TRIS are to be avoided for this, as they would interfere with the MALDI process if not removed.

The optimal amount for trypsin varies: in solution and with relatively large amounts of proteins (micrograms!) a ratio of 50 : 1 between substrate and trypsin is considered optimal. In gel-based digests usually 10–150 ng of trypsin are used, although the amount of sample protein is only 0.1 ng to about 2 000 ng. This large amount of trypsin is used because it has to diffuse into the gel, where it then digests the gel fixed protein leading to the release of the peptides in solution. These processes are helped by cutting the gel into small pieces and dehydrating/shrinking it with acetonitrile (ACN); upon addition of water based buffers containing trypsin, the gels expand very rapidly and the trypsin is soaked into the gel. Gel-based digestions of small amounts of protein are very challenging for several reasons. First, the amount of trypsin is critical. Not enough trypsin and there is not enough of a digest, too much and the digestion is inefficient as the immense surplus of trypsin will lead to trypsin 'jumping' on and off the proteolytic side, without performing digestions other than digesting itself, which in turn will deliver peptides that will mask the sample. Furthermore, peptides bind to any sort of surface or get lost on the liquid surface, never to be solubilized again, and in samples with small protein amounts this can lead to a substantial loss of peptides. Trypsin is very stable, so overnight digestions are the rule; these can, however, lead to the loss of preferentially large or hydrophobic peptides to the walls of reaction vessels. All these problems become apparent in gel-based proteomics once the amount of sample is below about a picomole (20–200 ng in a spot, depending on the protein size); so while it is possible to identify proteins from gels below this level it becomes increasingly difficult with lower success rates and higher workload involved (e.g. expect 80% success rate in automatic approaches for spots above 1 pmole, down to 10% success rate in manual approaches for less than 50 fmol). As sample amounts get smaller, staining procedures as well as destaining procedures (after the spot is cut out of the gel) become increasingly important as does the whole sample handling, including potential chemical modifications and the quality of all chemicals involved. Gel-based tryptic samples go onto MALDI target plates without much pre-cleaning; it is in principle possible to remove half a microlitre of a typical 20 µl digest solution a couple of hours into the digest and spot it onto a MALDI target (Section 3.1.1). This measurement usually provides the best coverage of large or hydrophobic peptides, which can easily get lost during clean-up or prolonged digestion. Clean-up can consist of repeated rounds of lyophilization or the use of self-made or commercial micro-reverse phase (RP) columns. One has to bear in mind that the micro-RP column approach delivers good uniform sample quality, often combined with the complete loss of some peptides and always with a great loss of sample. However, the signals usually look better, as signal suppressing contaminations (Section 3.1.1) are also removed.

Tryptic digests for MudPIT approaches are less challenging (at least for cellular lysates, serum or urine need optimization of digest procedures), as they are performed in solution with much better efficiency and recovery of low abundance peptides, and surface losses are prevented

by very abundant peptides. Here one has to ensure that all enzymes within the lysate are denatured first (typically by SDS or urea) before the sample is diluted to allow trypsin to be active (up to 2 M urea, 25% ACN, 0.1% SDS and even higher if milder detergents are also present). If the sample is not denatured first, enzymes might change its composition (e.g. proteases, phosphatases). Trypsin is added in a weight ratio of 1 : 20–1 : 100 (trypsin : sample) for 4–16 hours, with 8 hours being more than enough but 16 hours not being detrimental, in samples of about 50 µl typically containing 1–5 µg/µl of sample protein. It is also possible to use immobilized trypsin on porous beads, to avoid trypsin self-digestion peptides in the sample. TRIS can be used as buffer, as the sample will be run over several columns before it enters the mass spectrometer.

2.1.4 Overview of separation technologies

The analysis of the proteome has to start with the display of the proteins or peptides in some form. Given the

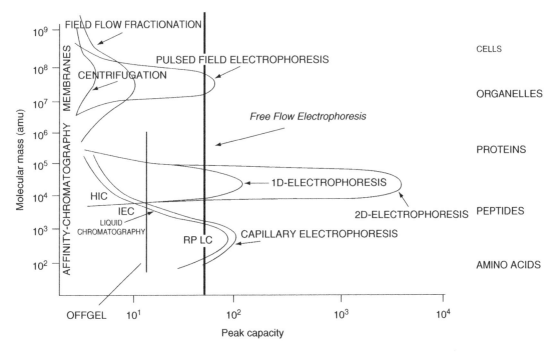

Figure 2.2 Separation methods in proteomics: comparison of mass, range and resolution. Different methods can be combined to enhance peak capacity/sensitivity. In the case of MudPIT, ion exchange chromatography resulting in 5–12 fractions is combined with RP LC and MS/MS is used as read-out. Since modern mass spectrometers can cope with 5–15 co-eluting peaks, the resolution of MudPIT is 25–180 times higher than the resolution of RP LC used in isolation. In GeLC a similar resolution is achieved by slicing a 1D SDS PAGE gel into 9–20 pieces and performing a tryptic digest followed by RP LC MS/MS on each of the resulting "fractions". OFFGEL electrophoresis can deliver up to 20 fractions for small molecules ranging up to about 10^6 Da. HIC = hydrophobic interaction chromatography. Reproduced and modified with permission from Kellner *et al.* 1998. © 1998 Wiley-VCH Verlag GmbH & Co. KGaA.

Table 2.4 Comparison of separation methods. Asterisks are an indication of strength on a scale of 0–5.

Method	Robustness	Resolution	Versatility	Sensitivity	Compatibility with MS	Throughput
2D SDS PAGE	***	*****	***	***	***	****
2D HPLC	***	****	*****	*****	*****	****
FFE	**	***	****	****	****	***
GeLC	****	****	*****	*****	*****	*****

huge number of proteins, we need elaborate separation technologies for this purpose. These separations are then followed by staining and/or mass spectrometric analysis. Figure 2.2 gives a general overview of the separation methods available for proteomics.

Traditionally, proteomic analysis always involved gel electrophoretic separations of intact proteins or even protein complexes. In the last decade this approach has been enriched, supplemented and overtaken by a plethora of approaches either avoiding gels in the separation of proteins or even omitting protein separation completely and rather working with peptides derived from the proteins and separating these in gel-free approaches. Methods have been developed to isolate protein complexes/organelles or complete cells prior to proteomic analysis, evidently again avoiding the usage of gels. In general we will distinguish between gel based and gel-free methods. Most of the gel-free approaches are often summarized as hyphenated methods, based on the form of the separation devices (columns and think tubings). Of course each of these separation methods works more or less well with either further separation methods or approaches for protein/peptide analysis, like the 'top-down' approaches, where a complete protein is fed into an MS to be analyzed, or the 'bottom-up' approaches, where peptides derived from a protein are analyzed in the MS.

Generally gel-based approaches work best with bottom-up analysis, while gel-free approaches are used either to feed into gel-based approaches, or are completely detached from gel-based approaches. Gel-free or hyphenated methods work with both bottom-up and top-down approaches.

Why is there no optimal separation technology? Well, the tasks are tremendous, and the requirements can be subdivided into the fields of reproducibility, throughput, cost efficiency, resolution (peak capacity), (protein amount) capacity, the loss of low abundance proteins (as well as sensitivity) and of specific classes of proteins (or versatility), and compatibility with previous isolation

methods (or robustness), as well subsequent analysis in terms of throughput and sensitivity.

Each method shows a unique blend in terms of the above requirements that will be discussed in the following chapters; a general overview is given in Table 2.4. Since we are comparing apples with pears, it is important to notice that the expressions may have different meanings in different separation technologies. Resolution for 2D SDS PAGE is for proteins, in the case of 2D HPLC for peptides, and in the case of free flow electrophoresis (FFE) for both. In terms of throughput, while in 2D SDS PAGE 2 000 protein isoforms can be visualized and compared/quantified in one experiment, they cannot be easily identified in later experimental stages; in 2D HPLC 1 000 proteins can be identified, but no easy comparison with a different sample can be made from these data.

Gel-based technologies/2D gels

From the point of view of analyzing proteins by mass spectrometry, it is not straightforward to have to extract them from a gel, so why are gel-based technologies used so often in proteomics? Gels can deliver a highly defined environment for protein separation, counteracting effects of diffusion and allowing sensitive detection. The widely used polyacrylamide gels (Figure 2.7) can be created virtually free of any charges, so they are optimally suited for charge based separations, like isoelectric focusing (IEF) in the first dimension of classical 2D electrophoresis. Polyacrylamide gels show only weak interactions with proteins (in the presence of SDS) and hence will not absorb most of the proteins. We can expose gels (and the proteins in them) to a variety of different chemical treatments; they mainly stay inert and allow detergents, reducing agents or denaturing agents to be used during separations. Polyacrylamide gels can be cast in different percentages (equal to different mesh widths) to suit the separation of large or small proteins, down to oligopeptides. The combination of denaturing agents and reproducible mesh width in gels allows a highly reproducible and stable separation of proteins and polypeptides by apparent size, the second dimension in classical 2D electrophoresis. After separations proteins are usually 'fixed' to the gel, for staining or storage. This fixation is often irreversible on the protein level but not on the peptide level, so that peptides can be extracted from the gels after staining and detection. Gels are compatible with a wide range of protein detection methods, like detection of radioactivity, staining by colloidal coomassie, silver staining, or fluorescence detection, to name just the most popular. For methods that need better access to the proteins like antibody reactions, proteins can be eluted from gels by Western blotting onto membranes. This can also be used in proteomic approaches for better MS analyses (e.g. in the molecular scanner approach). In a typical 2D experiment, proteins are irreversibly fixed into the polyacrylamide gel and only peptides can be extracted by digesting the proteins in the gels. This process can be very efficient, although in most instances a big proportion of the peptides is lost on the way to the mass spectrometer during gel extraction and peptide clean-up. This is exactly were the biggest advantage of non-gel-based technology lies, as the peptides are transferred without losses into the mass spectrometer (see Chapter 3).

2D gels

While there are several forms of electrophoresis that can be used in proteomic approaches, we will mainly talk about SDS PAGE and high resolution 2D gel electrophoresis (Figure 2.3). Other gel-based separation technologies, such as coomassie brilliant blue (CBB) or Schegger gels, which can be used for the separation of protein complexes by replacing SDS with CBB and thus running them under native conditions, will only be mentioned briefly. There are several well established procedures for 2D gel electrophoresis. In proteomics the combination of IEF under denaturing conditions in the first dimension and SDS PAGE in the second dimension (high resolution 2D electrophoresis) is most widely used. Other 2D gel approaches use non-reducing SDS PAGE as first and reducing SDS PAGE as second dimension or other more specialized combinations.

High resolution 2D gel electrophoresis, a term established for the combination of IEF and SDS PAGE, was invented independently by O'Farrel and Klose. Its principles are still used by most researchers today, except that

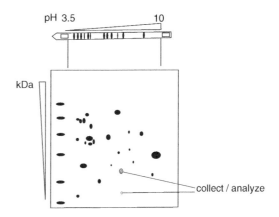

Figure 2.3 High resolution 2D gel electrophoresis overview. The size of most gels for proteomic applications is 20 × 20 cm. This allows the detection of about 2 000 spots from 100–200 μg of protein (up to 5 000 μg for preparative gels). Typically proteins from around 10^6–10^7 cells are loaded. Protein detection limits are well below 5 ng, but often limited by mixing/overloading with more abundant proteins. Under ideal conditions down to 1 000–10 000 proteins per cell can be detected (using radioactive detection methods), but analyses by MS are usually not as sensitive.

immobilized pH gradients (IPGs) for the first dimension have mainly replaced the mobile gradient formation of the original approaches (compare Figures 2.6 and 2.7).

The great success of high resolution 2D gel electrophoresis comes from the immense resolution power and relative robustness of the approach; more than 4 000 protein isoforms can be separated and displayed in a single experiment, migrating as clearly separated and roughly round 'spots'. These spots can be subsequently analyzed by a range of methods, including Western blot and MS analysis. High resolution 2D gel electrophoresis is highly reproducible and improvements over the years have made it accessible to everyone, not just specialists. Herein lies the reason for the bad reputation of irreproducibility that people sometimes connect with 2D gels: not everyone capable of running some 2D gels is used to high throughput systems, so experiments are performed slightly differently, and this results in different looking gels, apparently not reproducible.

The high resolution of the 2D gels is a result of using two orthogonal methods, which means that the separation in both directions depends on different, unconnected principles; the charge of a protein and its isoelectric point (pI) (separation principle for the first dimension) is totally

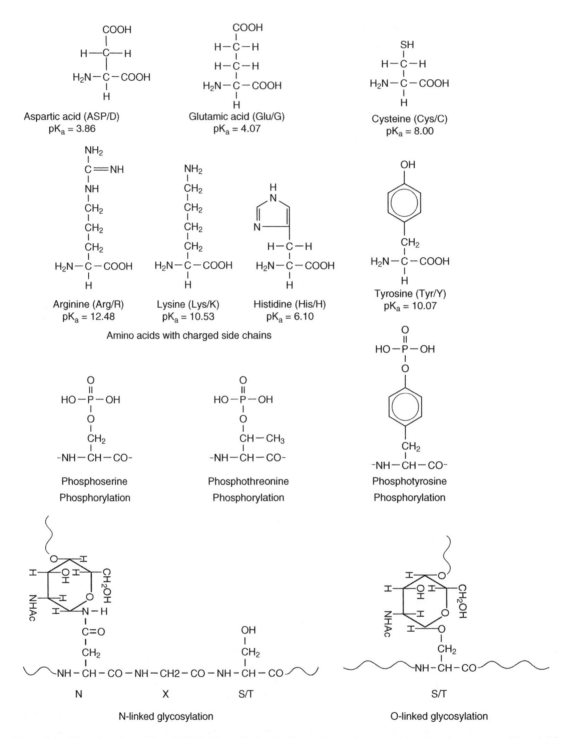

Figure 2.4 Charged amino acids and PTMs. The pK_a for the first and second proton of the phosphate group is 2.8 and 5.8, respectively. All depicted side chain modifications lower the pI (i.e. introduce negative charges); other charged modifications are sulfonation and deamidations. Oxidations can be encountered in vivo or can be artefacts occurring during sample preparation.

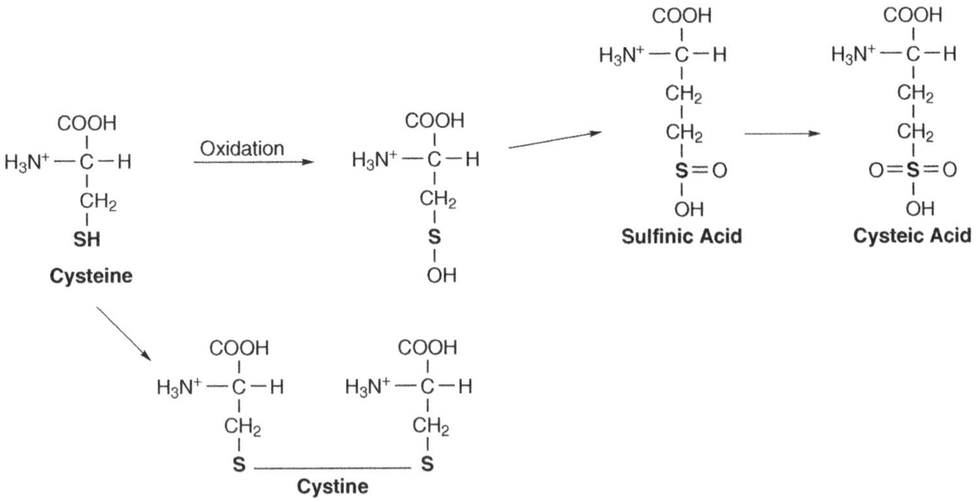

Figure 2.4 *(continued)*

independent of its (apparent) size (separation principle for the second dimension).

2.2 GEL-BASED SEPARATION

2.2.1 Separation by charge/isoelectric focusing

All proteins contain charged and potentially charged residues: the N-terminal amino group, the C-terminal carboxy group and a variety of charges introduced by amino acid side chains and PTMs.

The charged amino acid residues are shown in Figure 2.4 together with their pK_a values. For our purposes the pK_a can be regarded as the pH at which the residue carries no net charge and hence appears to be neutral.

PTMs that can introduce charges are also shown in Figure 2.4 with their pK_a. Please note that all the indicated pK_a values are approximations; since the pK_a is influenced by the structure and the neighbouring amino acids in a complete protein the 'real' pK_a can vary, typically within a range of about 0.4 pH units.

Groups with a pK_a smaller than 7 are called acidic, groups with a pK_a larger than 7 are called basic. The pH at the pK_a is also called the isoelectric point, from the Greek *isos* for neutral; at its pI a group is neutral and does not move in an electric field. If the pH is smaller than the pK_a the proton of the acidic/basic group will be predominantly dissociated, if the pH is higher than the pK_a the proton will be predominantly bound to the acidic/basic

group. At a neutral pH (pH = 7) it follows that acidic groups are negatively charged and basic groups are positively charged. For the whole of a protein the situation is similar; the charges of all the potentially charged groups are added up, and the pH at which the result is zero is the pI of the protein. At this point the protein might well be charged, it is just that it carries as many negative charges as positive charges; as a whole it is therefore neutral and will not migrate in an electric field. What happens if the pH of the surrounding medium is different from the pI of the protein? If it is lower, the protein as a whole is positively charged; this results from more acidic groups becoming neutralized and more basic groups becoming more positively charged. The further away the pH of the surrounding medium is from the pI of the protein, the stronger this positive charge will become. The story changes when the pH of the surrounding is higher than the pI of the protein; the protein becomes negatively charged overall (see Figure 2.5). How are all these basic principles used to separate proteins for proteomics? Well, the proteins show (theoretically predicted) pIs roughly in the range of pH 2.5–12, so ideally we would create a pH gradient within that range on a carrier medium, (say) a gel, and add our proteins. As soon as we apply an electric current (Figure 2.5) our proteins will migrate according to their charge; negatively charged proteins migrate towards the anode (+ pole), positively charged proteins towards the cathode (− pole). The closer the proteins get to their pI the less they are charged. When they reach their pI,

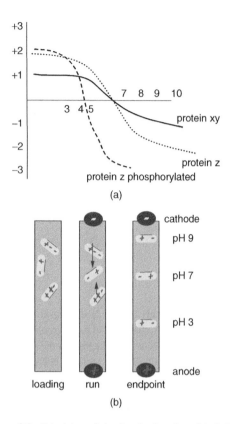

Figure 2.5 Principles of isoelectric focusing. (a) Relation between pH of the surrounding medium and charge state of proteins. Proteins xy and z have the same pI, but different charge densities. Phosphorylation of protein z adds negative charges and shifts the pI to a lower value. (b) To initiate an IEF run samples are loaded first. According to their pI and the position of loading, different proteins will have different charges. Once a voltage is applied they will migrate due to these charges. However, the charges change with the position in the gel; eventually they reach their pI, all charges are neutralized and the proteins stop migrating.

form. The same applies to all other charged modifications listed earlier in this chapter. This is one of the reasons why high resolution 2D electrophoresis is so popular and adds to its tremendous resolution power.

We mentioned before that we 'only' need to generate a pH gradient in a gel to allow for IEF. Just how we do this will be dealt with in the next two chapters, as there are two methods to generate this pH gradient; either the pH gradient is formed during the separation run of the proteins (mobile pH gradients) or it is cast and stable in the gel before the run (immobilized pH gradients).

Mobile pH gradients

In principle, this application is very simple: a mixture of different buffering substances (called ampholytes, from the Greek *amphoteroi* meaning 'both', in this case positive and negative), with different pK_a values is generated and added to the buffers needed to cast an SDS PAGE gel in a glass tube, about 3 mm in inner diameter and 20 cm long. When a current (see Figure 2.5) is applied across the gel rod, the substances will migrate according to their pK_a: the most acidic ones (charged negatively at anything but very low pH values) will migrate fastest towards the anode (+ pole) and overtake everything else on the way there, the slightly less acidic ones are next in line, in the middle of the rod we will have neutral substances with as many positive as negative charges at pH 7, at the cathode (− pole) will be the most basic substances (charged positively at anything but very high pH values), followed by less basic to neutral substances down to the middle of the rod. Once this pH gradient is established, our proteins will follow it and be separated according to their pIs (Figure 2.6).

What is the difference between these ampholytes and our proteins, when they both behave in very similar ways according to the above overview? Principally they are the same, the differences are in the details. First, it is possible to generate a molecule with, for example, a pI of 6 in many different ways. A protein with a molecular weight of 45 000 Da will be made up of about 410 amino acids. About 100 of these might contain potentially charged side chains, most of them will be charged very weakly at any pH between 2 and 10. So we have a molecule where at any given time no more than, say, the equivalent of four full charges pull a massive molecule of 45 000 Da (or one charge per about 11 000 Da), not a very impressive

they stop moving and stay there for hours and days. IEF even fights diffusion, a natural movement of molecules that would broaden the peak or spot of proteins in gels; as soon as the proteins diffuse away from their pI they get charged and are forced back to the position of their pI.

In IEF the proteins are separated due to their charges, which means that not only different proteins but also charge-isoforms of the same protein can be separated (see, for example, Figure 5.22); hence a phosphorylated variant of a protein can be separated form its unphosphorylated

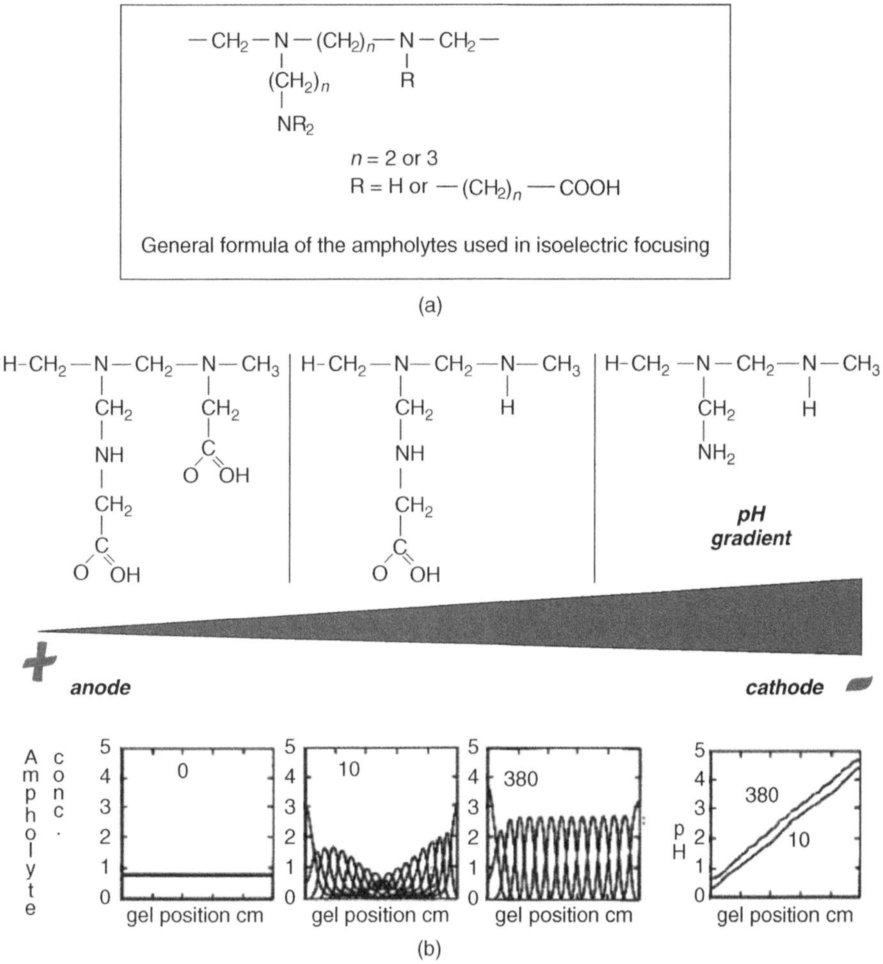

$$-CH_2-N-(CH_2)_n-N-CH_2-$$

(CH$_2$)$_n$ R

NR$_2$

n = 2 or 3

R = H or $-(CH_2)_n-COOH$

General formula of the ampholytes used in isoelectric focusing

(a)

H-CH$_2$—N—CH$_2$—N—CH$_3$ H-CH$_2$—N—CH$_2$—N—CH$_3$ H-CH$_2$—N—CH$_2$—N—CH$_3$

CH$_2$ CH$_2$ CH$_2$ H   CH$_2$ H

NH   C   NH   NH$_2$

CH$_2$ O OH   CH$_2$ **pH gradient**

C   C

O OH   O OH

anode   **cathode**

Ampholyte conc.

0	10	380	380 / 10

gel position cm   gel position cm   gel position cm   gel position cm

(b)

Figure 2.6 Mobile pH gradients: (a) shows the typical structure of ampholytes as they are used for IEF. According to substitution of R and the value of n, their pI is changed and they align themselves into a pH gradient during an IEF run. The positional variation for a simulated run containing only six different ampholytes is shown in (b). Note that in reality there are hundreds or thousands of different ampholytes. For the quality of the run it is very important that the pIs of all ampholytes (and thus their buffer capacities) are spread out evenly over the pH range of the separation. pH gradients are established shortly after the initiation of the run and stay stable for long enough to allow the proteins to follow on the pH gradient until they reach their pI.

pulling power. A typical ampholyte might have the same pI of 6, just like our protein, but it might only have a weight of 200 Da (Figure 2.6). It might contain only four chargeable groups, but these will allow it to be charged from −2 to +2, according to the surrounding pH, resulting in up to one charge per 100 Da, which means 110 times more force or pulling power across the gel for the ampholytes compared to the protein! Both, ampholytes and proteins, will act as buffers, both will have the best buffering capacity at their pI, the protein might be able to buffer four positive or negative ions away, the ampholytes only two. This is countered by a massive difference in the amount of molecules per experiment. Typically there will be about 1 pmol of the protein but about 1 μmol of the ampholyte: a thousand times more, that is an about 500 times higher buffer capacity for the ampholyte compared to the protein. There might be about 500 major different ampholyte species with maybe 300 different pI, but there will be typically thousands of different proteins with more than 1 000 different pI in the experiment.

Several conclusions can be drawn from the differences between proteins and ampholytes: the pH gradient that is ultimately formed depends on the sample that is separated. A mixture of proteins will most likely produce the same gradient as another (e.g. lysate from bacteria or human cells) as the charges/buffers within the sample are statistically spread. Pure or enriched proteins will distort the gradient. This can influence the position of proteins in the gel, or destroy the whole separation as it can create zones of uneven conductivity and under extreme circumstances a physical disruption of the gel during the run. As the separation progresses, the pH gradient might change slightly as degradation products of proteins influence the run. Also contamination or impurities in the gel/sample can influence the run, leading to the possibility of introducing run-to-run differences.

Immobilized pH gradients (IPG)

This technology was the breakthrough for the use of 2D gels by a wide community. While mobile pH gradient gels can be temperamental to run and sensitive to the nature of the sample or chemical contaminations, IPG gels are much more robust, reliable, versatile and stable. The IPG is fixed within the gel and is actually cast when the gel is cast. The pH gradient is formed by a mixture of a handful of Immobilines, which are covalently attached to modified acrylamide (see Figure 2.7).

By mixing the Immobilines a reproducible pH gradient can be achieved. This gradient can always be made exactly the same today as it can be made in 10 years from now, something not always given with ampholytes since they consist of hundreds of components and production batches are slightly different and manufacturing processes might change.

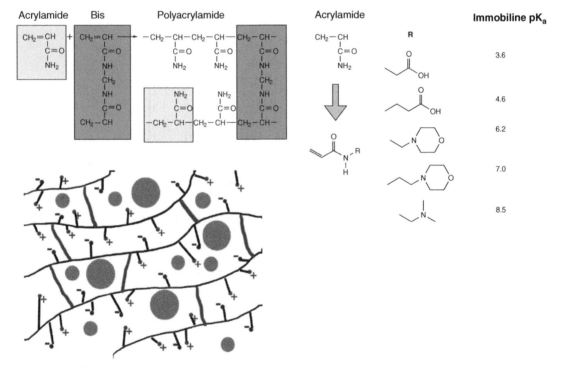

Figure 2.7   Immobilized pH gradient electrophoresis. Immobilines are used together with bisacrylamide to form a 3D mesh through which proteins have to be forced to migrate. One of the protons of the standard acrylamide is substituted by an Immobiline specific residue, giving it a specific pK_a. This is the negative logarithm to the base 10 of the pK, and thus can be injterpreted as the pH at which the specific Immobiline has its best buffer capacity. Additional Immobilines to those depicted are in use today, typically a total of 11. The Immobiline specific residues define the pH of the surrounding medium for the proteins during IEF. The Immobiline specific groups cannot migrate during an IEF run as they are polymerized into the gel.

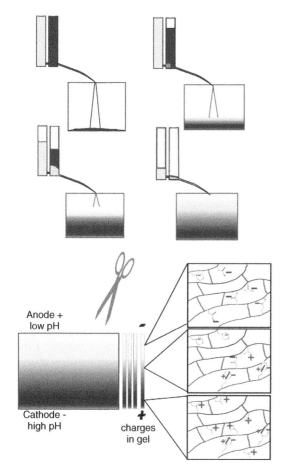

Figure 2.8 Casting IPG gels. See text for detailed explanation. The lower right panels show the charges immobilized in different regions of the IPG gel.

For IPG gels two or more mixtures of different Immobilines with acrylamide/bisacrylamide are combined to cast a gel (Figure 2.8). The pH gradient of the final gel spreads between the pH ranges of the pHs of either of the mixtures, while the pH maxima and minima result from the pure 'high pH' mixture or the pure 'low pH' mixture, respectively. Mixing of more than two different Immobiline solutions may deliver a more even or a stronger pH gradient or a 'non-linear' pH gradient with a flatter region for better separation. Gels with such gradients can be purchased ready made from different suppliers. In the lab usually only two different Immobiline mixtures are used, and tried and tested recipes are available which work perfectly well. For the casting of an IPG gel two modified

acrylamide mixtures with different Immobilines are used; a more acidic (light) and a more basic mixture (heavy) in separate cylinders that are connected (Figure 2.8). These solutions are mixed while pouring the gel.

The solution in the front cylinder (dark, basic) is poured first, and after a short time the cylinders are connected so that the solutions can mix while being poured into an assembly of a plastic support in between 2 glass plates. The front solution contains more glycerol, is therefore heavier and sinks to the bottom faster; a weight gradient forms that is also a pH gradient. The gel is poured to about 0.5 mm thickness onto a plastic support in between two glass plates. Towards the end of the casting process the solution in the front cylinder (heavy, basic) is exhausted. This means that the top of the gel is cast exclusively from the light acidic solution. The polyacrylamide/ Immobiline gel is allowed to polymerize onto a plastic support base and then washed several times to remove unpolymerized acrylamide and the glycerol. The gel is dried and ready to be cut in 3 mm wide strips, rehydrated and used for IEF. Samples are hydrated in the gel or loaded on top of it during IEF. Every point of the gel has a different pH with a corresponding buffer capacity, depending on the mixture of modified acrylamide.

Although this procedure seems complicated, immobilized pH gradient gels are very simple to handle. The runs are highly reproducible because after polymerization the gels are washed. This means all ions and unpolymerized components of the gels are removed. In mobile pH gradient gels the run starts shortly after polymerization and any impurities influence the gel run. While it is fair to say that mobile pH gradients are used for the biggest possible 2D gels (40 cm × 30 cm × 0.75 mm) with the highest resolution (up to 10 000 spots), these results are only achieved by experts. Immobilized gels in standard commercially available sizes of 20 cm × 0.3 cm can separate about 2 000 spot easily from complex samples. The handling of immobilized gels is very simple. The gels are cast on a plastic support and frozen away after washing and drying; you can literally tie a knot in such a gel and still use it for a run later. Mobile pH gradient gels are cast in glass tubes. If you treat these glass tubes incorrectly it is very difficult to press the gel out after the run and in any case the gel 'sausage' breaks very easily during manipulation and transfer on top of the second dimension gel. Drying of the spots on plastic supports also makes them very versatile: you can rehydrate them in any buffer you want for the run.

By changing the volume of the rehydration buffer the percentage of the gel can be varied from about 3% to about 1.5%. Mobile pH gradient gels have to be at least about 3% acrylamide/bisacrylamide, otherwise they would fall apart too easily during manipulations. Bigger proteins and proteins with weak charges run naturally better in lower percentage gels. One would assume that membrane proteins would benefit from the lower gel percentage of IPG gels; in reality they run better on mobile pH gradient gels. This might be due to the fact that in mobile gradient gels they are always surrounded by ampholytes which keep them mobile. In IPG gels a small amount of ampholytes is still used to allow better solubilization/migration of the proteins, but a well run mobile pH gradient gel has still the edge for membrane proteins and also for basic proteins (pI greater than pH 7). The latter is possibly due to the generally shorter run time of mobile pH gradient gels (typically about 8 hours compared to about 17 hours).

IEF is rarely used alone in proteomics; rather, it is used together with SDS PAGE as first dimension in high resolution 2D gel electrophoresis. Recently it is used very successfully as a pre-fractionation method in the form of OFFGEL electrophoresis. Different implementations of this method are possible, but usually a standard IPG strip is divided into 10–20 zones, with liquid reservoirs above each of the zones. Proteins, or more often peptides, are separated by gel-based IEF but accumulate in the liquid reservoirs on top of the gel zones in a very robust and reliable fashion and can be easily collected for e.g. typtic digestion or RP LC.

2.2.2 Separation by size using SDS PAGE

SDS PAGE is a very robust and reliable high resolution separation technology for proteins/peptides. In principle, proteins are denatured by SDS to disrupt any 3D structures (Figure 2.9) and thus any protein function. SDS has a hydrophobic tail which interacts very strongly with the peptide/protein backbone in a sequence independent constant ratio of about one SDS molecule to about two amino acids (1.4 : 1 on a weight/weight basis). This denatures the proteins and gives them a very strong negative charge, due to the hydrophilic sulfate group of SDS (Figure 2.9).

Moreover, the proteins thus acquire a constant charge per length, very important for their further separation. If we now put these proteins in a chemically inert mesh (polyacrylamide) and apply a constant force (electrical field) to pull them through this mesh we can separate

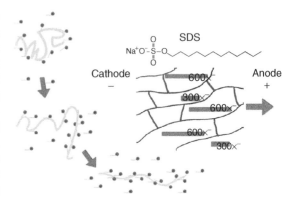

Figure 2.9   Basic principles of SDS PAGE. See text for detailed explanation.

them easily based on their size (number of amino acids); see Figure 2.9. Using marker proteins it is also easy to assign them apparent molecular weights, based on their migration relative to the marker proteins and to the dye 'front' of the gel. These apparent molecular weights are accurate to within about ±8% for most proteins. However, some proteins retain some short stretches of non-denatured amino acids or SDS cannot bind in the usual stoichiometry because they are phosphorylated. Each single phosphorylation might make a protein appear bigger ('shift' it) by about 2 kDa and multiple phosphorylation may cause it to 'smear' over a 10–20 kDa region of the gel. Some proteins migrate very aberrantly altogether, and there are proteins that migrate at 80 kDa, being only 20 kDa in size. More relevant for our application, different proteins are affected in different ways by variations in the composition of the mesh (gel): on a 10% gel a protein might appear slightly below a marker protein while on a 12% gel it might appear slightly above the same marker protein. This is quite common and affects the appearance of 2D gels. So it is important to bear in mind that gels only give us an apparent molecular weight.

Polyacrylamide gels have been introduced in this book before; for IEF as the first dimension in 2D gels they are used usually at about 3% of acrylamide (forming long chains) and the 'crosslinking' bisacrylamide, typically in a ration of 40 : 1 by weight. In SDS PAGE higher percentages are typically used, ranging from about 7% to about 20% (Figure 2.10).

These different percentages allow proteins/peptides of a different size range to be separated efficiently (see Figure 2.10). It is also possible to use gradient gels,

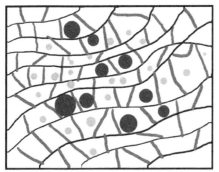 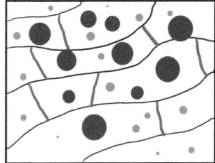

Gel acrylamide concentration	Linear separation range (kDa)	Useful separation range (kDa)
8%	35–92	32–200
10%	25–68	20–150
12.5%	19–55	15–100
15%	12–43	8–65
20%	8–20	6.5–25

Figure 2.10 Separation of protein in linear SDS PAGE gels depending on total acrylamide concentration (30% acrylamide, 0.8% bisacrylamide mixture). Note that values have to be treated with care and used only indicatively. Depending on gel size and running conditions especially proteins at the smaller end of the scale might migrate with the dye front. Special methods exist for peptides below 10 kDa, which allow separation of peptides down to 500 Da. Peptides smaller than 10 kDa migrate in a sequence dependent manner.

typically from 7% to 15%. These allow a wider range of proteins to be separated. For 7–15% the useful size range would be about 15–200 kDa. However, it is a lot more laborious to cast the gradient gels reproducible enough for 2D applications. We mentioned above how most proteins show slightly varying changes in migration behaviour at different polyacrylamide 'mesh' percentages. These changes in the migration patterns in 2D gels from one gel to another are enough to put many people off using the gradient gels. All possible problems in reproducibility of gel runs get enhanced in gradient gels (see below).

While the principles of SDS PAGE are straightforward there is a lot to be said about its diverse implementations in proteomics/protein analysis, though here we will just mention some of the most relevant aspects.

SDS PAGE in one-dimensional separations

SDS PAGE for 1D separations normally utilizes a stacking gel in combination with a separating gel. The separating gel is cast first and takes up about 85% of the whole gel length (or height, when run vertically). This gel determines the separation characteristics of the gel, that is, gradient or not and separation range determined by percentage acrylamide/bisacrylamide. Once polymerized, a stacking gel is cast on top of the separation gel. The pH and ion strength in the separation gel are higher than in the stacking gel and the stacking gel is of low percentage and usually polymerizes fast (10 minutes) as opposed to the 45 minutes desirable for the separation gel. The stacking gel contains pockets for sample loading, offering a sample volume of 10–60 μl to be loaded, depending in the size and thickness of the gel. SDS PAGE sample buffer usually contains TRIS, SDS, glycerol (to make the sample sink into the loading pocket), DTT/β-mercaptoethanol and bromophenol blue as a marker for the ion-front so we know when to stop the run and how to calculate the relative mobilities. The differences in ion strength and pH result in a focusing of the proteins at the border between the two gel types, allowing for higher loads and sharper bands. Mini-gels are loaded with 1–20 µg of proteins, while larger gels (20 cm long) can be loaded with 1–200 µg, depending on the sample composition and the goal of the analysis. The SDS covered (and hence highly negatively charged) proteins are forced through the gel by a electrical field, typically between 3 and 30 V/cm, resulting in running times from 90 minutes for mini-gels (6 cm long) to 7 hours or even overnight for longer gels (20–40 cm). For a fast run the current is usually limited (e.g. 50 µA per 20 cm gel), resulting in a low voltage at the beginning and a high voltage at the end of the run (e.g. 70–700 V for a 20 cm gel). These fast runs require efficient cooling of the gels, and for best results it is important to keep the field strength low at the beginning to allow the proteins to enter the separation gel properly. It is even possible to load larger volumes than the loading pocket can fit; up to 2.5 times the pocket volume can be loaded by letting the sample run slowly into the stacking gel, pausing the run as soon as the blue front reaches the bottom of the pocket and reloading. Once the bromophenol blue front has reached the end of the separation gel the run is stopped, so SDS PAGE is not an equilibrium method like IEF. Depending on the purpose of the separation the gel can now be stained or the proteins visualized in any number of ways (see next chapter) or the gel can be further processed for Western blotting.

SDS PAGE is a very powerful and robust separation method and is often combined with LC peptide separations in the GeLC approach. Here a complex sample is resolved by SDS PAGE, and the resulting gel lane is chopped into 10–20 pieces of equal protein complexity or gel volume. The proteins in each gel slice are digested by trypsin and separated by reverse phase liquid chromatography (RPLC) online coupled to a mass spectrometer via electrospray ionization (ESI). GeLC does not have the same problem of losing less abundant proteins that we encountered in 2D approaches, as the protein concentration in every step is high, due to the more abundant proteins present in each gel slice. Although proteins might be present in more than one slice (especially when gels are overloaded!) the resolution of SDS PAGE for proteins is much higher than that of LC methods.

SDS PAGE as second dimension in high resolution 2D gel electrophoresis

When SDS PAGE is used as second dimension for 2D gels, the proteins from the IEF gel have to be transferred

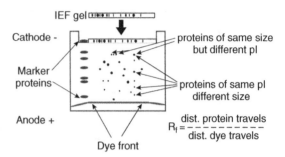

Figure 2.11 High resolution 2D gel electrophoresis. Proteins are fist separated according to their pI using IEF, and after equilibration to SDS containing buffers the IEF gel is placed on top of an SDS PAGE gel. Size markers can be used to determine the apparent molecular weight of 'spots' by comparing their Rf value to the Rf value of the marker proteins.

into the gel as efficiently as possible (Figure 2.11). For this purpose the buffers have to change from the 'urea world' of denaturation into the 'SDS world' of protein denaturation/solubilization. After the first dimension (IEF) the proteins are in either gel rods of 2–4 mm diameter or (more practical to handle) plastic-supported ultra thin gels (3 mm × 0.5 mm) of low percentage (3–4%). These are equilibrated into a TRIS buffer containing SDS and a slightly lower amount of urea (6 M). The urea concentration has to be lowered because SDS gels do not run very well at high urea concentrations. At the same time the proteins are usually reversibly reduced with DTT or DTE. Then this reversible reduction is made permanent by incubation in iodoacetamide. Although the pH of 6.8 (which is needed as the IEF gel doubles up as "stacking" gel for the second dimension) is not optimal for these reductions, cysteine residues are still transformed quantitatively into carbamidomethyl cysteine. This transformation is important to identify peptides that contain cysteine by mass spectrometry. Without this transformation, all of the cysteine residues would get oxidized to any number of higher oxidation states (up to 3× oxidized to cysteic acid or +48 Da, methionine is usually only oxidized once, +16 Da; see Figure 2.4), each giving the peptide a different mass and making its identification by mass spectrometry difficult (see Chapter 4). Now the first dimension gel is placed on top of the second dimension separation gel, acting as a stacking gel. It is very important at this stage that both gels make a good contact without any air bubbles between them. The quality of the separation gel surface (and hence its proper and complete polymerization) is also very important for the same reason. Ultra-thin IPG gels can easily be slid in between the glass plates used to cast the separation gel (e.g. 0.5 mm first dimension gel, 0.75–1 mm second dimension gel). Rod-type IEF gels are difficult to place on top of second dimension gels: they break easily and are typically thicker than the second dimension gel. This can be addressed by casting the second dimension gel right to the top of the glass plates and/or cutting the first dimension rod in half. With all types of first dimension gels it helps to cover the gels up with agarose and press out any air bubbles. The agarose (1%, low iso-osmotic, freshly made up in SDS running buffer) hardens and allows the gel to stay in place and helps to avoid any new air bubbles finding their way between the gels. The first dimension gel is equilibrated in SDS/TRIS buffer pH 6.8, while the second dimension gel has a pH of 8.8. Together with changed ionic strength, this creates similar conditions to discontinuous SDS PAGE, where the proteins get focused as they slow down at the border between stacking gel and separation gel. After the ion (blue) front has migrated well into the separation gel, the run may be interrupted to remove the first dimension gel, which can help to get cleaner looking gels and overall better results. Transition from first to second dimension gels is a critical step; the proteins have to transit from a 4% into a 7% or even 12.5% gel. If the gels are 'squeezed' together the electrical resistance may vary across the gel, leading to different efficiencies in protein transfer and smeared or disappearing spots. While all proteins will ultimately leave the IEF gel, they might not do so at the same time: they smear across the gel and fail to form spots and remain undetected. If the technique employed is not efficient this may lead to the preferential loss of 'problematic' proteins, which are either large, do not bind as much SDS as others or are very hydrophobic (or a combination thereof; integral membrane proteins).

Two-dimensional electrophoresis is still very popular, despite more recent developments in high throughput protein separation based on LC. This is due to its overall very good reproducibility, extremely high resolving power and relatively easy access ('cheap' instrumentation). This is all helped by the immense expertise that has been built up over the years and its huge versatility with a diverse set of analysis methods (see next chapter), including a wide range of protein stains, blotting/immunological detections and the possible combination with mass spectrometry. The position of a protein in a gel can be diagnostic of several PTMs (degradation, ubiquitination, glycosylation

and phosphorylation, to name but a few). This will be evaluated further in Section 5.4. Differential proteomics using 2D gel electrophoresis has been very successful for all these reasons. However, it also has major drawbacks:

- 2D gel electrophoresis does not work for all proteins;
- 2D gel electrophoresis is not easy to quantify absolutely (but easy to use if only relative quantifications are needed);
- linkage to mass spectrometry is not very efficient in terms of sensitivity;
- there are always high losses from sample application to accumulation of proteins in a 'spot', with a loss of 20–90% of the whole protein amount used being typical, so on average about 1 mg of protein has to be loaded to recover about 200 μg in all the spots on the gel.

In any case, 2D gels display a complex pattern of spots that needs to be analyzed (see Section 2.3.2). If there is not a very complex spot pattern (below about 200 spots), it is questionable whether it is really necessary to perform a 2D gel experiment; a simple SDS PAGE might give far better results with less protein loss along the way.

2.2.3 Other gel-based separation approaches

Schegger gel electrophoresis, or CBB electrophoresis (also called blue native), is in principle similar to SDS PAGE but with SDS replaced by coomassie brilliant blue (CBB), a dye that stains/binds predominantly to tyrosine. CBB is used to give the proteins a strong negative charge, in a reaction very similar to staining the proteins. Since CBB is not a detergent, the proteins remain essentially in their native states. The charge transfer is also not as efficient as with SDS: the forces upon the proteins during electrophoresis are much smaller. All this together makes Schegger gels the gel-based method of choice to analyze protein complexes, typically multi-protein components of up to millions of daltons. As a very practical touch, the proteins run stained through the gel and slices of a gel can be put on 'normal' SDS PAGE gels, thus showing at which size and in which complexes smaller proteins migrate. The method is not quite as robust as SDS PAGE. However, there are very reliable protocols that will deliver good results. Sensitivity is somewhat less than normal gels, and ideally you want to load several milligrams and not tens to hundreds of micrograms as in SDS PAGE. After the run proteins are visible due to the stain and are not fixed in the gel. The method is easily combined with MS analyses to identify components of the complexes.

Gel-based isotachophoresis (ITP) can be used for small molecules as well as peptides and proteins. The samples have to be salt free, and the ITP does not have to be stabilized by gel at all and can be performed in solution. The resolution is a lot lower than SDS PAGE, but higher sample concentrations can be loaded and a coupling to MS is possible.

Protein separation on gel-containing chips is (at the moment) purely analytical (proteins cannot be isolated) and follows similar principles as SDS PAGE. While the whole set-up is relatively expensive for a gel system, the single chips can be cheaper than gels, and samples are run within minutes. Sensitivity and resolution are comparable to 'normal' gels, however only a minute sample amount is needed for the actual separation and the rest gets lost during sample application.

2.3 VISUALIZATION AND ANALYSIS OF PROTEINS/PEPTIDES IN GELS

After gel-based separation, proteins are mostly kept within the gel and need to be fixed and visualized in the gel. Without fixation proteins would diffuse rather fast. Within an hour or so the spots are not only not as sharp as they were, and if the gel is not in between glass plates but in a liquid the proteins will diffuse right out of the gel! So most staining procedures incorporate protein fixation as a first step. This is usually achieved by acid and alcohols and is irreversible; basically the protein denatures in the presence of the acrylamide 'mesh', and the coils are not easily untangled (except for some very small peptides/proteins below 15 kDa, which are not fixed very well by standard procedures). Imagine the precipitation process of a fried egg or cheese: irreversible. This means that the complete proteins cannot be retrieved from the gel after fixation. Of course if the protein is digested into peptides remarkable recovery rates can be achieved (over 90%). Different stains have different advantages, disadvantages and sensitivities and will be explained in detail below.

Gels can be dried (ideally between cellophane sheets) and stored for years. If dried at room temperature the procedure does not influence analysis by mass spectrometry, except for faint spots where fresher is

always better. If gels are kept wet, they need to be fixed or the proteins will diffuse out of them and they will perish even if kept cool (fungi) if not kept in alcohol which needs regular refreshing (evaporation). Drying of gels removes the zinc stain and most fluorescent stains, so has to be avoided for these stains.

2.3.1 Visualization of proteins in gels

Coomassie stains

Coomassie is a dye originally developed to stain wool. It binds preferably (and in a reversible reaction) to tyrosine residues in proteins. The binding is linear over a wide range of protein concentrations (about 2.5 orders of magnitude) and is nearly independent of protein sequence (but not tyrosine content). This means that with a margin of error of about 50–150% one can deduce the absolute amount of protein from standard proteins, either in an assay after reading the optical density (OD) at 595 nm in solution or in gels by comparing the intensities of the bands/spots of interest with those from standard proteins of known concentration. However, using different standard proteins (e.g. IgG or BSA) one can get readings as much as 50–150% different! Coomassie stains are also quite specific to proteins (unlike, for example, silver stains!), which means they give 'clean' results for proteins even if RNA or DNA is present. However, ampholytes are stained perfectly, which is why IEF gels (carrier and IPG) need to be treated (washed) extensively before staining. This also applies to silver stains (see below).

There are two types of coomassie stain, the more conventional stain with solubilized CBB and the colloidal coomassie stain.

In the conventional stain the dye is dissolved in acidic water/alcohol and stains the proteins by binding within a fixed gel followed by a de-stain in a more dilute fixing solution. The longer the staining the higher the sensitivity, up to the limit of detection (see below). This staining takes usually 20–60 minutes and the de-staining 2–6 hours (with several changes of solution), depending on the amount of protein and the thickness of the gel. The staining solution can be reused several times and a lot of solution can be saved by putting any kind of sponge in the de-stainer, but only if the proteins are not meant to be analyzed by mass spectrometry, as reused liquids may induce unwanted chemical reactions and sponges contain polymers that interfere with mass spectrometry.

Staining can be done very rapidly (2 minutes) in the microwave, as can de-staining (10 minutes). Avoid over-heating the solution and do not attempt this for high sensitivity or mass spectrometry work due to the chemical modifications and degradations that can be induced!

Colloidal stains take a bit longer. In principle, the colloidal coomassie staining solution is a concentrated solution that needs to be diluted before use. The staining procedure can take up to two days for ultimate sensitivity. Solutions can be prepared from scratch, but affordable and reliable commercial products are available and the speed of the whole procedure can be increased significantly.

Depending on factors such as gel thickness and band/spot sharpness, coomassie stains can detect proteins from 10 to 50 ng upwards in gels, with the colloidal stains being the more sensitive ones.

One of the most positive aspects of the coomassie stains for proteomics is that the author knows no other gel stain that is more compatible with mass spectrometry, not even fluorescence stains. However, coomassie stains are not the most sensitive detection methods. As a rule of thumb, if you can detect a protein in any coomassie stain you should be able to identify it by MS or MS/MS. If you can see the band/spot easily on a photocopy of a scan, you should even be able to attempt to identify PTMs.

Coomassie stained gels can be stored wet for a while (weeks) and dry for years. If they are dried at room temperature and acids removed beforehand, they can be used after years for successful MS analyses of the more abundant spots.

While coomassie is regarded as a good quantitative stain, it must be said that, based on the chemistry of the stain, different proteins on a gel will show a differential staining behaviour (Figure 2.12), and saturation as well as intensity of the stain are amino acid sequence dependent. This is a consideration that applies to all gel-based stains.

Silver stains

There are several principles of silver stains, but they all rely on differences in 'electrochemical potential' to trigger the formation of molecular silver precipitates from a solution with silver ions. The process is highly sensitive and can occur spontaneously and is thus much harder to control than the aforementioned coomassie stains. Also silver staining involves a lot more washing and incubating steps, thus involving hands-on time of several hours as opposed to several minutes with

coomassie stains. However, the overall time can be a lot shorter than some coomassie protocols (3–6 hours). Moreover, silver staining is not specific to proteins (it is similar to the photographic process triggered by light). Once the precipitation of silver is initiated in the gel it can be difficult to control and the reaction can proceed unevenly throughout the gel. The sensitivity and intensity of the silver stain may also depend on contaminations within the gel (e.g. RNA and DNA or ampholytes). From all this it follows that silver staining would not be the first choice for quantitative measurements. When used as a quantitative method the range of protein amount per spot in which reliable results are obtained is rather small, about one and a half orders of magnitude. Furthermore, different proteins show different tendencies to stain, that is, the intensity and colour (membrane proteins can tend to be green) for each protein vary much more than, for example, for coomassie stains. Silver stains are not as well compatible with MS as other stains. However, there are some modifications and chemical conversions that address this situation, for example omitting formaldehyde (lowers sensitivity a bit) drying of gels at room temperature and solubilization of silver (de-staining) before extraction of the proteins for MS analysis.

All this suggests that we cannot recommend the use of silver staining, but the opposite is the case: it is highly recommended, as long as you are aware of the (potential) pitfalls. Silver stains are one of the most often used stains for 2D gels as well as all other gels. They are very sensitive: depending on gel thickness and spot/band size, as little as 0.1 ng of protein can be routinely detected. It is very cheap and there is no need to rely on expensive equipment. The use of good quality reagents is all that is necesssary. The stain is permanent and strong, so it gives good signals for cutting out spots/bands and lasts for years on dried gels. Most steps are easily automated and several gels can easily be stained in parallel, swimming on top of each other.

Apart from the more intensive chemical treatment and the silver in the gel another reason why it is harder to identify proteins from silver stained gels than from coomassie stained ones is that there is usually far less material; a faint silver stained spot may contain only 1 ng and even a well visible spot only 50 ng, while with coomassie by comparison one can expect to have 20 and 200 ng, in weak and reasonably strong spots, respectively. Also consider that 1D gels are often used

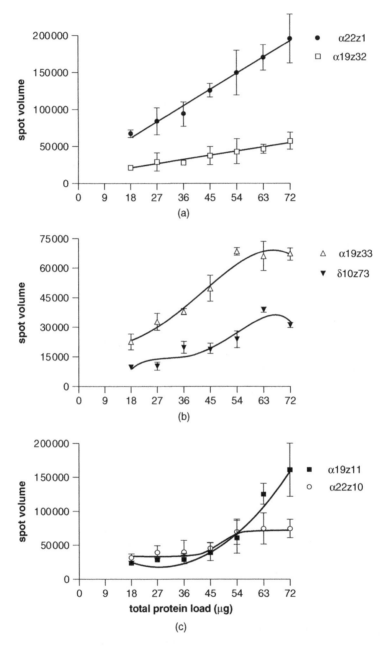

Figure 2.12 Linearity of coomassie stains. Identical lysates were loaded in different protein amounts on 2D gels. Images of coomassie stained 2D gels were digitized and spot volumes analyzed. Different proteins show different behaviour regarding linearity and stain intensity at saturation. Notice, for example, that α19z33, δ10z73 and α22z10 reach saturation intensity well below the intensity at which other spots still increase in stain linearly. Standard deviations from three repeat experiments are shown. Every protein thus needs its own standard curve, and an identical intensity in between two spots does not mean equal amounts are present. However, comparative analysis can be performed; it is just not clear (without an analysis like the one shown) where the linear range for each protein is. The same applies to calculations of how much total protein was loaded; if either BSA or IgG is used as a standard, the difference in total protein amount loaded can be 50–150%. Reproduced with permission from Consoli & Damerval (2001). © 2001 Wiley-VCH Verlag GmbH & Co. KGaA.

to estimate the stain sensitivity; if a protein is efficiently transfered during all 2D steps and migrates as one spot in a 2D gel (typical for highly soluble marker proteins), the sensitivity is higher in a 2D gel than in 1D gels, as the spot has a smaller area than the band.

Zinc stains

A stain that can be useful for some applications is the negative zinc stain. Its sensitivity is somewhat similar to the colloidal coomassie stain. However, proteins are not permanently fixed in the gel. The stain shows the background (empty gel) in foggy white, while proteins remain clear. The stain is only visible on a wet gel and is difficulty to scan or photograph. It is also more difficult to cut out weak spots under visual control; the silver stain is best in this respect. It is not easy to produce a uniform stain across the gel, a problem that can be overcome by using (expensive) kits for the stain.

The fact that proteins are not fixed in the gel has to be taken into account when cutting proteins out for tryptic digests and MS analysis: following standard procedures (for fixed proteins) a considerable amount of protein will be lost during washes of the cut-out gel piece in preparation for the digest. However, it is easy to fix the protein in the gel piece, wash and then proceed as usual. If you want to elute the complete protein from a gel, then this stain comes into its own.

Fluorescent protein stains

Fluorescent protein stains for gels became popular in the late 1990s. Initially they were very expensive, offered somewhat lower sensitivity than a good silver stain (despite companies' claims to the contrary) and were difficult to handle in the gel analysis process. This is due to the fact that the stain can only be seen with a relatively expensive reader, manual spot excision is not possible and spot cutters need to be able to either 'see' the stain or allow the alignment with landmarks and hope the gel does not 'warp' during the treatment. So when it comes to determining where exactly on the gel the spot of interest is (e.g. in order to cut it out) one can get a little lost: gels warp when handled and no screenshot or printout of a scan is a true and exact reproduction of a gel. A 0.5% error on a 20 cm gel can result in a 1 mm deviation, enough to miss your spot of interest and get a contaminating spot instead. There are a variety of stains available with different protein interaction characteristics

and different wavelengths for excitation and read-out. It is important to make sure that your reader is compatible with these characteristics, and that other characteristics are compatible with your procedures (e.g. some read-outs are not compatible with the auto-fluorescence of plastic-backed ready made gels, some fluorophores bind to glass, so you cannot use glass vessels which have advantages over plastic in MS compatibility). Fluorescent stains are sensitive, the stain itself is easy to perform and the linear signal range can be very large, much larger than that of the gels themselves or any other stain.

One of the most exciting developments in fluorescent stains (apart the reduction in price over recent years) is the development of 2D fluorescence difference gel electrophoresis (DIGE). Here two or three samples that are to be compared are stained with different fluorescent stains and separated as a mixture together on one gel. The fluorophores are covalently attached to the proteins. They change the proteins' charge state (making them more acidic) and there are different strategies to cope with this effect. On the positive side, once the labelling procedures are worked out for particular samples and strategies, less protein is needed for the analysis (but not for MS identification . . .) and most importantly, gel to gel variations are reduced drastically, so less technical repeats and gels in general of the same samples are needed, Also, a stoichiometric labelling can make more proteins accessible to 2D analysis, as proteins are notoriously difficult to separate reliably at pH higher than 7–8, and these proteins are shifted to pH 4–5 by the covalently attached stains.

There are now also fluorescent stains available that stain only glycosylated proteins on 2D gels, allowing a closer look at this very important and complex form of modification.

Radioactive detection strategies

In functional proteomics it can be very interesting to look at a subgroup of proteins, for instance newly synthesized proteins and proteins with PTMs. For this purpose radioactive in vivo labelling of the proteins is a very versatile tool. By pulse labelling with methionine or a mixture of methionine and cysteine (typically about 30 minutes to 2 hours) all newly synthesized proteins are labelled. When analysing the proteins resulting from such an experiment on 2D gels, all the radioactive 'images' of each separated protein spot can be compared to the total protein content of a sample which could be detected by any of the earlier mentioned methods, e.g. silver staining. Pulse-chase experiments allow the protein degradation half life of proteins to be estimated, an important, not strictly genetically encoded determinant of protein abundance. Phosphorylation of proteins can be observed very conveniently by in vivo metabolic labelling with ^{32}P- or ^{33}P-orthophosphate. Metabolic labelling with radioactive isotopes is also a very sensitive way to detect proteins. For example, in mammalian cells it is on average possible to detect 'spots' that are only present in about about 1 000 proteins per cell with radioactive labelled amino acids using a wide pH gradient (e.g. pH 4–7) and standard procedures (and proteins well behaved during 2D electrophoresis . . .). The sensitivity of metabolic labelling with ^{35}S amino acids is about three to five times higher than the protein detection by silver stain. If a protein is highly phosphorylated (or has a high turnover rate of phosphorylation) as little as about 200 proteins per cells can be detected in this way, due to the higher specificity of the label and the very strong signal of ^{32}P, but not ^{33}P.

Not only is metabolic labelling the most sensitive protein detection method, it can also be the detection method with the biggest dynamic range of linear protein detection in 2D gel electrophoresis. The linear range in which a weak radiation emitter like ^{14}C or ^{35}S can be detected is about 5 orders of magnitude, and therefore larger than the potential of 2D gels to present proteins in a linear fashion, which is about 4 orders of magnitude.

For special reasons (proteins of interest lack methionine/cysteine, or for a sequence-independent label, that is, for absolute quantification) it might be desirable to metabolically label proteins with ^{14}C of ^{3}H derivatives. These labels are very expensive and less sensitive than ^{35}S labels, since the weaker radiation of ^{14}C and especially ^{3}H is absorbed to a great extent in the gel, even if dried. Radioactive detection can be combined with non-radioactive stains, but in the case of coomassie and especially silver stain the radioactive signal can be weakened by the 'cold' stain; in the worst case weak signals from a low energy radiation source have to penetrate a layer of metallic silver.

While metabolic labelling has many advantages, it comes with some disadvantages as well: it is expensive, special labs are needed, cell incubators and the cells have to be manipulated before the experiment, which can be difficult (organs or whole animal studies) or prone to artefacts (starvation conditions to increase label uptake; see below).

A typical set-up for detection of proteins with radioisotopes would start from cultured cells. These cells are washed several times and then grown for a short period in a medium that lacks the compound used for labelling or any indirect source for this compound. Such starvation can last from 10 minutes to an hour, depending on the nature of the cells and the duration of the labelling. Without this step the efficiency of the labelling and the sensitivity are greatly reduced, since the labelled compound has to compete with unlabelled ones to be incorporated in the protein. Then the labelled compound is added for between 20 minutes and 24 hours, usually for about 2 hours. Cells are washed and 2D gel analysis is performed in the usual way. Regardless of stain, the gels are fixed and washed to remove any remains of label not fixed to proteins and to reduce the background. Radio labelled gels can be exposed wet, except for labelling with weak emitters like ^{14}C and ^{3}H, but the signal gets stronger and the bands/spots much sharper when the gel is dried. The radioactivity incorporated into the proteins can be detected using films or so-called phosphorimagers. The sensitivity of both methods is about the same. However, it is much more convenient and faster to use phosphorimagers. Using a phosphorimager the signal is already digital, which is important for the gel analysis. This also means it is difficult to get a precise print-out if one wants to cut out an otherwise invisible protein from a gel for MS analysis (compare also Section 'Fluorescent protein stains').

Phosphorimagers are about 3–10 times more sensitive when compared to the best films. The sensitivity of films can be boosted by several methods: infusing gels with 'fluorescing' substances (e.g. salicylic acid, but the bands get less sharp, useful for all but ^{32}P and problematic in combination with MS and protein stains!), using fluorescent screens in exposure boxes at $-80\,^{\circ}$C and pre-flashing films to make them more sensitive. Even with all these methods phosphorimagers are always faster and have a higher dynamic range of detection (about 4 orders of magnitude) compared to 3 orders of magnitude for films or 2 orders of magnitude for pre-flashed films. Exposure times can range from hours to two weeks for all but ^{3}H and ^{14}C, where they can reach up to six months; not a very practical approach in most cases. With films it is usual to take several exposures in order to deal with the low dynamic range of detection. It is important to get similar exposures for all experiments and also to take into account that some radio nuclides have short half lives. ^{32}P has a half life of 14.4 days, so the first two weeks of exposure will be the strongest.

Using differential shielding methods, it is also possible to detect two radioactive species at the same time: exposing a gel with two films and covering the gel with aluminium foil can deliver the ^{35}S signal and the ^{32}P signal differentially. It is also easy to overlay the signals from films: fluorescent stickers allow for perfect alignment. The same can be achieved with radioactive ink in the phosphorimager. Conveniently small paper strips (5 × 5 mm) are marked with ^{14}C ink which is dried under the cellophane sheet in the corners of the gel.

2.3.2 Analysis of 2D gel images

In most proteomic experiments using 2D gels the questions one wishes to answer are related to either absolute or relative quantification of proteins ('spots') or differential analysis between different samples (=gels). With typically about 2 000 spots per gel and a dozen or more gels to compare the task of image analysis is a much bigger one than actually performing the gel runs! An exhaustive analysis is usually not even attempted and only certain aspects of the experiment are analyzed – typically only a question such as 'which spots show a difference of more than 50% in intensity in at least three out of four experiments?' and not 'which spot show any statistically significant difference?' or 'which spots can be grouped together (clustered) based on different criteria?'.

It pays to invest at the beginning of a project in some test runs to get a feel for how much information can be extracted from the gel images one is able to get from a certain sample: the worse the gels look, the less information can be extracted and the more gels need to be run, again increasing the amount of gels needed per analysis. And any kind of image analysis on a 'bad' gel takes a lot longer than from a 'good' clean gel with high resolution.

Digitization of gel data

No matter how a gel has been stained or how the proteins have been detected, all gel images need to be digitized for further analyses. Sometimes the data are generated in digital form already (e.g. phosphorimager data, fluorescence). While this is straightforward in most cases, the digitization will determine how much information you can extract later from your data.

Phosphorimager data are usually saved in generic file types for compression. Transforming these into TIF files can lose a lot of the quality of the original data. Resolution is usually not affected but the dynamic range can be cut quite drastically. It is worth using the phosphorimager file format for the image analysis, if your program permits this. If not, why not ask the manufacturer if they can include the feature in their software? There is a good chance they might. Resolution is another important question. Resolution on a phosphorimager can be measured in 'pixel size' while other sources like scanners use dots per inch. Within a given analysis the resolution has to be the same, so if you compare phosphorimager data with silver stain data you might have to go through a image converting software (e.g. Photoshop) to get all images to the same resolution. In practice something like 100 μm pixel size and 150 dpi works usually fine even for weak spots. Increasing the resolution further just increases the file size and analysis time. It is advisable, however, to take the highest possible bit rate for your image depth (8–24 bits per pixel are used). The bit rate determines how many discrete values of intensity a single pixel can have. It is zero for a blank signal and 256 for black if the bit rate is 8, and 1 024 for a bit rate of 10 and 65 000 or 16 million for 16 and 24 bits, respectively. If you imagine that the background gives an intensity per pixel of about 10 than with a bit rate of 8 your strongest spot (and hence also the highest increase in spot intensity) is a factor of 50 stronger than your weakest spot. With 16 bits your (theoretical) dynamic range increases by a factor of 100 and with 24 bits by a factor of more than a million. The highest useful bit rate will depend on the settings of your phosphorimager (notice that when transforming files you might lose image depth) and the quality of your gels and scanning equipment for stains. For silver, coomassie and zinc stains you must have a high quality flatbed scanner with a wide useful OD range (2.5 is good) in transillumination mode. Even cheap and unsuitable scanners offer a high enough resolution – remember that 150 dpi is good enough and 300 dpi is more than enough. However, only quite expensive semi-professional and professional flatbed scanners offer a wide OD range of operation and allow you to make the best use of the dynamic range of your stain. As an alternative to scanners, some phosphorimagers may be ideal for scanning stained gels. For best results you should be able to put a wet gel on the scanner. As long as it is not dripping wet, there is no damage to be expected for the scanner. If you must, put the gel in a transparent plastic file, with the edges cut open to allow easing in and out of the gel. Remove air bubbles by gently pressing them out.

Image analysis

There is nowadays no excuse not to use computer assisted image analysis of 2D gel images as there are a wide range of programs available, for every budget, operating system and project size. While some programs are easy to use and deliver fast results for simple questions, others are more refined and have more capabilities with some even working automatically. You can imagine that these different capabilities and software philosophies are also reflected in a price range that spans nearly 2 orders of magnitude. Also think at the beginning of a project about whether the person (biochemist?) performing the experiments can also spend weeks in front of a PC doing a complete analysis of his work. Even with highly automated systems, the results are not better than those derived from good software relying on manual (computer assisted) analyses, and the hands-on time can be comparable as automated results still require manual evaluation.

How does software compare gels?

There are two different approaches in principle: comparing pixel by pixel and performing spot detection after the gel image comparison; and performing spot detection first and then comparing the spots and spot patterns. Both methods deliver good results, but the advantage of the first is that you do not even have to perform a spot detection to get decent, if not complete, results. Spot detection and exact quantification from 2D gel images are quite a complicated task and, depending on the software used, may involve a lot of user input, more so than image comparison.

Comparing gels pixel for pixel is a straightforward task for a human operator. Obvious spot patterns from two gels presented in different colours can be assigned easily (Figure 2.13) and manually overlaid, thus making them seed spots. In one of the two gels the areas surrounding the seed spots are than 'warped' to make them more similar. Warping can be done by several mechanisms (algorithms) and generally aims to improve an initial overlay into a more perfect overlay. In the initial overlay only the most obvious and often strongest spots are matched (typically actin, in the middle portion of the gel), while in later overlays some 50–100 seeds from all regions of

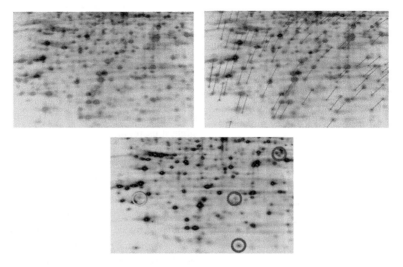

Figure 2.13   Image analysis of 2D gels. Digitized images of lysates from two different signalling conditions (Raf kinase on/off) in mammalian cells are compared in a software supported approach (Delta2D, Decodon). 'Raf on' is blue in the original and 'Raf off' is orange (dark and light shade of grey, respectively). Without spot detection, similar looking spots joined by the operator (lines in top right panel) and gel images are warped over each other. The resulting picture (bottom panel) shows identical spots in black and differential spots in blue or orange (indicated by dark grey and light grey circles, respectively). Note that this identifies differential spots without gel quantification or even spot recognition, with about 30 minutes of hands on time per gel pair.

the gels are matched. If the overlay appears like one gel with just some spots in slightly different colours, the first overlay phase of gel image comparison (20 minutes to 2 hours per gel pair if done manually) is ready and just needs controlling. Every seed spot gets a certain vector (the amount and direction it has been warped to match the other gel). If these vectors are like a cornfield in the wind they are consistent and most probably right. 'Wrong' or outstanding seeds are removed and it is checked that the seeds are distributed all over the gels to make sure every 'match' is reliable and supported by spot patterns in the vicinity. This last criterion is sometimes difficult to meet for the corners of the gels and can of course also be controlled by protein identification via MS. The only pitfall might be that it is relatively easy to see in MS if you are dealing with the same protein, but it is virtually impossible to ensure that two spots have the same PTM pattern, which of course is very important and will influence the migration pattern on a 2D gel for many PTMs.

Spot detection is the process in which the computer decides which pixels belong to the same spot. Spot detection is absolutely essential for a complete computer assisted comparison and quantification of gels. A pixel is (depending on your datafile resolution) about 0.1 mm in diameter, and a spot can range from about 1.5 to 5 mm, equivalent to several hundred to thousands of pixels. All software has an algorithm to detect which pixels belong to a certain spot. All these algorithms have in common that our eyes and brains can do a better job, but when comparing just two situations with four gels each, you would have to assign and keep the data for about 6 000–8 000 spots in your memory. When you think about more realistic and larger projects it is easy to see the advantages of letting the computer help you with the task. An example of a popular spot detection algorithm is the analogue of describing pixels as 'holes' in the gel image. The more intense a pixel the deeper the hole, and the more pixels next to each other show a high/similar intensity the wider the combined "pothole". Now imagine rolling a ball across your image. If the ball is only one or a few pixels in diameter it will fall in every scratch and noise of the image. If your ball is roughly the size of a spot it will roll over small artefacts and only fall into real spots and only those will be combined by the software and recognized as spots. Taking special care of oval forms of some spots you can develop a set of parameters to describe a spot: pixel

size (area), intensity of all pixels, intensity distribution within a spot, circularity, position of the centre and sharpness, for example. These data sets of all spots on a gel can then be compared to other gel data sets. Here the software can compare vectors (length and $x-y$ directions) and try to minimize them. Assuming most spots are unaltered, the software will try to minimize these vectors to overlay the gels.

A very different approach is to get away completely from spot detection for image comparison. The software compares each pixel's intensity and tries to find a vector group with minimal length and maximal similarity and 'warp' one of the images so it perfectly fits the other. However, for quantification pixels still have to be merged into spots afterwards. All software is helped immensely in the process if the human operator makes some initial alignment of obviously identical spots and/or crops the images so that identical spots are in near identical positions on the gels (Figure 2.13).

Most software offers fully automated algorithms with varying results. They all work well on the companies' example gel images, so the onus is on you to test the software on your own images (with sample overloads, streaks, etc.) before you spend a lot of money.

The experimental strategy has to take the above mentioned points into consideration. Also, it is clear that every result has to be reproduced several times to be valid and the more often the original experiments are repeated (and the better the gels!), the smaller are the changes in protein abundance that can be detected. At the same time more spots get excluded form the analysis, as they might not be detected in all experiments. Typically changes of $\pm 30\%$ in spot intensity and about ± 2 mm in the spot position in relation to other spots can be detected with a reasonable number of biological and technical repetitions of the experiment. Typically, if one wants to compare two situations (A and B) about six gel images from well separated gels should give very good results. This represents three independent experiments (biological repeats, see Figure 2.14; note that more technical repeats might be needed to get really well separated gels).

There are different possible strategies for comparing all the gels. The easiest strategy starts by comparing two gels of one experiment as pairs (one gel A and one gel B, run in parallel) and combining the data in a mastergel. This mastergel is a hypothetical gel, containing all the spots encountered in both gels with information as to

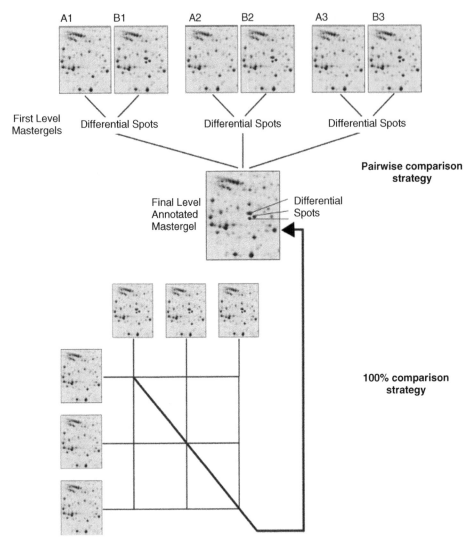

Figure 2.14   Image analysis strategy for gel-based proteomics. For detailed explanation of the pairwise comparison and the 100% comparison strategy, see text. The gel pairs (A1/B1), (A2/B2) and (A3/B3) are results form biological repeats comparing situation A with situation B.

which of them are differential. The same is done for the remaining four gels, so that three first level mastergels are generated. The information from these is then combined into a final mastergel, which now contains annotations for all spots from all comparisons (e.g. changes in two out of three experiments or in three out of three). Another approach is the pairwise comparison of all gels of situation A with all gels of situation B. Since all the data are stored and easily retrievable one can check all differences and combine them in a mastergel. This 100% comparison is surely the best, but also the most labour intensive comparison possible.

The following is a brief description of the kind of challenges and results you can expect from your gel comparison experiments. Some examples of real life results of qualitative (which spots/proteins are there at all?) and differential proteomics (also in comparison to other methods) are given in Chapter 5.

In a typical experiment with mammalian cell culture material, in which you aim to compare two situations, it is usually sufficient to carry out four independent experiments for a good quantitative analysis. It is always better to have more gels, but you soon reach the point of diminishing returns, whereby to improve your data a little you need to put in a lot of effort and more and more spots will not be observed on all of the gels. To obtain eight good gel images from two situations (four gels for each) it might be necessary to perform some more experiments: single gels might not run sufficiently well or the conditions of the experiments might not be reproduced sufficiently. On a typical gel there are 2 000 spots; this results in 2 000 potential indicators of slight changes in temperature, cell density, serum content or speed of sample preparation, so there are bound to be some differences in gels from identical conditions. As a rule of thumb, if a single gene/protein is changed between two cell lines about 20 spots will be noticeably different (i.e. intensity changes of more than 30–40% or changes in position greater than 4 mm). If the culture conditions change dramatically (e.g. stress/starvation) you can expect hundreds of spots to change. Similar samples should yield about 80% identical spot intensities or positions; this leaves room for up to 400 changed spots. Similar estimates can be made for bacteria, plant or fungi, perhaps with some 30% fewer spots for bacteria.

Different cell lines or yeast/bacterial strains can be difficult to compare because there are not many identical spots. Samples from the same cells but different sub-cellular compartments can also vary considerably and be impossible to compare directly. In this case it is still possible to do image (or sub-proteome) comparisons by comparing fractions of diminishing similarity and end up comparing fractions that are so different that they could not be compared directly. One typical example is to compare a whole cell extract to a nuclear extract and then the nuclear extract to an extract of nucleoli or the nuclear matrix. While this is relatively straightforward to do, it would be impossible to compare whole cell extracts with, for example, nucleoli in a reliable fashion. Of course to make sure one has done the right comparison it is very helpful to test whether identical spots were overlaid by MS analyses. After these comparisons one can transfer the spot information (position, intensity, any changes during treatment) from the nucleus/nucleolus and nucleus/nuclear matrix comparison to the whole cell gel image/mastergel. From the combined data one can estimate the purity of the fractions, the enrichment factors and, most importantly of all, whether a protein is found in a sub-fraction because of artefacts or for biological reasons. This decision is very important when it comes to refining lists of proteins that are contained in certain biological samples, such as mitochondria. Finding all the proteins present in a sample (and its artefacts) is one of the two main driving forces in proteomics; the other is to compare the protein composition of samples after differential treatment (e.g. plus or minus presence of a drug) or from very similar samples (normal cells versus cancer cells).

2.3.3 Analysis of proteins from gels

Every method that can be used for SDS PAGE gels can also be used for 2D gels, often with additional and better quality results, albeit with additional costs and workload. It is therefore advisable to reflect on whether it is really necessary to perform a 2D gel analysis, or whether a 1D approach can deliver the desired answers.

It is beyond the scope of this book to discuss in detail every possible way proteins can be analyzed after SDS PAGE, so we will focus first on the mass spectrometry based approaches that are likely to deliver answers for many applications in combination with 2D gels and then go on to the more traditional protein analysis methods that are likely to give a wider audience valuable answers within the context of proteomic projects.

Mass spectrometry

For a thorough introduction to the mass spectrometry of biomolecules, see Chapter 3. Here we will focus on some basic principles of sample preparation and what to expect of MS analyses in connection with 2D gels and, to a lesser extent, SDS PAGE gels.

The most successful method in the analysis of proteins from gel-based proteomics is the digestion of proteins to peptides within the gel followed by peptide analysis using MALDI-ToF (time of flight) MS (peptide mass fingerprint Figure 2.15). The main aim of these MS analyses is to generate data that can be compared with 'simulated' experiments in the computer; hundreds of thousands of sequences from DNA databases are translated into proteins, 'digested' by trypsin and the data analysis by MS is simulated. If the experimental data match the data set of a simulated experiment from a specific sequence,

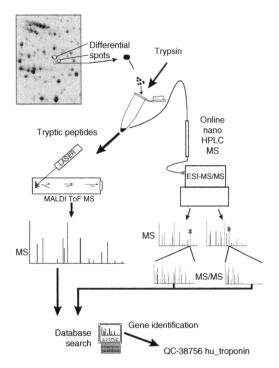

Figure 2.15   Principle of MS and MS/MS based identification of gel separated proteins. For detailed explanation see text.

the protein is said to be 'identified' in the database, or matched to a certain sequence by database searching. In another approach, which can generate more and better data, the tryptic peptides are resolved and concentrated by nano-HPLC (Figure 2.15). The separated peptides are eluted directly (online) into a mass spectrometer using ESI. Here MS spectra of each of the eluting peptides are generated, the mass spectrometer chooses certain peptides to undergo further fragmentation and the fragment masses are measured (MS/MS). Again the experiment is simulated with all known DNA sequence and matches to the real MS/MS data sets are identified.

Apart from these basic approaches there are a lot of others that can be used for refined/extended analyses (e.g. GeLC MS or search for PTMs; see Chapter 4).

For most MS based analyses the proteins of interest have to be digested into peptides with compatible size and composition to get meaningful results.

For this, the spots of interest, as identified by image analysis, have to be cut out of the gel. Often a preparative gel loaded with a 5–10 higher amount of protein is used,

so that a higher amount of the spot of interest is present, but these preparative gels can also result in more chemical contaminations and neighbouring protein spots contaminating the spot of interest. The samples intended for analysis should be cut right out of the middle of the spots, and the area cut out should be no larger than approximately 1.5 mm in diameter. This area from the middle of the spot contains most of the protein of interest and results in the least contaminations from other proteins and, very importantly, from the gel matrix; contaminations from the gel can be detrimental to further analysis. Keeping the gel thin (0.75–1 mm) further reduces these contaminations.

It is very important to stress that for the gel only very pure ingredients should be used and an optimal polymerization needs to be achieved. Most acrylamide solutions lose quality after four to six weeks. They are still perfectly usable for applications outside proteomics, but in proteomics the deterioration influences the apparent percentage of the gel, resulting in changed migration pattern, and the accumulation of unpolymerized acrylamide and bis-acrylamide. These can then attach to proteins and peptides, reduce the yield and change the mass of peptides so that they are no longer useful for database searches and in fact diminish the significance of 'hits'. Also, any contaminations from the gel matrix carried over to the mass spectrometers will severely diminish the ionization and need to be kept to a minimum.

The gels can be cut with a pointed scalpel or a dedicated sharpened hollow tube of a fixed diameter. For both tools the gel should be placed on a glass plate (a surface that is easy to clean) to avoid cross-contaminations.

If you are working on more than several dozen samples a day or even high throughput analyses, you might want to invest in an automated spot cutter or even a digestion robot. These machines have to be set up carefully if you want them to be as good as your hand-cutting – in particular, sensitivity is often lost with these machines.

Using high quality reagents, a tiny amount of the digest can be spotted directly on a MALDI target, which is then inserted in to a MALDI ToF MS and analyzed (see Figure 2.15). A typical digest volume for a single spot would be 20 µl, and 0.5 µl of this solution can be loaded on a target after 3 hours of digestion. Alternatively, digests can be carried out overnight before peptides are either cleaned up using micro-RP columns in pipette tips or by repeated freeze-drying (optimally under optical control down to 1 µl) and re-suspension in 5% acetonitrile. If you have a high quality sample in copious amounts of protein on the gel (picomoles or a CBB spot), all the above methods will deliver very similar results. Under less favourable conditions (weak silver stained spot, low femtomole amounts or contamination) you will see differences in the results depending on your preparation method: direct samples from the digest show many peptide peaks that might disappear with the other methods, predominantly large masses or hydrophobic peptides (see Section 2.1.3).

There are several preparation methods for samples on the MALDI target, the most popular of which is the dried droplet technique, because it works with a variety of matrices and is simple, fast and reliable. The matrix, usually alpha-cyano-4-hydroxycinnamic acid (CCA) for peptides and 2,5-dihydrobenzoic acid (DHB) or sinapinic acid for longer peptides and proteins (see Figure 3.12), is simply added to a small droplet on the target and mixed by pipetting it up and down; typically 0.5–1 µl of each solution is added on a target.

In principle, a spot cut out from a typical 2D gel (or, even worse, a band from an SDS PAGE gel) will never contain only one protein; it will usually be a mixture of several proteins. However, with a high resolution gel in most cases the protein responsible for the stain (and hence the detection of the spot) will be the most abundant and hence the easiest to identify. Given this situation and about 1 pmol of the protein of interest in a spot, MALDI ToF MS will result in a spectrum with some 5–40 peptides for the peptide mass fingerprint (PMF) analysis (Figure 2.15). This should give you a good chance of identifying the main protein and account for anything between 10% and 80% of the ions in the spectrum and between 10% and 60% of the theoretically expected peptides in the protein. For a more confident identification, further analyses with mass analyzers having MS/MS capabilities can be performed (see Chapter 4).

It is important to bear in mind that the identification of a certain protein in a spot means exactly this: this protein is present at this position in the gel. It does not prove that:

- this is the main protein of the spot;
- this protein is responsible for any differential staining;
- this protein is not also present somewhere else on the gel;
- this protein is not present in the spot as an artefact (e.g. as a smear from horizontal or vertical streaks).

If the same protein identified in a spot of interest can be identified from surrounding areas, in particular from strong, seemingly unrelated spots, it is very likely that its identification in the "spot of interest" is merely a contamination. If the identified protein is well characterized, Western blot is a very good idea to confirm the hypothesis that the identified protein is indeed at this position (and perhaps others, indicating changed PTMs) or is present in the samples in a changed amount.

It is also important to remember that MS is a very powerful tool in combination with database mining to identify a protein in a gel; just exactly which protein of a highly conserved family it really is can be more difficult to establish (see also Chapter 4).

Using SDS PAGE as a powerful pre-fractionation approach, GeLC can identify many proteins using RPLC MS/MS. Here a lane of an SDS PAGE gel is cut into 10–20 slices, each containing up to more than 100 proteins that can be identified by MS/MS. The proteins in the gel are digested similarly to single gel bands with the addition of chopping the gel into small pieces. However, MALDI ToF MS cannot cope with this mixture of peptide from so many proteins. Instead the peptides are separated, concentrated by RPLC and then injected via ESI into a mass analyzer capable of MS/MS analyses (Figure 2.16). MS/MS isolates single tryptic peptides and fragments them in the mass analyzer. The masses of the resulting fragments are amino acid sequence dependent. Together with the mass of the original tryptic peptide this data set can be searched against the simulated data sets from huge DNA databases with hundreds of thousands of entries. This can than identify which proteins were likely to have been present in the gel slice. This method can even detect quantitative information on proteins differential in gel slices from similar samples, even if a stain of the gel would not have been able to show any differences because of 'overcrowding' and lack of resolution. Even detailed information on PTMs of proteins in the original gel can be derived (Chapter 4).

Analysis of proteins from 2D gels by traditional methods

Proteins separated by 2D gels can be analyzed in principle by any traditional method more typically associated with 1D SDS PAGE. Usually these methods work even better due to less background and higher purification of the proteins of interest.

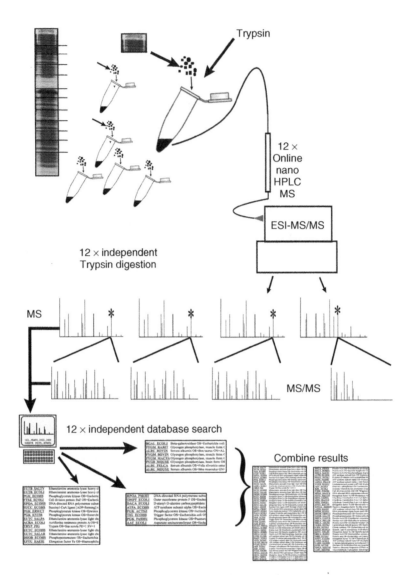

Figure 2.16 Principles of GeLC MS/MS analyses. SDS PAGE gels are cut into 10–20 (e.g. 12) slices with equal protein content rather than size, and characteristic proteins/bands are used as markers to allow reproducible divisions. Each slice can give rise to tens to hundreds of identified proteins. The combined data are curated (e.g. is a protein present in more than one slice? Is it a meaningful degradation or result from a PTM or just a 'smear'?) before quantification of MS data can be attempted (Chapter 4). Hundreds to thousands of proteins can be identified from a single gel.

Complete 2D gels (or sections thereof, to save on reagents like antibodies) can be blotted onto membranes for a variety of further analyses like Western blots. The result of Western blots from 2D gels are usually clearer than from 1D gels because there is less unspecific protein covering the spot of interest. So higher dilutions of the first antibody can be used. However, if a whole gel is blotted certain problems may arise. While small blots are conveniently blotted in a semi-dry way, the blotting efficiency for large gels (e.g. 20 × 20 cm) is not very good; a lot of heat is generated and a lot of current is needed. Better results (but with more cumbersome handling) can be achieved with wet-blotting tanks. These tanks allow up to three full sized gels to be blotted at once in about 3 hours. The amount of antibody needed for a Western blot of a complete 2D gel can be prohibitive (20 ml diluted antibody solution is needed, as opposed to about 5 ml for a small 1D gel). This is the reason why often a small portion of a 2D gel for Western blot is cut out. The rest of the gel can be stained to make sure the overlay does not cause any problems and proteins detected by the antibody can be correlated exactly (to 1 mm tolerance) with spots from stained 2D gels. This can be important if Western blots are used to corroborate findings from MS analyses. Often a protein has several isoforms and only one is detected as changed in an experiment involving mass spectrometry. The Western blot can show (but rarely quantify) all other isoforms on the 2D gel. Western blot is usually more sensitive and specific than a typical MS analysis of spots from a 2D gel (below 0.1 ng with a good antibody, compared to 5 ng or so for MS analyses), thus Western blot and MS analyses supplement each other very well. Western blots of 2D gels can be stripped of the first antibody and then reused to detect another antigen. This can be repeated several times (starting with the weakest antigen/antibody pair) and is very useful, since in combination with staining the unblotted proteins still in the gel, the exact position of several proteins/spots can unambiguously be determined from one experiment. The blot can also be stained using reversible stains like ponceau S (before blocking the membrane), and scanned in. This together with labelling the blot by pencil or fluorescent labels allows a very accurate overlay. There are disadvantages in only blotting part of a 2D gel: it is easy to cut out the wrong region since some proteins migrate several pH units away from their predicted site (especially after PTMs), and then it is easy to miss isoforms that might be relevant for the interpretation of results. These additional spots might not be the original spot of interest, but related to it and far away on the gel from its original position. It is important to remember at this point that spots are often composed of several proteins. The larger the gel and the narrower the pH range, the bigger the chance of only getting one protein per spot. However, even the identification of a protein from a particular spot by mass spectrometry and Western blot does not prove that it is this component of the spot that, for example, is responsible for the changes of intensity of this spot in differential proteomic approaches, and independent, MS based quantification may be needed for a strict proof.

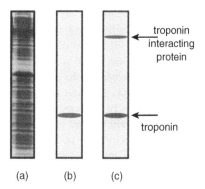

Figure 2.17 Principle of Far Western blot for protein/protein interactions. A sample containing, for example, complete cellular lysate (prey) is loaded on an SDS PAGE gel stained with coomassie (a) or blotted onto nitrocellulose membrane (b, c). The membrane in (b) is incubated with the antibody to troponin, which is detected in this classical Western blot at its correct size at 35 kDa. Blot (c) is given the Far Western blot treatment; proteins on the blot are fully denatured, then partially renatured and incubated with purified troponin (bait). After washing, the blot is tested for the presence of troponin. The original troponin band from the lysate appears, along with a weaker band where the purified troponin has interacted with a cellular protein on the blot. Variations of the Far Western blot include loading of purified proteins (e.g. different constructs to analyze the binding domain) or incubation with fluorescence labelled or radioactive bait.

In Far Western blots the proteins from 1D or 2D gels are blotted to a membrane and then incubated with other proteins to find interacting partners (Figure 2.17). These other proteins are typically well characterized and pure preparations are used. Their binding to specific partners is detected by probing the Western blot (now called Far Western) with an antibody for the soluble binding partner. Once the exact position of binding is characterized, the proteins (on identical run gels) can be digested on the membrane or from gels and identified by MS and database mining.

Proteins can also be analyzed by MS after 2D gel electrophoresis and blotting to membranes. In theory this should allow for very sensitive detection and highly efficient digestion since the proteins no longer have to be extracted from gels. Some labs prefer this method to in-gel digestion, but of course there are drawbacks. The blotting efficiency may vary (due to bubbles in the tank or blotting chamber), stains on membranes are not as sensitive as stains on gels, and blotting membranes can get saturated with proteins, so the quantification is better done in gels. Blotting membranes can dry and change their characteristics during analysis, and basically the more manipulation steps the higher the chance of artefacts. Peptides after the digest can rebind to membranes, which is why detergents are included in the digestion, which in turn can have negative effects on later steps. Interestingly, full length proteins can be analyzed from membranes using MALDI ToF MS with infrared lasers or by extracting the proteins by precipitation in acetone. All these methods work best with small proteins and rather large quantities, which explains why they never became mainstream technologies. It remains to be seen if they can be reliably integrated with top-down proteomics approaches, but the most recent publications show that massive parallel protein digestions can be very efficient on membranes, leading to enhanced detections of less abundant proteins.

An interesting approach is the development of the molecular scanner. In principle, a standard 2D gel is run and then blotted in a special way (see Figure 2.18). The blotting membrane contains immobilized trypsin, so that the full length proteins from the 2D gel are digested during the blot and the resulting peptides are blotted onto another membrane. Then the peptides are analyzed by MALDI ToF MS directly from the membrane. While technologically challenging, with recent developments in all areas necessary for success needed, it seems a

workable technology. However, a standard MALDI ToF MS would need about five days to scan the whole blot of a $20 \times 20\,\mathrm{cm}$ gel at $100\,\mu\mathrm{m}$ resolution, producing an immense data set, containing very important information but ultimately also much redundancy (e.g. from areas that contain too little peptides for a detection or 'empty' areas).

2.4 GEL-FREE SEPARATION TECHNOLOGIES

2.4.1 *Principles of HPLC*

Hyphenated technologies became a mainstay in proteomics from 2000 onwards. At that time it became possible to separate very complex peptide mixtures by 2D nano-HPLC coupled online to ESI (electrospray ionization) MS with considerable sensitivity. In this chapter we will explore the basic principles of this new breed of HPLC, with column diameters of $25\text{--}100\,\mu\mathrm{m}$.

It is very interesting to compare the flow rates of nano-HPLC to the rate of sample consumption in a standard offline nano-ESI source, which is about $50\,\mathrm{nl/min}$. At this flowrate, ESI is at its most sensitive; the efficiency of the formation of charged peptides is supported by starting off with small droplets, and they do not get much smaller than being sprayed from a $1\,\mu\mathrm{m}$ tip at $150\text{--}300\,\mathrm{nl/min}$ flow rate (see Chapter 3), which is typical of nano-HPLC.

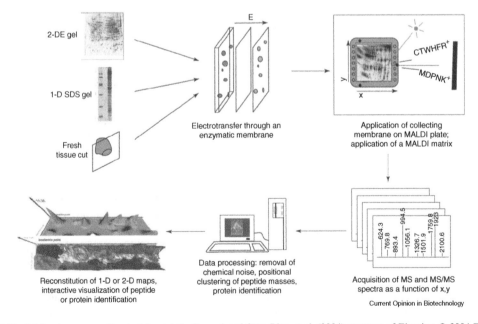

Figure 2.18 Molecular scanner. See text for details. Reproduced from Binz *et al.* (2004) courtesy of Elsevier. © 2004 Elsevier.

Standard HPLC systems operate at flow rates of several millilitres per minute; it is impossible to evaporate all the solvents efficiently into tiny droplets small enough for optimal ionization. So there are two main reasons why classical HPLC is not very sensitive in a proteomic setting; the proteins/peptides get diluted in columns that are far too large, and then most of the eluting peptides are not ionized but get blown away from the entrance of the mass spectrometer by some kind of 'turbo' spray installation. These settings for normal HPLC work well for many analytical tasks, because if you have micro- or even nanograms of a (small) substance it does not matter that you throw 98% of it away before mass spectrometry, since the latter is several hundred times more sensitive than typical HPLC detection by UV/visible light. By going down from typical column diameters of 4.6 mm in standard HPLC to 50 μm in nano-HPLC the sensitivity is increased by a factor of about 8 000. This is in large part due to sample concentration combined with separation; if a peak elutes over 30 seconds, in nano-HPLC this means its elution volume is typically 100 nl – in other words, some 200 fold higher concentrated than in a typical sample. Despite its advantages (like lower solvent consumption, high back-pressure in tubing after column and low flow inhibits problems with bubble formation) nano-HPLC has some specific challenges compared to standard HPLC, as we will see below, but it also solves many problems in proteomics and is now a central technique in sample separation.

Chromatography is a separation principle in which the analyte always has the choice of either binding a stationary phase or moving on with a solvent. The name 'chromatography' derives from early experiments, where the green pigments of plants were separated and developed into their own colours, the green being made up of yellow, red-brown and green pigments. Although the early experiments were done using paper as a stationary phase in upright tanks, the principles are still the same in HPLC (Figure 2.19).

The advantage of a tightly packed columns and narrow connective tubing in HPLC, which results in high back-pressure of the column, is a high speed separation. Apart from the obvious benefit of speed, this ensures that the analytes do not have much time to diffuse, resulting in sharp peaks and high resolution separations. An HPLC run is measured in minutes; paper chromatography takes hours. In all chromatography methods the stationary

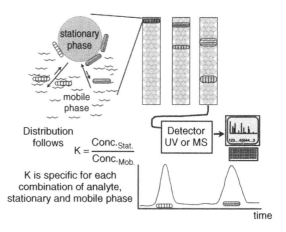

Figure 2.19 Principles of chromatography. Two analytes (e.g. two peptides) are in an equilibrium between stationary phase and mobile phase. In LC the stationary phase is the surface of small beads, inside columns. The mobile phase is a liquid and is changed during an LC run, thus shifting the equilibrium. Peptides elute as peaks form in the column, the longer the elution, the broader the peak. See text for further details.

phase is always surrounded by the mobile/liquid phase (in the very successful gas chromatography the mobile phase is a gas, the stationary phase is the liquid covering a long wound-up capillary). And in fact, there always has to be a clearly defined equilibrium between stationary and mobile phases, otherwise no clear separation is achieved, and the peaks 'smear'. So the longer the chromatography process takes the broader the 'peaks' of a separation will be. If the conditions during the run change (e.g. pH, density of material, or temperature, poorly mixed mobile phase) the equilibrium fluctuates and the peaks smear again. From this we can also see that chromatography is not an equilibrium method; we are exploiting volatile differences in affinities, so the column size, diameter and the running parameters in general have to be optimal and reproducible for consistent results of high quality.

In typical HPLC the high pressure is a result of trying to squeeze liquids through the tightly packed material that forms the stationary phase in the columns; the material is packed so tightly that solvents need to be pumped through the column at high pressure to achieve reasonable flow rates. The advantages of this tight packing are high resolution capacities and narrow peaks as well as high sample load capacities and reproducible separations (Figure 2.20).

IEC   RPC

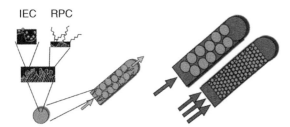

Figure 2.20 What goes on during HPLC. The performance of LC based separations depends on many parameters, critically on column packing. The tighter a column is packed, the higher its peak resolution capability. Tighter packing and long columns with small diameters result in huge back-pressure and thus HPLC systems are needed. The column material (beads) has to be chemically inert and stable at high pressures. The surfaces are coated with materials that determine the separation capabilities of the column; for example, charged surfaces for ion exchange chromatography (IEC) and hydrophobic surfaces for reversed phase chromatography (RPC). The smaller the beads, the higher the better the column performance, but the back-pressure gets higher as well.

It is also very important to ensure that structural folds of proteins or peptides do not influence their elution during LC. To achieve a reproducible elution it is essential that charges and hydrophobic residues are present on the outside of the proteins and peptides. Under native conditions peptides and proteins might associate which each other (in an artefactual way) as well as with the separation surfaces. Associations also influence how and which charges are present and accessible to the column material on the outside. Proteins pose more problems in this regard than peptides, as some proteins can show very strong 'unspecific' interactions with each other or even with the material which carries the interactive surfaces for LC. Thus both ion exchange and reversed phase chromatography (IEC and RPC) for proteomics are always used under denaturing conditions, where they have the strongest resolution power and concentrate peptides in sharp peaks eluting in reproducible time frames, independent of sample concentration.

Another important consideration for all chromatographic methods is to make sure there is no interaction between the stationary phase and the analyte. Even weak interaction would distort and smear analyte peaks. This is why all LC methods have higher resolutions for peptides than for proteins, where the weak interactions add up to strong ones. At the same time interactive groups needed

for interactions are spread all over a protein, so they cannot all make contact with the active groups on the columns at the same time. In the worst case the proteins are loaded on the columns never to be detected again, due to the summation of weak interactions with the column material. This can be because they actually do not elute from the column or in such a long outdrawn way that they are in the 'bleeding' background of the column. This effect, or better the absence of this effect from SDS gels, explains the huge popularity of gel-based systems; acrylamide does not retain proteins/peptides under denaturing conditions, but column material might. Hence it is much more popular to perform all separations under denaturing conditions; peaks form more easily, they are much sharper, recovery rates reach nearly 100% and the sensitivity is higher than under native conditions. Even under denaturing conditions a very important aspect of each chromatography is the wash step after the separation step: all non-specifically column–analyte interactions are removed, often by using harsh buffer conditions, and the column is clean for the next run. It is important to not mistreat the columns, as a sharp increase in, for example, unpolar solvents might bind proteins/peptides even more strongly to the column. If an analytical column used for a certain protein is then used for proteomics, some residue will bleed into the next run, depending on how 'roughly' the column was treated and on the concentration of the protein loaded. This memory effect may last for rest of the lifetime of the column. It may not be obvious during separations, but only after MS analysis and proteomic database searches. Columns are consumables and after 80–300 runs will have to be replaced. Factors that influence the lifespan of a column are particulate material that might clog up the column, denaturing or excessively swelling or shrinking the column material when buffer changes occur too rapidly. Nano-columns are somewhat more resilient to rough treatment than bigger columns, but letting them dry out will not do them any good.

In standard HPLC the connections and tubing before and after the columns do not contribute significantly to the back-pressure created. However, in nano-HPLC these parameters are slightly changed by capillary connectors/ tubes that create a significant amount of back-pressure because of their small inner diameters (10–20 μm). While this can add some 10–20 bar to the high pressures (typically up to several hundred bar) at low flow rates there are also advantages from this back-pressure in nano HPLC:

the back-pressure of the tubing after the columns is typically so high that there is rarely any problem with bubbles in the solvents after the columns. For proteomic applications there are different types of column materials, as will be discussed in the chapters on the different separation principles. However, they have in common that physically and chemically inert materials are used; these stationary phase material must withstand chemically the different solvents used in the separations, they must not swell or shrink strongly in these different solvents and they must not be compressed under high pressures or flow rates. The stationary material is usually based on silica, pressure resistant and chemically inert under typical HPLC conditions. Usually columns are filled with tiny spheres or beads of the stationary phase; the smaller the spheres the higher the quality of results because of the relatively larger surface area (larger capacity for separation) and more uniform packing (sharper peaks, higher resolution). Typically for nano-HPLC the beads are 3–10 µm in diameter, with column inner diameters from 50 to 100 µm for reversed phase column and 0.3 to several millimetres for ion exchange chromatography (IEC) and size exclusion chromatography (SEC). Column lengths are typically around 15 cm for reversed phase and from 1 to 15 cm for IEC and SEC. There are also porous beads; these allow higher flow rates and higher capacities because of their larger surface area and less resistance to the flow. However, they tend not to be so stable over time and show somewhat stronger changes of flow rate with solvent composition; so in proteomics, where the quality and reproducibility of the results are more important than the speed, they are not widespread in use, but as the quality constantly improves and demands on throughput increase they are surely being used more often. There are also so-called monocrystalline columns; here the inner surface of the column is coated with the separating material. The back-pressures are lower, flow rates are high, and run times very short, but again, where reproducibility separation and sensitivity for MS is paramount, they do not compete quite as well, partially because with short run/elution times the mass analyzers do not have the time for an extensive analysis (Chapter 3). Because of their low back-pressure they are used sometimes in applications after column runs, for example to digest proteins online with trypsin.

A complete HPLC set-up for typical proteomic applications is shown in Figure 2.21. A typical proteomic application is (one-dimensional) nano-HPLC reversed phase

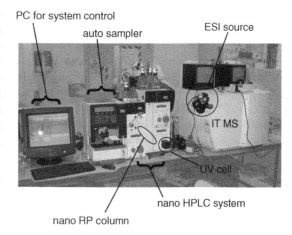

Figure 2.21 Components of a proteomic nano-HPLC set-up. For detailed explanations see text. The monitors on top of the IT (ion trap) mass spectrometer are for controlling the ESI spray needle position and function. Samples are stored in the cooled autosampler so that several runs can be performed unattended.

online coupled to ESI MS. We will first have a close look at this set-up and then switch to multidimensional nano-HPLC coupled to ESI MS or offline MALDI MS approaches further below.

A typical proteomic HPLC set-up is built around a computer to control all the components. Ideally the computer records/controls all run programs and parameters as well as internal pressures, and also collects the final data from the mass spectrometer (see below). Samples are usually first centrifuged to remove any particulate material and then loaded via an autosampler where several samples can be kept in special vials. The samples are injected in a programmed way into the sample loading valve. Modern set-ups can load amounts as low as 1 µl reproducibly, but a more typical value is 20 µl. Switching the sample loading valve puts the sample under the control of a system of high pressure pumps (at least two, up to four). These pumps deliver the sample onto a precolumn (to avoid the main column being clogged up with any particles) and, after another valve, the sample goes at a controlled flow rate onto the separation column(s). The mobile phase is a mixture of different solvents or buffers of different composition. This composition varies during the run and the pumps have to achieve a perfect mixture at very low flow rates (typically hundreds of nanolitres per minute flow over the column). From the final separation column (typically a reversed phase column) the sample passes through a high sensitivity capillary UV cell and than straight into the ESI interface of a mass spectrometer, typically capable of MS/MS analyses.

Now let us have a look at the individual components of the system. In older systems there may be two computers controlling the whole process, one for the HPLC and one for the mass spectrometer. Ideally the manufacturers of the mass spectrometer and HPLC system have collaborated to allow the whole system to be integrated into one software package on one computer. This makes changing the methods, finding faults and generally running the system on a daily basis much easier. The computer has to control which sample is loaded at what time. Usually you would run the sample with the lowest concentration first and inject the sample with the highest protein concentration last in a series of automated runs during a day or overnight. Often there is not enough sample for more than one run, in which case it might be desirable to run these samples first and while a human operator is around, just in case something goes wrong (e.g. a needle clogged up or connector leaking) so that the sample can still be saved or measured.

The computer controls the autosampler, a cooled device for storing samples before the runs, the needle that takes samples from the autosampler (repeatedly if needed) and the valves (typically at least two) of the HPLC system. It also controls the pumps of the HPLC and, if applicable, the flow splitter. With such a system, different methods can be programmed and run, the parameters being the amount injected into a sample loop, transported from there to one of several possible columns, where the eluates of these columns go, and so on. The composition of the solvents used at any given time is also controlled. Basically the sample is controlled all the way from the autosampler to the detector(s). Flow rates and pressure targets are also programmed in special method files. These often come pre-programmed and can be adjusted to specific needs, arising from different column materials and sizes. The computer also controls the flow splitter if applicable. Most systems cannot generate the low flow rates and high pressures needed for proteomic applications (typically below 300 nl/min and up to 180 bar) while guaranteeing proper mixture of the solvents. The way round this is to use a flow splitter: the pumping and mixing of the liquids occur at a much larger volume, then the flow is split, one part going into a long capillary (typically 1 m in length and 10–20 μm inner diameter) and from there into the waste, the rest going onto the system

and ultimately through the column. By changing the flow capillary (or any other parameter) one can exert a level of control over the flow going onto the column. Of course during a run the flow and pressure are determined mainly by the pumping speed of the pump and the back-pressure of the system: on one side the flow splitter capillary and on the other the system of columns where the sample goes. The system is set up so that most of the pumped liquid (90%) goes to the waste via the flow splitter capillary. There are only a few reliable flow splitter systems on the market, one problem being that the back-pressure of the system of capillaries and columns does not stay constant all the time; as the solvents change so does the viscosity of the pumped liquids and thus the back-pressure from the column. In addition, the material in the columns changes in different solvents (ACN/water/alcohols) so that it is difficult to achieve a constant flow rate in a flow splitter system during an HPLC run. Huge differences in flow rate can result in poor sensitivities or complete loss of ionization due to failure of the ESI process.

The sample and the solvents are delivered via fused silica capillaries. They are chosen for their inert chemistry and superior behaviour under high pressures; fused silica capillaries change their volume only slightly under pressure, are flexible enough to make variable connections and can be produced with inner diameters from 5 to 200 μm. The capillaries used for nano-HPLC applications typically have an inner diameter of 20 μm; much smaller and they cause too much back-pressure, much larger and the gradients or separated analytes mix inside the capillaries and destroy any attempt at separation. The capillaries are connected by various means, and these connectors are quite critical. If finger-tight fittings (with threads) are used they can create a large dead volume if not connected correctly. A dead volume of, for example, 1 μl would equate, at 300 nl/min flow rate, to a run time of 3 min on the column. In other words, if there were such a dead volume somewhere in the system during elution, the mobile phase gradient could be significantly disturbed. The elution of a (presumably mixed) 1 μl dead volume typically represents 10% of the whole useful RP HPLC separation volume (if the HPLC run is 90 minutes in total, the phase in which peptides elute is only 30–40 minutes), and would thus ruin the separation completely.

Another aspect controlled by the computer is the UV detector. While a standard UV detector is not sensitive enough for proteomic applications, it will show the mixing of the liquids at the beginning of a run

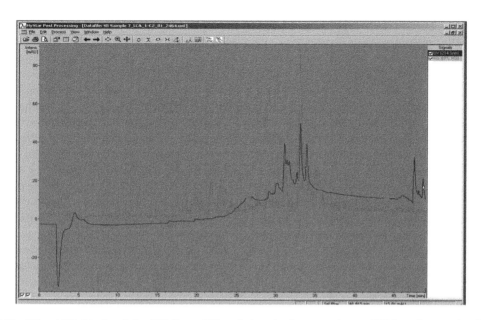

Figure 2.22 UV and MS detection during RPLC runs. This typical result of a proteomic nano-RPLC run from a gel purified spot of low abundance shows no signal for the peptides in the UV lane (214 nm, black line). The UV lane shows the start of the run (negative absorbency) and the 'dirt' eluting at the end of the gradient. The peptides are visible in the total ion current (TIC) lane of the MS (grey line). The TIC lane shows how several peptides co-elute and their summation signal is visible. Using MS/MS the mass spectrometer can still analyze single peptides. The acetonitrile (MeCN or ACN) gradient is indicated as percentage of buffer B. The true ACN content is slightly different as it results from mixing buffer A (5% ACN) with buffer B (90% ACN). Acetonitrile is also known as cyanomethane, ethanenitrile, ethyl nitrile, methanecarbonitrile and methyl cyanide.

and the contaminations apparent at the end of the run (Figure 2.22). Very sensitive UV detectors are now available, which do not interfere with the sample; the sample only has to pass a slightly longer path through capillaries to be measured in the UV spectrum (see Figure 2.21). These systems might even show some of the peptide peaks of gel isolated proteins (particularly when they 'ride' on one another), and they will show signals if complex samples are loaded (GeLC 2D HPLC), but not for all peptides for which MS data is generated. Even if the peptides are not always seen in the UV lane, having a sensitive UV detector can be very useful in running the system on a daily basis, in order to diagnose malfunctions of the system as quickly as possible. Additional interference free detectors can be luminometers, coating the capillaries inside with scintillating wax, used as radioactivity detectors, predominantly aimed at 'hard' emitters like ^{32}P.

Of course the main detector in a proteomic nano-HPLC system will be an ESI coupled mass spectrometer. The capillary of fused silica from the HPLC needs to be connected to a nano-capillary with a defined spray tip,

usually a capillary melted and pulled to produce a tip with a diameter of around 1 μm. This tip is often coated with a metal to allow an electrical connection between the spray needle and the ESI source. The shape of the ESI source is dependent on the manufacturer, but will almost always include some control as to where to place the tip and some optical controls to make sure you can see that you have placed the spray in the most efficient and stable position by a combination of optical control and control of the ions produced (see Figures 3.15 and 3.16). The mass spectrometer itself can be a relatively simple MS instrument but will most often be an instrument capable of MS/MS. Mass spectrometers typically coupled to nano-HPLCs for proteomics include ToFs, triple quadrupoles, quadrupole/ToF hybrid instruments, Pauli ion traps, linear traps, Fourier transform ion cyclotron resonance or orbitrap mass spectrometers (Section 3.3). A very successful and efficient mass spectrometer for nano HPLC in proteomics is the Pauli ion trap.

The computer needs to control the settings of the mass spectrometer, such as the measurement start time, the type of measurements made (e.g. for the ion trap, scan at medium resolution, perform MS/MS on the three strongest peaks, go back to scanning, ignore the previously scanned ions for future MS/MS, perform MS/MS on the next three strongest peaks, and so on) and when to stop the measurements. All these settings control events online, that is, during the run itself. Once the HPLC run is finished the data are stored, all the instrument settings and diagnostic informations are filed away together and the next run can start.

If the nano-HPLC is used to feed an offline MALDI MS, after the UV or ion detector the computer might have to control a spotting device. This can also be done manually, and the spotting device just spots pre-defined volumes, together with a matrix solution or on top of a pre-existing coating of matrix, on a moving MALDI target plate. This then has to be taken and measured in a MALDI ToF MS once the HPLC run is finished. As an added complication, it is possible to split the flow after the UV filter, to have part of the sample going onto a MALDI target or an autosampler for offline activities, while the rest of the material goes on to be measured by an online ESI MS. While in principle attractive, this post-column split can be technically demanding since in the low pressure environment after the column small changes in back-pressure (clogged ESI needle or matrix build-up on the spotter) can easily change the flow split. Flow problems in the low pressure part after the column in general are an example where a combination of UV and ESI MS detector is very useful; it allows a fast diagnosis as to what is wrong with the system when, for example, the MS signal is not as strong/good as expected. If (i) the UV signal is OK, the flow over the column is unaffected, thus the system up to the UV cell is working, but there might be a fault with the spray or the MS. If (ii) the UV signal is gone as well, the fault may be with the pump, buffers, columns or (most likely) with one of the connectors. It is inevitable that such a complex system will fail from time to time, but run times from a couple of days to a week or two without major disturbances should be achieved. Thereafter it should be a matter of an hour (or half a day at most) to get it up and running again (e.g. changing an ESI needle, tightening a connector or changing a pre-column).

Often a second computer with another set of software is now needed to extract useful data from the experiment offline, that is, after the run. The multiple peaks are calculated back to their real mass (singly-, doubly-

and triply-charged), the MS/MS data are joined for the different ions (e.g. MS/MS data from a doubly- and triply-charged ion of the same peptide are pulled together) and a new type of data file is generated (a text file) which can now be used for a database search to identify the proteins that were present in the run. Another objective might be to identify modifications of a peptide/protein that is known to be in the mixture or to compare several runs and perform quantifications of peptides labelled by stable isotopes (e.g. stable isotope labelling with amino acids in cell culture (SILAC) or the isobaric tags for relative and absolute quantification (iTRAQ) reagent). We will look at these steps of the analysis in Chapters 3–5 in more detail.

In a real set-up, there will be more complications, such as complex switching ports to allow switching between 2D and 1D HPLC, different columns (shorter/longer diameters, depending on separation requirements) or precolumns for sample concentration and filtration of any debris remaining after sample pre-clean-up and to protect the separation columns, which can last hundreds of runs.

Column ovens are often included, to ensure constant running conditions, as the viscosity of the medium changes with temperature. Increasing the column temperature to 42 °C, for example, decreases run times, sharpens peaks and reduces the overall pressure dramatically.

Reverse phase chromatography

In RPLC the stationary phase is hydrophobic while the mobile phase is more hydrophilic at first. To elute the analytes the hydrophobicity of the mobile phase is increased and the equilibrium for the peptides is gradually shifted from the stationary to the mobile phase. Because in most traditional chromatographic methods (before the 1970s and before RP became as popular as it is today!), the separation took place along hydrophilic surfaces (see Section 'Ion exchange chromatography'), this 'new' chromatography was termed 'reverse phase' chromatography. Similar to SEC and IEC, the stationary phase is mostly composed of inert beads. However in RP, the surface of these beads is covered with aliphatic chains (hence the phrase 'reverse phase').

On different columns and for different applications the length of the aliphatic chains, the working end of the RP, is variable. Denoted by C2–C18, the aliphatic chains are 2–18 CH_2 groups long (see Figure 2.23) and interact with

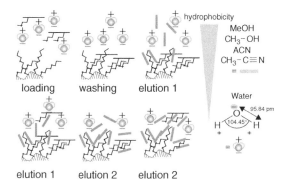

Figure 2.23 Reverse phase chromatography. Upon loading, peptides bind with their hydrophobic sidechains to the hydrophobic residues on the surface of the RP beads (in proteomic applications typically C8–C18). Unbound contaminations (salts!) are washed off and the amount of ACN in the elution buffer is increased. Peptides with low overall hydrophobicity elute under these conditions, while peptides with more hydrophobic sidechains stay bound, but migrate slowly down the column. As the amount of ACN in the elution buffer is increased even peptides with numerous or strongly hydrophobic sidechains elute from the column. See also Figure 2.19 and text for further details.

hydrophobic protein/peptide stretches. C8 and especially C18 are the most popular materials in general applications as well as in proteomics. Interestingly the same amino acids that provide charges for IEC are also the ones that provide a significant amount of hydrophobic side chains for RC chromatography: arginine and lysine. Since trypsin is used most of the time for digestion of proteins prior to separation, every peptide contains at least one arginine or lysine. This creates some minor problems when using RPLC as the second dimension in 2D HPLC after IEC (see Figure 2.27).

RPLC is by far the most popular chromatographic method in general applications and proteomics alike. It is the most powerful chromatographic separation technology for proteins and peptides in terms of robustness, capacity, dynamic range, resolution and reproducibility. A variation of RPLC is hydrophobic interaction chromatography (HIC). For our purposes it is simplest to assume that HIC is very similar to RPLC, except that the hydrophobic surfaces are a little less hydrophobic and differently shaped than the long RPLC aliphatic chains; they are either C2–C8 chains or a variety of aromatic rings (aryl groups). Not only is the hydrophobicity of HIC stationary phases lower than in RPLC material, the

density of hydrophobic groups is also significantly (about 10 times) lower, so the overall binding is much weaker as well as the binding capacity in terms of amount of analyte. The strength of binding to the column is lowered even further in HIC by using salt in binding/elution buffers (see below), so overall the interactions in HIC are of very low intensity compared to RPLC. Hence, HIC is best suited for larger peptides and complete proteins taking full advantage of their subtle differences in hydrophobicity, while RPLC has a much higher resolution and binding capacity and works best on peptides. In fact, proteins can bind so strongly to RP columns that they are impossible to elute quantitatively and bleed out of the column long after the original run. RPLC is always run under denaturing conditions and needs polar solvents to elute the analytes, most commonly methanol and the ubiquitously used ACN. These solvents have to be perfectly mixable with water but less polar than water, and, as can be seen in Figures 2.23, methanol and ACN fit the bill nicely. In RPLC and HIC the driving force for analytes to bind to the column is the exclusion of the polar solvent, either water or water and hydrated ions (mainly salts). The energetically unfavourable contact between hydrophobic and hydrophilic molecules/surfaces is minimized if the analyte binds to the stationary phase. For our purposes this simplified view is enough to explain why salt concentrations and hydrophobic solvents can combine to elute analytes; the more ions there are in an eluent the less hydrophobic the eluent has to be to interrupt a given analyte from the column. This is because there is less free water available to hydrate the column surface and the analyte. So the driving force behind the analyte binding to the column is smaller, hence the binding appears to be weaker, as it is more easily interrupted. In RPLC salt is seldom used to elute analytes; the interaction between the highly hydrophobic peptides and the reversed phase is disrupted by a prolonged elution at high ACN percentages (Figures 2.22–2.24).

RPLC is widely used in proteomics on its own or in combination with other separation methods; for example, it is the second/last dimension in 2D HPLC or other multidimensional separations.

A typical application for RPLC is the separation of tryptic fragments of a protein spot from a 2D gel or a band from a 1D SDS PAGE gel in order to feed them via ESI into a mass spectrometer, for example an ion-trap, triple quadrupole or quadrupole–ToF MS. In this case

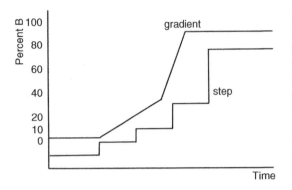

Figure 2.24 Elution profiles in liquid chromatography. See text for details.

the column is typically about 75 μm in inner diameter, 15 cm long and packed with C18 'material' with a bead size of 3–10 μm. A typical run would last about 60 minutes, with about 25 minutes during which peptides elute. The column is loaded with about 1–20 μl of the sample in a buffer containing an acid, typically 0.5% formic acid and 5–10% ACN. Formic acid is volatile, and therefore much more compatible with ESI (see also comments on ammonium acetate in Section 'Ion exchange chromatography'). The standard acid in RPLC (not connected to ESI sources) is trifluoroacetic acid, which interferes strongly with the ionization of analytes in ESI and tends to 'catch' most of the ions itself, but delivers superior RPLC resolution. One disadvantage of formic acid is its instability; solutions should be freshly made up from pure formic acid to ensure that it has not significantly evaporated. The acidic environment (pH about 2.5) and the presence of ACN ensure the denaturation of peptides and proteins as well as strong positive ionization of peptides in ESI. However, small, stable proteins such as trypsin stay enzymatically active even in 10% ACN (but not at pH 2.5). The interaction of analytes with RP material is strong, but still in equilibrium, so analytes can be eluted isocratically, that is, without changing the buffer and with the loading buffer running. However, sample peptides loaded at 10% ACN and eluted at this concentration would elute after a long time (depending on the column dimension) in very broad peaks. Hence, it is best to elute by the addition of more hydrophobic solvents. A step or a gradient elution can be chosen (Figure 2.24). For complex samples with dozen or more peptides, a gradient offers the best resolution with very good reproducibility. To achieve the

highest possible resolution, runs are typically composed of two phases: a shallow gradient up to about 20% ACN and a steeper gradient up to about 60% ACN. Gradients can also be expressed (more typically) in 'percent B'. For an RPLC run two pumps are needed, designated A and B. Buffer A contains the loading buffer and buffer B typically contains 80% ACN. So 60% ACN represents about 72% B (with 10% ACN in A). Peptides are typically eluted in 15–50% ACN, depending on the column dimensions and the flow rate (Figure 2.22). The more hydrophobic peptides will elute later in the run. Amino acids that make peptides more hydrophobic are shown in Table 2.5. Interestingly, the same amino acids that make peptides strongly charged for IEC/SCX are also the ones contributing significantly to their hydrophobicity. We will look at the consequences in Section 2.4.2 on the combination of hyphenated technologies.

Table 2.5 Physicochemical parameters of amino acids in peptides relevant to LC. Next to the physicochemical parameters shown there are also others – size, surface area, steric hindrance and hydrogen bonding – that are important. There are different ways of calculating or measuring these parameters and they do deliver very different results. The results shown in the table are most relevant to the situation of having these amino acids in a peptide during IEC or RPLC on C18 material and elution with ACN. Molecular weight also plays a role in RP, acting in the same direction as increased hydrophobicity. -- means low values, xxxxxxx means high values.

Amino acid	Hydrophobicity	Polarity
Gly (G)	x	--
Ala (A)	--	--
Val (V)	xxxx	x
Leu/Ile (L/I)	xxxxxx/xxxxxxxx	x
Ser (S)	xxx	xx
Thr (T)	xxx	xx
Phe (F)	xxxxx	x
Arg (R)	xxxx	xxxxxxx
Cys (C)	xxxxxxxx	xx
Met (M)	x	xx
Tyr (Y)	--	xx
Trp (W)	xxxxx	xxx
His (H)	--	xxxxxxx
Asn (N)	xxxxx	xxx
Asp (D)	xx	xxxxxxx
Gln (Q)	--	xxx
Glu (E)	xx	xxxxxxx
Lys (K)	xxxx	xxxxxxx
Pro (P)	xxxxxx	xxx

After the separation phase of the run there has to a regeneration phase; this is typically an increase to 100% B and a step to keep it there for several minutes followed by a steep descent to 5–10% B and an equilibration of the column to these conditions. Since these are also the loading conditions the column is now ready for the next run. If the column is not used for a while it should be kept in storage buffer (usually containing methanol). For a typical proteomic set-up it is best to keep the column running whenever possible. The nano-HPLC technology can sometimes cause problems with deposit build-up or tiny leaks. When the system is stopped for a while there may be problems in starting it up again: ranging from the HPLC pumps, going to different connectors and up to the ESI needle tip. A low flow setting over the weekend is better than switching the system off completely!

Ion exchange chromatography

IEC is typically used in two proteomic applications: as first dimension in shotgun proteomic approaches like MudPIT, also called direct analysis of large protein complexes (DALPC), or in a variation of this approach to perform something very similar to IEF on a column (chromatofocusing). While IEC is fully compatible with MS analyses if it is not the last step in any separation combination, it cannot be linked directly to ESI MS or offline MALDI MS easily. This is because ions that have to be used to elute analytes from the columns would interfere with ESI and to a lesser extent also with MALDI. Furthermore, IEC does not have the highest resolution of all chromatographic methods – this title goes to RPLC. Therefore, it makes a lot of sense to use IEC for pre-fractionation in proteomic experiments, followed by another method, preferably RPLC, which in turn is coupled online to MS analyses.

IEC can be used on native or denatured protein as well as on peptides. The basic principle is that the stationary phase consists of charges immobilized to a chemically and physically inert material inside the column, on the surface of small beads. Before a run the column is conditioned and loaded with the right charge (positive or negative) and the sample is loaded under conditions that induce a strong charge on the analytes (Figure 2.25). Typically the pH is low during sample loading and only weak buffers and low salt concentrations are used in order to make the analyte bind strongly to the column, as any salt competes

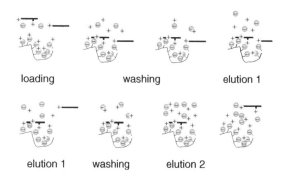

loading   washing   elution 1

elution 1   washing   elution 2

Figure 2.25   Strong cation exchange (SCX) as an example of ion exchange chromatography (IEC). Before using SCX the column has to be equilibrated under low salt conditions at a low pH (about pH 2.5). The flow-through of the loading/washing step is the first useful fraction. Although all peptides contain positively charged groups under these conditions, not all will bind due to sterical hindrance. Washing removes uncharged contaminants (e.g. polymers), which will bind to other LC columns, like RP. To elute proteins the salt amount is increased, typically in a step elution profile. Gradients can also be used, but the washing step is not possible and samples have to be collected offline for 2D LC applications. Peptides with weak or few positive charges elute first. Peptides with stronger or more positive charges elute at higher salt concentrations. SCX columns are usually shorter and have a larger diameter than RP columns.

with the analyte in binding to the stationary phase. Nevertheless, the flow-through is typically the first fraction of IEC, since some analytes will not bind no matter what the buffer conditions (uncharged aliphatic proteins/peptides). Changing the pH or increasing the ion strength of the solvent will elute the bound material from the column in a controlled manner.

In proteomics strong cation exchangers (SCX) are typically used as the stationary phase in the column. In SCX the stationary phase contains a strong negative charge. The peptides/proteins are charged by lowering the pH to about pH 2 using formic acid in low salt conditions and then loaded onto the column. Analytes are released from the column either by a gradient or a step elution (Figure 2.24). Step elutions of SCX columns are typical of proteomic applications, and each fraction (starting with the flow-through) is directed onto a reversed phase column to perform 2D chromatography (for more details see Section 2.4.2).

Typical steps for elution from an SCX column are NaCl concentrations of 10, 25, 50, 100, 250, 400 and 1 000 mM. Another salt frequently used for SCX step gradients is

KCl which, similar in its usage to NaCl, can be used at somewhat lower concentrations of 10–500 mM. Depending on the application, the salt steps can be made narrower or wider, resulting in more or fewer fractions to analyze. However, there is a limit to the fractionation ability of IEC, and too many steps will result in considerable overlap of analytes in between the neighbouring fractions, thus leading to an unwanted increased complexity, redundancy and dilution of signals. The maximal possible resolution is influenced by many factors, among them the nature of the analyte and the buffer (e.g. pH, ionic strength) it is presented in, the dimensions and shape of the SCX column, the quality of the material in the column, the flow rate and the ions used for elution. While in principle any salt will work to elute proteins/peptides, different salts will do so with different efficiencies and qualities. Ammonium acetate, for example, is typically used at concentrations of 25–500 mM. It has the additional advantage that it can be used for 'one-column' 2D HPLC approaches. Because ammonium acetate is volatile it can be injected into the ESI needle without clogging it up too much and residual ammonium acetate will not interfere with peptide detection, unlike NaCl, which clogs up the ESI needle in small amounts and interferes with the ionization of the sample.

While SCX is the standard in most proteomics application, strong anion exchangers can also be used. In fact, in practical terms it pays to use both in parallel to get the best coverage of proteins/peptides to be separated, as the most sensitively detected peptides with the two methods are complementary. Note that the many silica based bead materials can get chemically unstable at a pH higher than 7.

IEC columns can also be used to separate analytes according to their pI (chromatofocusing). This technology is sometimes used as an additional dimension to enable one to separate complex mixtures better and thus to analyze less abundant species. In anionic chromatofocusing the proteins/peptides are loaded under high pH conditions, charging all analytes negatively, so that they bind to the anion exchange column. Titrating a lower pH buffer onto the column for elution, the pH is slightly different across the column over time; at first the top part of the column has a lower pH, and after a while into the elution the bottom part of the column is the only part that still has a high pH. As soon as the pH on any part of the column reaches the pI of a certain protein/peptide, it loses all its charge and elutes from this part of the column to a part

further below, until it is eventually eluted, with the basic analytes coming first, followed by more and more acids ones. In practice, it is difficult to match the buffer capacities of column, sample and buffer and to make sure that all charges of the analytes are in contact with the column material. The separation also has to be performed under low salt conditions, as salts interfere with the pI dependent interactions. Although this approach has its practical limitations, the method does deliver important information about the analyte, its pI, which can be influenced by PTMs and thus be very informative, and can be used for example also for database searches.

Capillary electrophoresis

Capillary electrophoresis (CE) is not one of the major separation technologies used in proteomics today. However, there are several well established uses for CE in protein separation, the most important for proteomics being the use of CE in chip based approaches. CE is based on a variety of separation principles, using the analyte's size, charge and hydrophobicity for separation. The driving force for sample movement in a homogeneous buffer system is a very high voltage, and minute amounts of sample are loaded (up to 1 µl). CE is used heavily in nucleotide sequencing approaches, and the biggest drawback in practical terms for proteomics was for years the lack of stable connection to ESI. These problems have since been solved, and CE is faced with having to compete with the well established RPLC. Advantages of CE include minute sample requirements, low flow rates (high sensitivities), with stable buffers during elution (again aiding extreme sensitivity) and robustness against salts and common contaminations from biological samples (lipids, nucleotides).

Size exclusion chromatography

Size exclusion chromatography (SEC) has recently found its way into proteomic applications so it will be briefly discussed here. It does not have high resolution on its own, but is used in combination with other separation principles to increase the resolving power of a entire approach (see Section 2.4.2), typically by separating whole proteins prior to other methods.

The principle is very simple. A column is filled with spheres or beads of the same size which contain pores (Figure 2.26). When a sample is pumped through such a column the largest proteins cannot enter the pores and

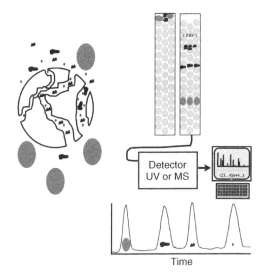

Detector
UV or MS

Time

Figure 2.26 Principles of size exclusion chromatography. SEC columns have to be long to allow for good resolution. While parameters such as dead volume (i.e. column volume outside the beads) and the quality/homogeneity of column packing influence the performance of every type of LC, in SEC they are absolutely critical. See text for further details.

are pumped only through the small volume in between the beads. They will elute first from the column. Very small proteins are able to penetrate into the pores. The total volume accessible to these small proteins is much larger than the volume accessible to the large proteins. At a given constant flow rate the small proteins will thus elute later. Any proteins of intermediate size might be able to access the larger pores or the beginning of the pores but will be unable to go deep into the beads; they will therefore be able to occupy an intermediate volume and, given the constant flow rate, will elute in intermediate times or elution volumes.

For proteomic applications SEC will always be performed in HPLC, but not necessarily in nano-HPLC, as eluting proteins will be used for further separations and the volume is not really an issue for this. SEC is ideal for changing the buffer of a sample, since small molecules will have a much later elution time than protein/peptides. While not as complete in removing salts/buffers from samples as dialysis, SEC certainly can be very fast in de-salting a sample with minimal sample loss.

In SEC, as for any other separation method for peptides and proteins, it is very important that all the analytes are in a defined state – either native or denatured. Separation

of proteins under native conditions using SEC for proteomics can be very useful: for example, to determine whether proteins are in homodimeric or other complexes. In the case of SEC, this would result in separation due to the increase in size of the dimers. However, it is nearly impossible to generate conditions under which all proteins are native (and soluble), and to preserve their association stages or even their PTMs. Under native conditions all enzymes, including proteases, phosphatases, kinases and glycosylases, should be active by definition. They will change the state of many proteins, which in turn will affect the proteins' structure, isoelectric point, size in solution and association stage. Inhibitor cocktails might go a long way to keep the situation stable for a given enzyme/molecule in a typical biochemistry experiment. However, in proteomics there are usually several thousand test proteins, and many of these will change PTMs.

Moreover, every association has a different rate constant – that is to say, every association is subject to constant dissociation and reassociation. Under stable conditions an equilibrium might establish itself for many proteins, but during many separations there are no stable conditions. So while the complexes partly decompose they cannot reform because their components are being separated. While this leads to a smear in the results when you look at one protein at a time, in a typical biochemical experiment under proteomic conditions you end up with difficult to reproduce and often confusing or meaningless and misleading results. This is the main reason why for most proteomic experiments aiming at quantitative or qualitative protein/PTM analyses denaturing conditions are used and special experiments are set up for association studies (see Section 5.5).

2.4.2 Combination of hyphenated technologies

All separation technologies that are connected by tubing are called hyphenated, and this term traditionally includes RPLC, IEC and SEC. These separation technologies are often combined to increase the separation power in proteomics experiments. As discussed earlier, a relatively complex starting material, a cell or tissue, contains several thousand proteins with several isoforms each resulting in even more tryptic peptides. Each of the hyphenated technologies can separate from a dozen to several hundred analytes into useful fractions, depending on the running parameters. Hyphenated technologies for proteomic applications are often used in combination with online ESI

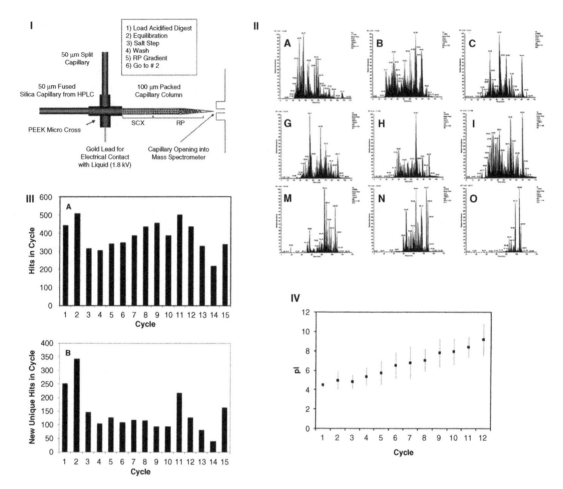

Figure 2.27 The overall performance of quantitative shotgun proteomics (MudPIT). (I) Detailed experimental set-up, including the flow split (to achieve low flow rates with standard HPLC pumps) the integrated SCX/RP column which is packed into the same fused silica capillary drawn into the spray needle for ESI, and the electrical contact for ESI. Such a set-up is 'home made' and, while it has advantages in terms of sensitivity, it is not as reproducible and technically more challenging. Most systems are organized differently (see text for details) but follow the same principles. (II) TIC lane (as an overview of some of the MS results) for some of the 12 RP runs following the SCX steps. Note that later runs contain more hydrophobic peptides, eluting later in the run. (III) The number of peptide hits in database searches with MS/MS data. The upper chart shows total hits, the lower chart shows unique hits. From a total of 250–500 hits per cycle, 50–350 are unique to one SCX fraction (one RP run), showing the amount of peptides eluting in more than one fraction. (IV) The pI of the identified peptides (calculated from their sequence) increases during the SCX run, indicating that Arg and Lys provide hydrohobicity and polarity under these conditions. The resolving power of this 2001 study is still on a par with that of many modern studies. Reprinted with permission from Wolters *et al.* (2001). © 2001, American Chemical Society.

MS/MS, which can analyze five to eight analytes in the time it takes a peak to elute from a column. Thus, in a typical MudPIT experiment (Figure 2.27), the theoretically possible resolution could be 8 (if about 12 different salt fractionation steps are used in IEC, allowing for some overlap) × 300 (RPLC) × 5 (MS/MS) = 12 000 peptides.

This impressive number is a conservative estimate. What it means is that not only can about 10 000 peptides be analyzed in one experiment; they will deliver meaningful MS/MS data, leading to the identification of peptides and proteins. This in turn typically results (in an experiment starting with a complex mixture like cell lysate or, even more so with, tissue lysate) in about 1 500 significant protein identifications (leaving several thousand peptides that cannot be clearly assigned). Even more impressive when one takes a closer look at the data, it becomes clear that multidimensional ESI MS/MS approaches can cope with low abundant proteins much better than gels; the rate of low abundant proteins identified in the samples follows the distribution of these proteins far closer than, for example, in gel-based approaches (see Figure 5.7). Another closer look at the data also reveals that the shotgun approach has no problem identifying membrane proteins, a type of proteins that is all but completely absent from 2D gels. This is because very hydrophobic proteins, such as transmembrane proteins, do not solubilize or focus well in gel-based IEF, they interact with the gel material and disappear from the analysis. Likewise, very small (below about 12 kDa) and very large (above 100 kDa) proteins are difficult to analyze with gel-based technologies. Here MudPIT has the very clear advantage that it does not discriminate against any physicochemical parameters of the 'parent' proteins; all proteins are digested into peptides and some or most of these will be just as easy to analyze if the parent protein was large, small or extremely hydrophobic. However, MudPIT does analyze the 'peptidome' rather than the proteome; we have no way of knowing how the peptides are really connected. If a certain protein has three phosphorylation sites with modified peptides found in MudPIT, we do not know whether there are actually proteins carrying three phosphorylations at the same time. This is information which we can often deduce from the full length protein's migration pattern in gel-based approaches (see Section 5.4 and the Webexercise part 3 for a discussion of this problem).

We will discuss the protein identification steps from MS/MS data on proteins in later chapters, but, looking at the impressive identification power of MudPIT, it is clear that there are more than 1 500 proteins in cell or tissue (perhaps 3 000–9 000 in analyzable expression levels in a cell line and an additional 6 000–8 000 proteins at very low levels of expression) and an even higher number of protein isoforms due to PTMs. So overall even the impressive resolution power of MudPIT is overwhelmed with protein and peptide diversity, in particular since it has to cope with an increase in the number of analytes when proteins are digested to peptides. There is also an overall loss of information when peptides are digested into peptides; this is particularly evident in the analysis

of PTMs or even the identification/quantification of proteins that share homologies and tryptic peptides (see Section 5.4 and Figure 5.8).

It must also be said that the inherent weakness of hyphenated methods coupled to ESI MS is the lack of an easy quantification of the data. While it is easy to identify most (if not all) components from simple mixtures, because of the redundancy with which every protein is detected, when the number of proteins in a mixture increases the redundancy decreases and not every component is detected in every run. With the inherent loss of redundancy it becomes even more difficult to quantify MudPIT data. As a rule of thumb, it is possible to quantify (and compare to other samples) proteins that are covered by more than five peptides; the more peptides are detectable, the more accurate the quantification will be. In real terms this means that, following established protocols and using 'off the shelf' equipment, one can expect to identify about 1 000 proteins in a typical multidimensional chromatography run of complex mixtures (e.g. cell lysate), but one is still a long way from being able to compare two different samples with each other. In other words, with the combination of hyphenated methods one can with relative ease find thousands of components in a given sample, but it is much harder to quantify them, make conclusions about the level of PTMs or indeed exclude the possibility that a certain protein is present in the sample. All these questions can be approached with more elaborate approaches such as iTRAQ, SILAC or immobilized metal affinity chromatography (IMAC) or in combination with gel-based approaches (see Sections 4.4 and 5.3). We focus in this chapter on the multidimensional chromatographies which are the foundation for iTRAQ, SILAC or IMAC and discuss their implementation in later chapters.

In a seminal paper, Wolters, Washburn and Yates (2001) used a combination of SCX and RPC to identify 1 500 yeast proteins in a set of simple experiments, 900 in a single experiment (Figure 2.27). This showed the expression of about 20% of all yeast genes (their model system) and is a larger number of identified expressed proteins than the whole yeast research community had produced using gels and MS in the decades before this experiment. Moreover, a very similar approach has become the mainstay of MudPIT used in hundreds of labs everywhere. In the original paper a biphasic column was used, containing IEC and RPLC beads in the same column. This set-up guarantees very high sensitivity and resolution, since it gets rid of a lot of tubing, and offers a cheap and elegant way of performing the analyses. Only two pumps are needed and many ordinary HPLC systems could be used to perform the experiments, columns can be 'home' made and incorporate the ESI needle tip. Among the disadvantages of this set-up are that the results are less reproducible since column to column variations are larger, and the whole set-up is less flexible. Everything that goes off the column has to go over the ESI needle as well, so KCl and NaCl decrease the sensitivity and clog up the ESI needles and one has to use ammonium acetate as eluent for IEC. Not everyone wants to make their own biphasic columns, and most manufacturers prefer to produce very reliable/reproducible monophasic columns. For all these reasons most labs use a set-up with two distinguished columns, four pumps, several automatic HPLC valves and ESI needles that are separated from the column, allowing for UV measurement of the peptides as well. An excellent account of how to use the biphasic column MudPIT approach is given in Fournier *et al.* (2007).

Many possible improvements to the above described 'classical' MudPIT approach have been tested. So far none have improved significantly on the original concept and none have been widely adopted, either because of being too complex, or diluting the sample or for other practical reasons. However, some of them have the potential to improve in the future or are useful for certain applications and will thus be briefly discussed here.

To increase the coverage of MudPIT one can run (in parallel) not only SCX/RPLC but also 1D HPLC and strong anion exchange (SAX)/RPLC. The same columns and eluents as in SCX can be used in SAX, but the pH is adjusted form 2.5 to about 8, which gets many negatively charged groups into the peptides. From this set-up you can expect to identify about 40% more proteins, while the workload, and sample amount needed, will go up by 250%. This is mainly due to the fact that no matter how good your search approach is, there are always some peptides/proteins that escape your measurement more easily in one method than in another, creating complementary peptides and a less than complete overlap between methods. For example, very hydrophobic peptides will always interact with the bead surface and recovery from SCX or SAX will be so low that they are not detected at all in 2D HPLC, but very well in 1D HPLC. As another example, some peptides will elute into two fractions from

the SCX/SAX and can thus be diluted to the point that they can no longer be detected. So in a 1D RPC run there is a higher chance of measuring these peptides, since there is no second column connected and the RPLC run starts with loading of the sample. Remember that in two-column 2D HPLC the salt of the first column gets into the second column, is washed away and then the run is started; some peptides may potentially be lost. In biphasic columns the detection of the first peptides from RPLC can be influenced/diminished by salt from the IEC steps. On the other hand, just repeating the experiment produces significantly better peptide coverage over the proteome. However, in line with what was said above, some proteins will not be detected easily (Figure 2.28) and performing successive MudPIT runs will increase the number of identified proteins by up to 50%, but the benefits will be very small after three runs. This is mainly due to the fact that the abundant proteins are measured over and over again, while just some of the lower abundant proteins creep up in later runs and the proteins that produce peptides that are unlikely to be measured (perhaps because they do not ionize so well), or are concealed under more abundant peptides, stay mainly hidden. Indeed, which of the peptides derived from lower abundant proteins (or other 'problematic' proteins) will actually be measured in each MudPIT run is a bit of a statistical event. This is not to say that MudPIT is irreproducible; it is an inherent problem mainly because the resolution of MudPIT is at least an order of magnitude lower than the sample complexity in samples such as whole cell lysate or tissues. Reducing the sample complexity allows measurements of low abundant or 'problem' proteins to be made much more redundant; if you have a chance to get MS/MS data on six peptides per protein, the chances of getting at least two peptides in every run are much higher than if you only ever had the chance to measure two peptides in the first place; if you only have a chance to measure two peptides, you might get no or only one peptide within a single set of MS/MS data. Two peptides with good 'hit' scores are a typical criterion for reliable protein identification (see Section 4.3). Abundance might not be the only reason why a protein is not identified; some very short proteins might only produce one or two tryptic peptides suitable for MS analysis; very large proteins with extensive homologies to smaller proteins will never be detected, as there is no way of telling if the peptides were generated from the large or the smaller proteins,

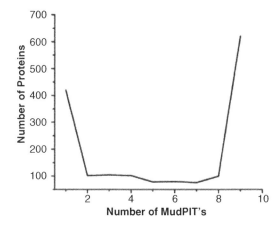

Table 1. Percentage of Proteins Experimentally Identified from 1, 3, 6, and 9 Combined MudPIT Runs Relative to the Results Obtained by Ghaemmaghami[28] at Different Abundance Levels, or Protein Copy Numbers per Cell

protein copy number per cell[a]	number of proteins[a]	% proteins identified from a different number of combined MudPIT runs			
		1	3	6	9
>1×10^5	80	96.3	97.5	97.5	97.5
(1×10^4)–(1×10^5)	536	71.6	79.3	83.4	85.6
(1×10^3)–(1×10^4)	2184	20.8	24.4	30.6	34.9
100–1000	1036	6.5	9.7	13.2	15.5
<100	32	0	3.1	6.3	6.3

[a]Results obtained by Ghaemmaghami et al.[28]

Figure 2.28 Benefits of technical repeats for MudPIT experiments. A yeast extract was prepared and analyzed in nine independent shotgun analyses, using 2D LC and an ion trap for MS/MS measurements. The graph shows for each protein in how many runs it was identified in the database searches. Most proteins are either identified in a single run or in all nine runs. The table summarizes the findings in relation to the expected abundance of proteins as deduced from a seminal study using fluorescent labelled proteins (see Figure 5.7 and Ghaemmaghami *et al.*, 2003). Repeating the runs under identical conditions does increase the amount of rare/low abundant proteins identified, but the returns for the enhanced workload (and sample consumption) are modest. Reprinted with permission from Liu *et al.* (2004). © 2004, American Chemical Society.

and smaller proteins are always preferred as 'hits' by the search software as the identification 'by chance' is much smaller (see also chapter 4.3). So basically a lower sample complexity allows a deeper penetration of the proteome using MudPIT-like approaches. This deeper penetration might result in less abundant proteins being measured or more PTM variations being characterized.

Pre-fractionation before MudPIT (e.g. by protein or peptide OFFGEL) can thus enhance the value of a proteomic study significantly. For this, a third hyphenated technology or a non-hyphenated method such as gel-based IEF fractionation or FFE can be combined with 2D HPLC.

One approach to reducing sample complexity is not to measure every peptide of a protein, but just a few. This can be achieved in approaches targeting 'rare' amino acids such as cysteine. Cys represents only 1.7% of all amino acids, whereas the statistical average is 5% per amino acid. This is of course for functional reasons, and the Cys content in proteins that need stabilizing by Cys bridges or in reaction centres is higher. Targeting Cys with chemical modifications allows most of the peptides to be excluded from the analysis (Figure 2.29). This in turn enables the analysis of less abundant proteins. The price to pay for this is that PTMs cannot be analyzed, as the peptide coverage over the complete protein sequence is low. On the positive side, using an approach such

as isotope-coded affinity tagging (ICAT) allows for a relatively simple comparative quantification of shotgun proteomic analyses (Figure 2.29).

SCX and RPLC are not completely orthogonal, that is, separation in the second dimension depends partially on the separation parameters used in the first dimension. This is mainly due to the fact that arginine and lysine are present in every peptide (due to trypsin specificity) and contribute to both the charge and the hydrophobicity of the peptides since they contain (at low pH) positively charged groups as well as hydrophobic aliphatic chains. Hence, the full potential of RPLC is not used as peptides do not elute all over the ACN gradient but tend to accumulate in the first half of the gradient in the low salt fractions, in the middle part in medium salt fractions and at the end (high ACN concentrations) in high salt gradient fractions (see also Figure 2.27). To address this issue several approaches have been tested. Hydrophobic interaction liquid chromatography(HILC) uses mainly hydrogen bonds for the interaction between peptides and stationary phase. The peptides bind strongly to the matrix in presence of a hydrophobic buffer (80% ACN), and this is how the samples are loaded on the column. For elution of the peptides a gradient starting with a high hydrophobicity and going towards a lower hydrophobicity is applied, which is the exactly reverse of the RP elution

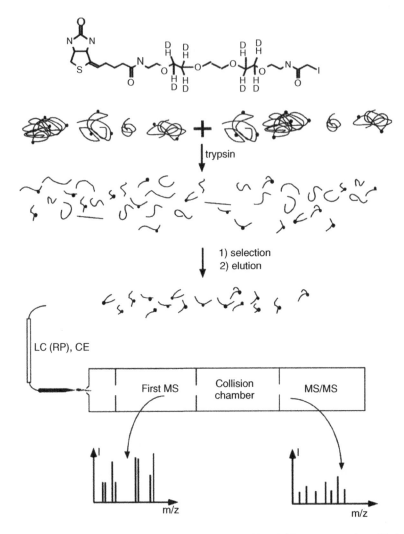

Figure 2.29   Targeting rare amino acids for sample complexity reduction. The ICAT reagent contains a biotin tag (left-hand end of the formula) and a iodoacetamide group (right-hand end of the formula). The iodoacetamide group targets Cys residues very specifically and efficiently, and attaches the whole ICAT reagent to every Cys-containing peptide. The reagent comes in two variants, light and heavy, which differ by 8 Da as four hydrogens in the linker region (middle of the formula) are replaced by deuterium (H or D, respectively, in the formula). After labelling with the ICAT reagent the peptides can be isolated using the high affinity and specificity biotin tag. To compare samples, lysates are separately digested by trypsin, labelled with either of the two tags (light, heavy respectively) and then mixed. The biotin-containing peptides are separated from Cys-free peptides and analyzed by MudPIT. Each pair of MS signals 8 Da apart is likely to represent the same peptide in light and heavy modified form. MS/MS is used to identify the peptide, the MS signal intensity is used for comparative quantification (see also Sections 4.4 and 5.3). Reprinted from Rabilloud, T. (2002), *Proteomics*, 2, 3–10, courtesy of Elsevier. © 2002 Elsevier.

gradients. The least hydrophilic peptides with the least hydrogen bonds to the stationary phase elute first (reverse order to RP). Although one would assume the system to be related to the elution order in RPLC, it is actually a lot more orthogonal to RPLC than SCX (Figure 2.30).

Although no more peptides can be detected in a single MudPIT experiment, they are more equally distributed among the HILC fractions. This allows better analysis of the peptides and, in particular, a higher rate of detection for phosphorylated peptides.

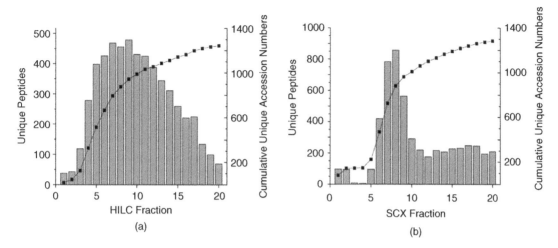

Figure 2.30 HILC as alternative to SCX in shotgun proteomics. One milligram of HeLa cell lysate was digested with trypsin and analyzed by shotgun proteomics, using either HILC or SCX as first dimension followed by RPLC and ion trap MS/MS analysis. *Bars* show the number of unique peptide identifications obtained in each of the 20 fractions and the black graph indicates cumulative unique accession numbers obtained across the gradient. While SCX and HILC show the same number of identified peptides and proteins/genes, the spread over the different fractions (i.e. higher number of unique peptides per fraction) is more even for HILC, indicating a better separation of peptides and a more orthogonal separation in dimension 1 and 2. The authors showed in the same study that phosphopeptide exhibits very good binding/elution behaviour in HILC. Reproduced from McNulty & Annan, *Mol. Cel. Prot.*, 2008; 7: 971–80 by permission of the American Society for Biochemistry and Molecular Biology.

2.4.3 Free flow electrophoresis

Free flow electrophoresis (FFE) was born out of the ambition to employ the very powerful gel-based electrophoretic separation technologies without using gels. This should in theory not only make the separations faster (there being no resistance to analyte migration by gel matrix), but also allow many separations that are not possible in gels because of the strong resistance of gel matrices to the movement of analytes or because of residual interactions between analytes and gel matrix (these cause all the difficulties of 2D gels, for example with membrane proteins). The strong resistance of analyte movement within gels can be partially overcome by decreasing the gel percentage. However, the lowest level at which PAGE gels are stable is about 7% for unsupported gels and about 2% for gels supported by (and cast upon) a plastic backing. In IEF these limits are fully exploited, since the IEF force is very low, in particular during the later stages of IEF, when most protein charges are neutralized. The largest proteins that can be analyzed efficiently by gel-based IEF are about 120 kDa. The largest proteins that can be analyzed accurately for their seize by SDS PAGE are about 250 kDa. The ability to analyze larger proteins in SDS PAGE than in IEF results from covering the proteins with the highly charged SDS molecule and thus generating a much higher driving force through the gel. However, some proteins are larger, and every protein works in a complex with other proteins and these complexes are severely damaged if not totally destroyed during gel-based electrophoresis. Large complex like active kinases (e.g., the Raf kinase is active in a multiprotein complex of 600 kDa) or even organelles and sub-cellular structures cannot routinely be separated by gel-based approaches, even when using Schagger gels, where SDS is replaced with CBB. Since FFE works in the absence of a gel matrix most of these problems can be overcome (and be replaced by a set of new challenges).

FFE is historically much older than the more recent attempts at OFFGEL electrophoresis (which still depend on the passage of analytes through gels) and in theory much more versatile. Unfortunately, it is also technically very demanding. FFE machines only recently evolved into more user-friendly units, and while the separation time might be as short as 20 minutes, it takes nearly a full working day to set up a run and clean the machine for the next run. FFE works with small ions such as salts, but also with larger ions such as peptides, proteins, protein complexes and even organelles and whole cells. The resolution of FFE is never higher than about 70 useful fractions, but the absolute resolution can be extremely high, for example 0.02 pH units between fractions in IEF. FFE can be run as a continuous process, so the amount of analytes that can be separated can be quite large: anything from 200 μg to 100 mg of protein in a day's work is typical for FFE. All these characteristics, together with a high reproducibility, establish FFE as good alternative for pre-fractionation in proteomics application. This pre-fractionation can be based on the isolation of sub-cellular compartments/organelles, proteins or peptides. Any of these components can be separated based on their speed of movement in an electrical field using certain buffer compositions, for example in zone electrophoresis (ZE) and isotachophoresis (ITP), or by virtue of its amphoteric nature in IEF. These different possibilities of size and separation method make FFE in its various guises a highly versatile separation technology with medium separation capacity.

FFE can be used with several electrophoretic separation principles, but we will focus on Free Flow IEF (FFIEF). Figure 2.31 shows the basic construction of an FFE machine. In principle, the gel is replaced by the laminar flow of a buffer. The composition of this buffer will be discussed later, but for IEF it has to contain ampholytes

to generate a pH gradient in an electric field and some components to stabilize this pH gradient and the separated proteins (basically by increasing the viscosity). This buffer is typically delivered by about five inlets and a wide range of buffers can be used to suit the different applications. On a full size machine the separation chamber is about $120 \times 500 \times 0.5$ mm (width × length × thickness). The buffer is pumped at a flow rate along the length of the separation chamber (typically 5–10 ml/min), creating a laminar flow (Figure 2.31). Electrodes are positioned perpendicular to this flow. Applying an electric field across the chamber results in the generation of a pH gradient perpendicular to the flow of the buffer. Establishing this gradient takes less than 20 minutes and then the sample is introduced at the beginning of the separation chamber. The speed of sample introduction must be matched to the media flow, in order not to disturb the laminar flow. In the first third of the separation chamber the pH gradient is established, and the sample follows along with the ampholytes until it is in the second third of the chamber where a pH gradient is formed. In the last third of the chamber the sample and the ampholytes move along a stable pH gradient and elute through 96 holes at the end of the chamber and are collected into separate fractions, each containing analytes with a different pI (Figure 2.32). For this fractionation to be efficient a

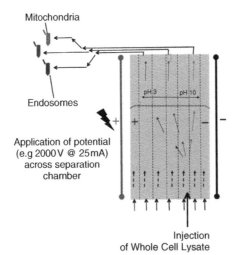

Figure 2.31 Principles of free flow electrophoresis. FFE can be used as shown in the figure for the IEF (FFIEF) of organelles from whole cell lysates. The pH gradient is formed in the first three quarters of the separation range. A total of 96 fractions can be collected, about 70 of which are useful in terms of pH range; the others 'buffer' against the electrodes. The left panel shows a run with each second inlet marked with a dye to demonstrate the laminar flow. Further details are given in the text.

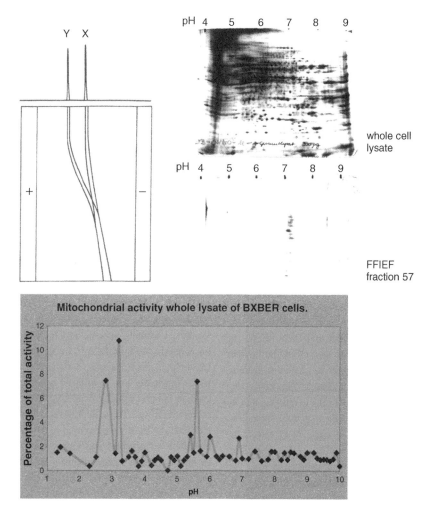

whole cell
lysate

FFIEF
fraction 57

Figure 2.32 Separation of proteins and organelles using FFIEF. FFIEF can be used for the pre-fractionation of proteins under denaturing conditions (upper panel). The right panel shows the results for 200 μg of cellular lysate either loaded directly on a 2D gel or after loading one fraction (no. 57) of a FFIEF run of the same lysate onto a 2D gel. In practice, it is useful to combine several FFIEF fractions and load them on matching narrow pH gradient 2D gels (see text). The lower panel shows the distribution of mitochondrial activity throughout the pH range after a native FFIEF run from lysates of mammalian cells. The mitochondria elute in one major fraction at pH 3 and several minor fractions at higher pHs. The distribution of the mitochondria is influenced by signalling events in the cells on very short time scales. Data generated in the authors laboratory by of Dr A. McConnell, University of Manchester, UK.

counterflow of separation medium is needed. This counterflow bypasses the separation chamber and enters at the top of the separation chamber. The counterflow basically helps to increase the flow and to direct it stably into 96 capillaries without increasing the flow in the separation chamber. The separation time (between loading and collection of the sample) is about 20 minutes. The longer the loading time, the more sample can be separated. Typically about 1 mg of sample can be loaded in about

30 minutes, making the entire run time about 50 minutes. The sample collection can be monitored by adding bromophenol blue to the sample. Loading the sample on the cathodic half of the separation chamber will result in the bromophenol blue stain disappearing in the first third of the chamber just to reappear in the second half on the anodic site, as a stable pH gradient is formed. Seeing blue/yellow drops outside the useful separated fraction range on the anodic side indicates that it is time to collect

the samples, and when the fractions near the anode become colourless the sample collection time is over.

Just like gel-based approaches, FFE can be run under denaturing conditions using urea and thiourea. However, the separation times become longer and the separation power under denaturing conditions is somewhat lower than under native conditions. Since it is desirable to have denaturing conditions in order to get reproducible results (see discussions of this in Sections 2.1.4 and 2.4.1), FFE is often used under suboptimal conditions. One field where native conditions are absolutely required is the isolation of protein complexes and organelles. While some protein complexes have been analyzed with FFE, it is in the separation of organelles under native conditions that FFE comes into its own; compared to any other method to isolate organelles, the separation is of high resolution and very fast. There is a vast body of work on microsomes and their maturation using FFE; the different subspecies can be directly separated in FFIEF as their surfaces change, and it is these surfaces that give the organelles their charge in solution and their pI. At the same time the roughness and size as well as the number of charges on the surface changes as different proteins associate with the organelles and change their conformation due to internal changes in pH. These changes can be exploited using ITP and ZE in FFE to separate organelles; a ruffling of the membrane or increase in size with the same charge will slow the migration of the organelles down and hence make them distinguishable in FFE ITP or FFE ZE. Changes in the charges of the organelle surface will also change the pI of the organelles, which can be exploited in FFIEF (Figure 2.32). Using FFIEF, mitochondria can be separated with an unprecedented resolution. As the sample is subjected to the electrophoresis for only 20 minutes, artefacts and degradation can be kept to a minimum, compared to hour-long separations of mitochondria using classical approaches.

The buffers used in FFE always have to contain reagents to increase the viscosity of the separation medium in order to allow for a stable separation. Glycerine or hydroxypropyl methylcellulose (HPMC) are often used for this purpose. Further typical components for IEF are ampholytes and non-ionic detergents like CHAPS to increase the solubility of proteins. In ITP and ZE other buffer components are salts needed for the separation. Most of these components can interfere with later steps in separations, like 2D gel electrophoresis.

Most components can be removed by either dialysis or (much faster) by acetone precipitation.

FFE can be used for the pre-fractionation of complex samples in several stages to supplement either 2D gel electrophoretic approaches or hyphenated methods. For example, complex protein mixtures can be separated by FFIEF into pH regions that match the pH regions of narrow pH range 2D gel systems. This allows a higher load of the IEF gels, since FFE not only concentrates the proteins in the desired pH region but also removes salts and other charges molecules (detergents, RNA/DNA or cell matrix components) that would otherwise limit the load of sample on the 2D gels. It has been shown that by using FFIEF to pre-fractionate, the number of spots on 2D gels can be increased nearly twofold and that the enrichment by FFIEF also allows the analysis of some membrane proteins by 2D gels. However, this application has not found widespread use, due to the complexity of FFE. FFIEF of proteins followed by RP HPLC ESI MS/MS can be used to characterize PTMs in top-down proteomics, enriching specific protein isoforms in separate fractions. Like OFFGEL pre-fractionation, this approach can help to regain some of the information on PTMs, which is usually lost in MudPIT approaches. FFE is also used to enrich peptides from tryptic digests of complex protein mixtures. These peptides are then analyzed by RP HPLC-ESI MS/MS and, similar to MudPIT, up to tens of thousands of peptides and thousands of proteins can be identified. An advantage of this approach is that the pI of the peptides is known for each RP run (from the pH of the corresponding FFIEF fraction) and this information can be used to generate more reliable peptide/protein identification by MS. By far the most successful application of FFE is the enrichment of cellular fractions and the pre-fraction of proteins in order to 'dig deeper' into the proteome by identifying more organelle-specific and low abundance proteins. Organelles and cellular compartments that can be isolated by FFE are lipid rafts, endoplasmic reticulum, mitochondria, granules, peroxisomes, membrane vesicles and endosomes, as well as whole cells (enriched from mixtures). FFE can also be applied in miniature (chip) format for sample clean-up.

2.5 VISUALIZATION OF PROTEINS/PEPTIDES FROM HYPHENATED METHODS

Analytes from hyphenated methods can be visualized by a variety of methods, non-destructive and destructive.

In contrast to gel-based approaches, stains are not very common. Since each of the components is usually present in high purity and concentration in a transient manner, components (or low complexity mixtures of analytes) are measured one at a time straight after separation in real time (online). Measurements after the separation (offline) are possible but rarely used. This means that a lot of data is generated in a relatively short time and needs to be handled in real time.

2.5.1 Optical detectors

Standard UV detectors for HPLC systems can be used for proteomic applications. They can measure UV/visible light from about 190 nm wavelength onwards. Peptides and proteins have a strong absorbance between 190 and 220 nm due to the peptide bond. Shorter UV wavelengths cannot be used because of the absorbance of solvents used in the separation, the absorbance of flow cell materials (e.g. silica) and the cost and useful output and life span of UV lamps. So mostly peptide/protein absorbance in HPLC elution is measured at 214 nm; at longer UV wavelengths mainly the absorbance of tryptophan is being measured. Tyrosine, phenylalanine and histidine have a strong absorbance due to their aromatic rings. Since not every peptide contains aromatic side chains and proteins vary greatly in their tryptophan content, measurements for these residues at 280 nm are not very often used. Also the overall absorbance is about four times lower than at 214 nm. Furthermore, typical contaminants such as RNA and DNA absorb strongly at 260 nm, adding to the guess-work of 280 nm measurements for peptides and proteins. In a typical nano-flow cell the actual flow cell is made out of fused silica and with such a flow cell, sensitivities can reach down to several tens of femtomoles of a single peptide or a protein digest, and absorbance units down to 0.002 maximum read-out can be measured reliably. With the best available set-up and detectors (nano-flow cell of low volume) it is possible to see some peaks in the UV detector lane from as little as 50 fmol of a digested protein. The peaks seen are not likely to be single peptides, but co-eluting peptides, increasing the signal. Also it is very likely that these peptides will contain tryptophan, as this increases the absorbance of the peptide bond strongly at 214 nm, and of course has a strong maximum at 280 nm. For complex samples it is perfectly possible that the visible peaks in proteomics samples are a mixture of peptides and contaminants (e.g. polymers or nucleic acids). In standard HPLC multiple wavelengths can be measured by diode array detectors. This is not possible in nano-HPLC, due to the limited information that can be extracted from the small peak volumes/intensities and the relative large flow cells necessary, which would distort/dilute the peaks. This information is also not crucial in proteomic settings, as the main detector is ESI MS/MS. So why bother with the UV lane in the first place? In everyday running of nano-HPLCs it can give valuable additional and reassuring information. If, for example, the UV lane shows peaks but there is no signal from the MS (or, more typically, a weak one) you can be sure something is wrong, after the UV detector. Typically the ESI could have a problem (misaligned spray or clogged needle). It is also reassuring to see the sample has gone on the RP column, which would be shown first by a huge change in optical density of the UV lane due to mixing of the solvents. If, for example, a connector or column was blocked, the whole flow might go in the waste, and not on the column. This is a result of using a flow splitter system in nano-HPLC; the flow is split according to back-pressures at the splitting capillary and the real sample pathway, and if one is blocked (even partially) the whole flow is re-directed to the other pathway. So ultimately a UV detection system is not strictly necessary, but good to have, to optimize the running of the system. The tubing diameter and length in hyphenated methods has to be kept small to avoid peak broadening and mixing. So in terms of sensitivity there are advantages to not using UV detection at all. The original reports on Mud-PIT did not use any UV detection. Of course the analyses were also performed by experts, and this is what proliferation of proteomic technology is all about – enabling mere mortals to perform very challenging analyses.

2.5.2 Mass spectrometers as detectors

The main detector for hyphenated proteomic separations is ESI MS/MS. Here we discuss the major aspects of the use of this technology for detection/analysis in hyphenated technologies; more details will be given in later chapters on ESI and mass spectrometry.

In a typical setting the analytes, having just passed the last column of a separation module (and hopefully also the UV flow cell), are sprayed out of a tiny capillary tip (typical diameter 1–5 μm) at atmospheric pressure into the orifice of a mass spectrometer. The capillary is typically gold coated to allow an electrical contact between the spray and the orifice of the mass spectrometer. The application of a strong electrical field is essential for a good spray formation and the ionization of the analytes. ESI for nano-HPLC is extremely sensitive in detecting ions for several reasons; the very high efficiency of ESI is the most important of these. For efficient ESI the buffer in which the analyte is dissolved has to evaporate in a very short time and physical distance, and this is best achieved at low flow rates with tiny needles; just the conditions used in nano-HPLC in proteomics. If one increases the flow rate/dimension of the HPLC this evaporation needs to be aided by turbo-fan generated streams of (inert) gas. Under such conditions (inner diameter of columns several millimetres to centimetres, flow rates up to millilitres/minute) the UV absorption signal may well be more sensitive than the MS signal.

So what exactly is the read-out from the MS measurement? Mass spectrometers are molecular balances. Here we only explore some of the basics of MS, as they apply to hyphenated approaches.

Mass spectrometers can only measure charged molecules in the gas phase. So for most proteomic experiments the peptides/proteins are charged positively by adjusting the pH as low as pH 2, forcing positive charges on the analyte before ionizing it. The principal method of ionization in hyphenated applications is ESI. MALDI is another popular ionization method, sometimes used for hyphenated approaches. MALDI can only be used in the offline mode, that is, the analytes are prepared for MALDI as they elute from the LC and are measured by MALDI based mass analyzers after the LC run (see Sections 2.4.2 and 4.3). A plethora of mass analyzers can be connected via ESI sources to the LC set-up for analysis (Chapter 3). Here it is sufficient to remark that these mass analyzers for most proteomic applications have to have MS/MS capability of some sort. This means that they can not only analyze the mass of incoming peptides, but can also chose a specific peptide (ion), fragment it in the mass analyzer and measure the masses of these fragments. In principle this enables the identification of the sequence of this peptide in one of the extensive DNA databases, containing hundred thousands of sequences from a variety of species.

Having passed the ESI, the peptides/proteins are devoid of their hydration and the mass analyzer can measure their exact mass, typically in the 'positive mode', with the analyte carrying one or more protons. If a peptide with mass 800 Da carries one proton it will be measured as having a

mass of 801 Da. If the same peptide carries two protons it will have a mass of 802 Da. While the peptide carrying one proton will be detected at 801, the peptide with two protons will be detected at $(800 + 2)/2 = 401$. Here we see that mass spectrometers cannot measure masses in dalton, but rather in mass/charge ratios, so the exact expression of the weight in the first example is 801 m/z or 801 Thomson (Th), as the unit is also sometimes called. In another example (right panel in Figure 2.33), a peptide with two protons, is detected at 801 Th or m/z. This time the peptide is doubly-charged; the measured m/z derives from the peptides real mass of 1 600 Da, plus the mass of two protons, divided by 2 as the ion carries two charges. How do we know if a mass at 801 Th derives from a peptide of 800 or 1 600 Da (devoid of protons)? For this we need to know

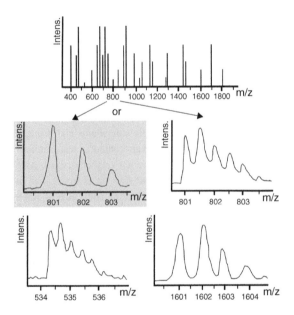

Figure 2.33 MS data for peptides from LC. A typical ESI MS spectrum from a moment in time during the elution of an RP gradient (e.g. signal integration from 20.3 to 20.8 minutes into the run). The signals in the upper panel represent single ions. Looking more closely at the spectrum, the signal at 801 m/z could look like the detailed spectrum in the middle panel on the left or right. In this case the spectrum on the right was observed, we detected a doubly-charged ion at 801 Th (or m/z), as indicated by the distance of 0.5 Th between the isotopic peaks. This proves the ion resulted from a peptide of 1 600 Da. This peptide can often be measured in the same spectrum in its triple charge state at 534.33 Th and as a singly-charged ion at 1 601 Th (lower panel). For more details see text.

the isotopic distribution pattern. This will be explained in much more detail in Section 3.1. Suffice it to observe here that all peptides and proteins contain many carbon atoms. About 1% of the naturally occurring carbon comes in form of the stable (i.e. non-radioactive) isotope of carbon 13. The other 99% are made up of carbon 12. This means every single peptide and protein is made up of a mixture of carbon that has a mass of 12 Da to 99% and a mass of 13 Da to about 1%. This explains why we see a mass spectrum as in Figure 2.33 for a peptide with the monoisotopic mass of 800 Da (panel in gray box Figure 2.33); if, by chance, every peptide in a given molecule is carbon 12, the mass of the peptide is 800, measured as 801, m/z because of the one proton attached to it. At 800 Da the peptide will be made out of about six to nine amino acids and, more to the point, contain about 44 C atoms, and 1% of these are carbon 13. So while most peptides will contain no carbon 13, some will contain one atom of carbon 13, some will contain two and some even three or four. Since we measure a huge number of individual peptides to generate the total spectrum we will end up with the statistics correctly presented in our measurements and thus see isotopic peaks ranging from 801 to 803, and maybe even 804 at a very low intensity. A peptide of 1 600 Da will be made up of 12–18 amino acids containing nearly 90 carbon atoms. Now the signal intensity for the monoisotopic peak (the first one with the lowest mass, consisting exclusively of carbon 12) is not even the strongest. Statistics predict that peptides with one carbon 13 atom are the most abundant isotopic species (see Figure 2.33).

Since all these isotopic peaks are 1 Da apart from each other, we can use the isotopic distribution to determine the charge state of a peptide; if the isotopic peaks are separated by 1 Th, the peptide is singly-charged as we know that carbon 12 and carbon 13 are 1 Da apart. If the peptide carries two protons it is doubly-charged, the isotopic peaks are still 1 Da apart, but appear on our measurement as being 0.5 Th apart due the fact that the ions are doubly-charged (Figure 2.33). Conversely, if we see that the isotopic peaks separated by 0.5 Th we know the peptide is doubly-charged. If the isotopic peaks are separated by 0.33 or 0.25 Th the peptide is triply- or quadruply-charged.

Almost all mass spectrometers used for ESI LC in proteomics are capable of producing MS/MS data, and indeed this is essential for protein identification in MudPIT. MS/MS will be dealt with in later chapters

in more detail and there are differences in the way the MS/MS data are derived and in the quality of the data between different mass spectrometers. In MS/MS the instrument first measures all ions within a certain m/z range, then isolates a single ion/peptide or protein and transfers energy into this ion (Figure 2.34). The energy transferred leads to the fragmentation of this ion and the masses of the fragments are recorded.

The initially isolated and fragmented ion is called the precursor or parent ion, and the fragment ions are called daughter or product ions. Peptides do not just break at any point; similar to a vase with handles, you can piece the vase back together even if it is shattered, particular if it breaks at a "weak" point, like the handles. Of course you might miss the finer points (of vase and peptide), but most of the time you would be able to distinguish a vase from (say) a beer jug. Since there are certain rules governing the fragmentation of peptides and proteins, going back to the amino acid sequence, the fragments can be used to identify the structure, and sometimes even the exact amino acid sequence (including PTMs) of the peptide or protein in question. Some mass spectrometers even have the capability to choose some of the fragment ions and perform and measure further fragmentations performing MS^n. MS/MS is a very powerful tool and a good high quality data set can identify the presence of a peptide with absolute accuracy; usually the detection of two peptides is sufficient to detect the presence of a protein in a mixture of several thousand proteins and tens of thousands of peptides. A detailed account of data analysis from MS/MS experiments in given in Chapters 3 and 4.

The mass analyzer is constantly measuring during an LC run, typically producing one measurement cycle in about 0.1 to 2 seconds. What we see in Figures 2.33 and 2.34 is thus only the picture of a single measurement during the elution of an LC run. Often (for improved sensitivity and resolution) several measurements can be summarized. Of course it only makes sense to summarize the spectra from times when a certain peptide is coming down from the LC, (say) from 10.4 to 10.8 minutes into a run (Figure 2.35); summarizing longer time periods mixes the signals of different peptides. Most of the time mass analyzers are set up to cycle between MS and MS/MS analyses, so either of these can be summarized over a period of time. Summarizing the entire run is pointless, as there would be ions at nearly every single m/z range. However, there are several ways to summarize the MS data of a

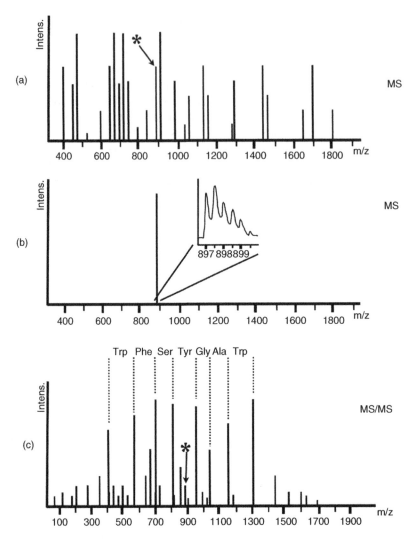

Figure 2.34 MS/MS data for peptides from LC. Before MS/MS can be performed an overview spectrum is needed in most cases to identify which ions are present ((a), MS) and to choose one ion for fragmentation (marked with an asterisk in (a)). Depending on the mass analyzer used, this does not necessarily result in a second MS spectrum showing only the selected ion ((b), MS). However, mass analyzers with MS/MS capabilities can single out a parent ion from the other surrounding ions. Usually a doubly-charged ion is chosen for fragmentation in ESI MS, as the fragmentation of doubly-charged tryptic peptides results most of the time in useful and informative spectra (see the inset for charge state determination in (b) and (c) for MS/MS). Different mass analyzers use different methods to transfer energy to the parent ion, but usually some form of collision with gas is involved (collision induced dissociation). In the resulting MS/MS spectrum there is often some residual non-fragmented parent ion present (marked with an asterisk in the MS/MS spectrum). Most fragments are singly-charged, and thus fragments can have a larger m/z than the parent ion, which is chosen in the form of its doubly-charged ion. The ideal fragmentation conditions for different peptides vary slightly in a sequence and charge state dependent manner: too much energy and only very small, not very informative fragments are generated; not enough energy and no fragmentation at all is observed. Often the fragmentations occur in the same place alongside several amino acids of the peptide backbone; an ion series is created. The difference in mass between the members of an ion series is exactly the mass of an amino acid, thus parts of the sequence can often be deduced. To identify a peptide from MS/MS data the spectra can be interpreted. However, this can be time consuming and error prone, so uninterpreted spectra can be used for database searches, and peptides of matching size and predicated fragmentation patterns are found by statistical methods instead in automated approaches. A single LC MS/MS experiment can create tens to thousands of MS/MS spectra. See text and Chapters 3 and 4 for more details.

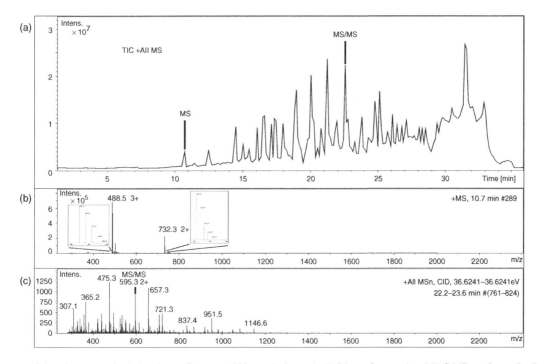

Figure 2.35 Summary of MS data from LC runs. (a) This graph shows the TIC lane of a complete RPLC MS run from a Pauli ion trap mass analyzer. All MS intensities are summed up. Note that the mass spectrometer also performed many MS/MS measurements, whose intensities are also summed up in the "All MS" TIC lane. The TIC signal is simply integrated over these time points, as there are no MS data generated during MS/MS measurements. Some instruments show the lack of data during the MS/MS measurements, and the signal intensity in the MS TIC lane goes to 0 for each MS/MS cycle. (b) Here we see the MS signal from one scan (number 289) at 10.7 minutes (position indicated in (a)), where only one peptide eluted. The peptide is measured as a doubly-charged and a triply-charged ion simultaneously. The insets show the three or two isotopic peaks per m/z unit. The insets show the data as a simple line graph, not as a raw spectrum. In a line graph features like peaks are processed into a single line. This reduces the amount of data by more than 80%, keeping hard drives clear and speeding up processing times significantly. Peptides for MS/MS analysis during the run are chosen automatically – there would be no time for an operator to make these decisions during the run (besides, they are needed to analyze data all the time). However, if needed the operators can scrutinize the data after the run, as shown in (c). The operator chose to look at the MS/MS signal of the doubly-charged peptide at 596.3 Th (the unfragmented precursor ion is labeled in the graph), eluting together with other peptides in the TIC lane peak around 22–23 minutes (see label in (a)). The spectrum shown is a summary of all MS/MS spectra of this ion, ranging from 22.2 to 23.6 minutes. This summation of all MS/MS spectra between scans 761 and 824 enhances the intensity of the signal significantly; while a single MS scan can deliver good data (see (b)), several scans need to be added up to get strong data onf the MS/MS fragment ions, as the signals get weaker when going from MS to MS/MS measurements. Also note that several MS scans were measured between spectra 761 and 824, and that each scan gets a consecutive number: scan 761 might have been an MS/MS scan, while 762 was an MS scan followed by another MS/MS scan, but this time with perhaps a different parent ion.

whole run, the most popular being the Total Ion Current (TIC) lane. All the data from the mass analyzer come with intensities. In print-outs TIC lanes (and all other MS data) are always normalized to 100%; that is, the range of intensities is set so that the monitor or paper is fully used. This is not how it is measured. Inside the mass analyzer is a detector that produces a voltage that correlates with the amount of ions of a certain mass reaching it at any given time. The units for these intensities are arbitrary and do not really mean very much. However, the signals over the entire mass range for one measurement cycle (e.g. 1.5 seconds) can be summarized. If this summarized signal intensity over the entire mass range is plotted over time, it results in the TIC lane (Figure 2.35). The mass analyzer can do a similar summary for the integrated signals of all peptide ion signals resulting from MS/MS experiment over the entire mass range and the entire length of time, resulting in the MS/MS TIC lane. These TIC lanes as well as single spectra can be displayed as the run proceeds, in real time. A complete summary of all the data is also available, in the form of a contour plot (Figure 4.13). However, this rather complex data view is only possible after the run and after some intensive data crunching.

Thus, besides seeing the TIC lane as it is acquired, there is also the possibility to display other MS and MS/MS data online during the measurement. For instance, the sum of the intensity of the MS/MS ions can be displayed, showing where the instrument acquires (acquired) MS/MS data. This is a much better indicator of data quality than TIC as the mass analyzer is usually set to perform MS/MS acquisitions according to several quality criteria. For example, the intensity of an ion has to be above a certain threshold level before MS/MS is attempted. In addition, a certain list if ions can be included or excluded for MS/MS purposes or the instrument might only chose the strongest ion within a range of 4 Th for MS/MS analyses, to ensure it will not perform MS/MS on several isotopic peaks of the same peptide. This shows us basically how much more versatile data dependent measurements are (taking the quality and/or quantity of the MS signals into account for planning the next measurement step) compared to 'ingnorant' data independent measurements (i.e. switching between MS and MS/MS of the strongest peak all the time). More complex aspects of data dependent acquisition will be dealt with in Chapters 3.

The MS/MS total intensity lane tells us online if there is any quality data coming out of the instrument. Also the instrument might display the current data being acquired (MS or MS/MS spectra); this will be a fast changing display, since the data acquisition rate is typically very fast, around a second for a spectrum, depending on instrument and measurement parameters. Not that there is not much point in watching the data from an LC run accumulate. The occasional check will tell us if everything is OK; if not, there is not a lot one can do for the current LC run. If the needle is not in position or clogged up or broken at least the next run can be 'saved' or its start delayed until the problem is remedied.

Offline analysis of the huge amount of MS and MS/MS data generated online is of paramount importance no matter what the intention of the experiment at hand. It is ideal to use another computer for the offline analysis, as this ensures that measurements are not interrupted, should the software crash or should the multitasking on the computer not proceed as planned. It is also a good idea to make sure the computer controlling the LC MS is not connected to the internet, with all the attending security and anti-virus software issues.

After the separation the raw data (or spectral data of 'line spectra'; see the example in Figure 2.35) are automatically analyzed to extract, for example, peptides from it. For this, a charge state might automatically be added (it being very important to decide whether the mass of a peptide detected at 401 Th is 800 or 400 Da; see the example in Figure 2.28) and all the measured MS/MS spectra from different peptide charge states might be connected (e.g. the spectrometer might have produced MS/MS data of the same peptide at 801 Th as doubly-charged and at 1 601 Th as singly-charged ions). Some assignments might be more difficult. For example, if in MS/MS data a peptide was measured at 25 minutes at 801.5 Th and at 26 minutes at 801.4 Th; are these peptides identical or are they different peptides that co-elute by chance? This is usually not interpreted, as to avoid wrong conclusions. It is more efficient to take decisions like this after a database search using the uninterpreted MS/MS spectrum; both peptides might match to the same data set from a database independently, thus confirming their identity. In practice, both sets of data may well derive from the same peptide but return different 'best hits' (see Chapter 4). It is important not to interpret the data too much before they are used in a second step to search in databases for matching peptides/proteins. In approaches for differential proteomics the sample might contain ions that should be separated by, for example, 6 Da in a SILAC experiment (see Chapters 4 and 5). The software will automatically search for such pairs, which might be separated by 2, 3 or 6 Th depending on whether the ions are triply-, doubly- or singly-charged and of course also see whether the charge state and distance correspond with user defined and MS/data quality dependent boundaries of accuracy. MS intensity data can also be used to quantify peptides in the LC run. In a first approximation we can assume that by comparing the peak volume, that is, the intensity at a certain m/z integrated over its typical LC elution time of 30 seconds, we can compare the relative abundance of closely related peptides. By 'closely related' we mean that they share the same amino acid sequence with the same PTMs, but different isotopic compositions (like a pair of SILAC peptides). Changes in amino acid sequence and PTMs can lead to changes in elution patterns from the separating columns, as well as to different ionization/detection efficiencies in MS as well as differences in transition throughout sample handling and hence hamper any such direct quantification. Furthermore, the ionization might be affected by experimental settings (e.g. experiment to experiment variations of the quality of the spray, differences in chemical noise and contaminations with salts or polymers) or limitations of the particular mass analyzer. All these problems make a direct absolute quantification of MS signals from hyphenated experiments impossible. However, relative quantification is possible and by including internal standards an absolute quantification is also possible. Since we need to know a lot more details of the MS background to deal with these questions we will discuss them further in Chapters 4 and 5 after we dealt with the basics of mass spectrometry of biomolecules in Chapter 3.

From all the above, it is clear that MS is a very powerful detection/analysis method for all hyphenated approaches, and that a lot of interpretation and careful adjustment of the measurements (not least the choice of mass analyzer) is needed to achieve optimal results.

2.6 CHIPS IN PROTEOMIC APPLICATIONS

There are several areas in proteomics where chips have been developed to replace more traditional aspects of sample preparation, separation of analytes or interfacing with MS. Most of the more mature products are used as an alternative for gel-based or gel-free separation and the methods employed are basically very similar to hyphenated technologies, which is why we discuss chips in proteomics at this point. The definition of a chip is somewhat loose, and methods such as surface enhanced LASER desorption ionization are described as chip based, as they use MALDI targets with differential binding/ionization properties (Sections 3.2 and 4.2.2).

Often chips for protein separation are marketed as being similar to gel-based approaches, just very quick (minutes), while sensitivity is similar and sample consumption minimal (microlitres, sub-picomolar concentrations). More often than not companies do not tell you what principles the separation is based upon, and more often than not it is capillary electrophoresis. CE has a reputation of being somewhat temperamental, but in the chip environment it is very stable, since all the parameters making it potentially temperamental are kept

constant, such as buffer composition and temperature and internal connections. Most chips come with one separation column only, built into the chip, so the runs have to be performed sequentially, also because there is only one detection device (fluorescent or UV) and the cost of the reader would be higher if there were more than one. Speed and high throughput are achieved because a single run is very short, a couple of minutes, compared to 30–40 minutes for an RP run or over an hour for a typical gel system.

Another flavour of separation chip technology are LC ESI chips, which replace the columns and ESI source of hyphenated approaches and can be interfaced to compatible mass analyzer. By changing the chip it is easy to switch between, for example, 1D and 2D LC. This development delivers similar results to the more traditional hyphenated approach, without the hassle of setting up different columns and taking care of the ESI interface, albeit at a price, making chips more popular in industrial settings.

Chips are also used in 'lab on a chip' approaches, which can save a lot of work in preparing samples for MS analysis. These approaches are not widespread enough at the moment to justify their detailed discussion in this book, but in a few years they may come to play a more important role in proteomics, as they have the potential to standardize procedures, for example in clinical proteomics.

Microarrays can also be used to fish for proteins/ peptides or to analyze protein/protein interactions. This can be done by fluorescence detection of interaction or by surface plasmon resonance, where the additional mass of an interacting partner can be detected by measuring low incidence light scattering. Both methods are available now to perform measurements on a proteomic scale and an example in protein interaction studies is shown in Figure 5.23.

Ultimately the use of chips has the potential to allow more sensitive analyses and minimize loss at different stages of proteomic analysis.

REFERENCES

Binz, P.A., Mueller, M., Hoogland, C. *et al.* (2004) The molecular scanner: concept and developments. *Curr Opin Biotech*, **15**, 17–23.

Consoli, L. and Damerval, C. (2001) Quantification of individual zein isoforms resolved by 2D electrophoresis: genetic variability in 45 maize inbred lines. *Electrophoresis*, **22**, 2983–2989.

Fey, S. and Larsen, P.M. (2001) 2D or not 2D, *Curr Opin Chem Biol*, **5**, 16–33.

Fournier, M.L., Gilmore, J.M., Martin-Brown, S.A. and Washburn, M.P. (2007) Multidimensional separation-based shotgun proteomics. *Chem Rev*, **107**, 3654–3586.

Ghaemmaghami, S., Huh, W.K., Bower, K. *et al.* (2003) Global analysis of protein expression in yeast. *Nature*, **425**, 737–741.

Herber, C.G. and Johnstone, R.A.W. (2003) *Mass spectrometry basics*. CRC Press.

Kellner, R., Lottspeich, F. and Meyer, H.E. (eds) (1998) *Microcharacterization of Proteins*, 2nd edn, Wiley-VCH Verlag GmbH.

Kohlheyer, D., Eijkel, J.C.T., Schlautmann, S., van den Berg, A. and Schasfoor, R.B.M. (2007) Microfluidic High-Resolution Free-Flow Isoelectric Focusing. *Anal Chem.* **79**, 8190–8198.

Liebler, D.C. and Mam, A.J. (2009) Spin filter based sample preparation for shotgun proteomics (author reply). *Nature Meth*, **6**, 11, 785–786.

Liu, H., Afrane, M. , Clemmer, D.E., Zhong, G. and Nelson, D.E. (2004) Identification of *Chlamydia trachomatis* outer membrane complex proteins by differential proteomics. *Anal Chem*, **76**, 4193–41201.

McNulty, D.E. and Annan, R.S. (2008) Hydrophilic interaction chromatography reduces the complexity of the phosphoproteome and improves global phosphopeptide isolation and detection. *Mol Cell Proteomics*, **7**, 971–980.

Millioni, R., Miuzzo, M., Sbrignadello, S., Murphy, E., Puricelli, L., Tura, A., Bertacco, E., Rattazzi, M., Iori, E. and Tessina, P. (2010) Delta2D and Proteomeweaver; performance evaluation of two different approaches for 2DE analysis. *Electrophoresis*, **31**, 1–7.

Patterson, S.D. and Aebersold, R.H. (2003) Proteomics, the first decade and beyond. *Nat Genet*, **33** (Suppl.), 311–324.

Rabilloud, T. (2002) Two dimensional gel electrophoresis in proteomics: Old, old fashioned, but it still climbs up the mountain. *Proteomics*, **2**, 3–10.

Richardson, M.R., Liu, S., Ringham, H.N., Chan, V. and Witzmann, F.A. (2008) Sample complexity reduction for two-dimensional electrophoresis using solution isoelectric focusing prefractionation. *Electrophoresis*, **29**, 2637–44.

Simpson, R.J. (2003) *Proteins and Proteomic, a user manual*. CSH Laboratory press.

Westermeier, R. (2004) *Electrophoresis in practice*, 4th ed, Wiley.

Wilkins, M.R. (1997) *Proteome research: new frontiers in functional genomics (Principles and practice)*. New York: Springer-Verlag.

Wilkins, M.R., Williams, K.L., Appel, R.D. and Hochstrasser, D.F. (eds) (1997) *Proteome research, new frontiers in functional genomics*. Springer.

Wolters, D.A., Washburn, M.P. and Yates, J.R. III (2001) An automated multidimensional protein identification technology for shotgun proteomics. *Anal Chem*, **73**, 5683–5690.

Zischka, H., Braun, R.J., Marantidis, E.P., Buringer, D., Bornhovd, C., Hauck, S.M., Demmer, O., Gloeckner, C.J., Reichert, A.S., Madeo, F. and Ueffing, M. (2003) Improved proteome analysis of S. cerevisiae mitochondria by free flow electrophoresis. *Proteomics*, **3**, 906–916.

Analysis of Peptides/Proteins by Mass Spectrometry

3.1 BASIC PRINCIPLES OF MASS SPECTROMETRY FOR PROTEOMICS

Mass spectrometry is an ever more popular analytical tool in many areas, from the point of a mass spectrometrist mainly defined by the substance class that needs to be analyzed. Proteins and peptides fall into the class of biomolecules, which also contains RNA and DNA. Indeed, many of the methods we discuss here can also be applied analogously for the analysis of nucleic acids. In terms of proteomics this becomes interesting when the interaction between proteins and RNA/DNA is to be analyzed; the same instrumentation and very similar ionization and separation processes can be used as well as identical data analysis approaches. Also metabolomics has partially similar approaches; some ionization principles, mass spectrometers and separation technologies used for metabolomics can be used for proteomics and vice versa; one just has to optimize settings to account for the smaller size of the compounds (usually comparable to peptides up to three amino acids), specific MS/MS behavior and the generally more diverse nature of the compounds. One advantage is that there are fewer metabolic compounds than protein species (around several thousand versus over a million). So much of what is said in this chapter may also be useful as an introduction to mass spectrometry for nucleic acids or metabolites.

Let us remind ourselves of the main aim of mass spectrometry for proteomics. For most organisms either we have genomic sequences, extensive mRNA, DNA or protein sequence databases, or we know of an organism or group of organisms that are similar in their genetic sequences. Mass spectrometry is thus used to generate lists of masses that can be compared to the existing pool of databases, which can sometimes be full of redundancies, faults, omissions and sequences that do not exist as proteins. This can be done in different ways (Figure 3.1); if we use whole proteins as a starting point for our mass spectrometric analysis we speak of top-down proteomics, since we start with the whole protein and fragment it during our analysis, measuring the masses of the fragments and comparing them to expected database entries.

Top-down proteomics works well on proteins with known sequences but unknown posttranslational modifications (PTMs), which we can deduce from the mass of the fragments if the quality of our mass spectrometric data is good enough. If we take proteins, or mixtures of proteins in the form of whole cells, and cut them into peptides before we do mass spectrometric analyses, we speak of bottom-up proteomics – 'bottom-up' because we start with something small (peptides) and try to rebuild the complete protein including PTMs from our experimental data and comparisons with the sequences from the above mentioned databases. Most proteomic studies are bottom-up; until very recently we simply did not have the ability and the mass spectrometers to perform top-down proteomics on anything but small, very abundant and easy to isolate proteins. If we have a pure protein (or a simple mixture of two to five proteins, one of which is predominant) a simple measurement of the fragments masses is enough to identify the protein from the databases using peptide mass fingerprinting (Figure 3.1). This approach might even allow us to annotate the odd PTM by bottom-up proteomics. This approach is based on the fact that we know the rules that lead to the cutting

Introducing Proteomics: from Concepts to Sample Separation, Mass Spectrometry and Data Analysis, Josip Lovrić
© 2011 John Wiley & Sons Ltd

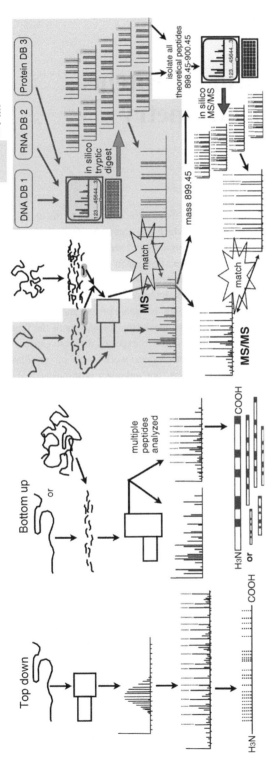

= PMF

Figure 3.1 Mass spectrometry as tool for proteomics. In top-down proteomics a purified protein is analyzed by a high resolution/performance mass spectrometer, typically resulting in a multitude of isotopic peaks and charge states. The protein ions are then fragmented in the mass analyzer, and by careful comparison of fragment masses with masses expected from the (known) sequence of the protein the presence of mutations or PTMs can be exactly localized. In bottom-up proteomics a protein (or a mixture of proteins) is digested by trypsin into peptides around 10–15 amino acids in length. These mixtures of peptides are analyzed by MS or MS/MS. Often there is a peptide separation step before the mass analyzer. From the data derived by MS or the additional fragmentations of the peptides in the mass analyzer (MS/MS) the presence or modification of proteins can be deduced. This information is always based on the identification of peptides, and often around 20–60% of the protein sequences are reliably covered by the peptides analyzed. Proteins can be identified by peptide mass fingerprinting (areas shaded grey). For this purose single proteins, often isolated from gels, are digested by trypsin and the masses of the resulting peptides are deduced from MS measurements. Sequence data from DNA/RNA databases (DB) are translated into protein sequences, and all these sequences are transformed into peptides that would be expected to result from trypsin treatment. For each sequence entry in any database a tryptic peptide mass fingerprint is thus generated (in silico tryptic digestion). All these theoretical spectra are compared on the fly with the experimental data, and the closest match is used to correlate a protein from an experiment with a database entry; the protein is identified. Protein identification by MS/MS can work on single proteins or even very complex protein mixtures as samples. All proteins are digested with trypsin, and the masses of all peptides identified by MS. Each peptide is identified on its own from MS/MS data. First all theoretical tryptic peptides from all databases that match the experimental mass to within a given experimental error (e.g. ±1 Da) are extracted. For each of these peptides a list of fragments expected to be generated by fragmentation inside a mass analyzer is generated (in silico MS/MS). The peptide from the experiment is identified by comparing the real data to all these spectra and finding the closest match. MS/MS data can generate ion series which result from peptide fragmentations in the same positions relative to the peptide backbone, just at different amino acid positions within the peptide sequence. Such fragments can be used to deduce the peptide sequence peptide outright (de novo sequencing), even if it is not present in any database. De novo sequencing needs good quality data and more sophisticated data interpretation, compared to MS/MS identifications from database searches of not-interpreted spectra.

of our protein to peptides of 5–30 amino acids; usually it is the sequence specificity of trypsin, which is used in the overwhelming majority of proteomic approaches. If we know the masses of each of the fragments we can generate a list of them – the peptide mass fingerprint (PMF). If we compare this to a PMF list generated from every single entry of the above mentioned sequence databases, we can generate a close match. Using statistical methods we can assure ourselves that the match is relevant; we say that we have identified a protein by PMF analysis. However, in mixtures of proteins this approach does not work; there are too many possibilities. For protein mixtures we need MS/MS approaches. Here we isolate single tryptic peptides in the mass spectrometer, measure their masses, fragment them and measure the masses of the fragments as well – two consecutive mass spectrometric measurements hence the name tandem MS (or MS/MS or MS^2). Peptides fragment in mass spectrometers following certain (not absolutely strict) rules. Knowing the rules, we can again compare our experimental data for each peptide with theoretical fragments of each peptide that (within a margin of error) were identical to the peptide mass before fragmentation. With this approach we can identify the presence of a protein in a sample by the presence of one of its peptides. However, as the data might still be ambiguous, more than one peptide is needed in practice to be sure about the presence of the protein. In order to identify PTMs we might need data of very high quality. Finally, it is even possible to deduce the sequence of peptides from MS/MS data; we can even use such data to correct and extend the (RNA and DNA) databases. However, this so-called de novo sequencing is rare, as high quality data are needed and often we do not have enough protein or time to generate them. All these analyses are explained in detail in Chapters 4 and 5, but it becomes clear that the analysis can be more meaningful if the data are of high quality. Just what 'high quality MS data' means and how they are generated is the subject of this chapter.

3.1.1 Introduction to mass spectrometry of peptides and proteins

The revolution in mass spectrometry of biomolecules has been one of the major drivers in the development of proteomics, and without it proteomics as we know it today would not be possible.

While MS has long been a very sensitive detection and analysis tool for small or volatile compounds, it only became possible to analyze biomolecules such as RNA, DNA, peptides and proteins in small amounts in the early 1990s. The problem was that biomolecules have a very low tendency to evaporate; their gas pressure is extremely low. MS always works with the analyte in the gas phase, better even when these analytes are ions. So for a long time the major problem for the use of MS was how to gently transfer massive molecules into the gas phase and ionize them. Vaporization and ionization have to be achieved without breaking the peptides/proteins into small fragments, which would be devoid of useful information. Matrix assisted LASER desorption ionization (MALDI) and electrospray ionization (ESI) became available around the same time for this kind of sensitive biomolecule analysis. In the early days of proteomics, MALDI equipped mass spectrometers were much more sensitive (albeit not very accurate at the time and not very good for MS/MS) than ESI equipped machines, but the sources for ESI MS became ever smaller and the flow rates lower. With the advent of nano-liquid chromatography systems and offline sprays, both ionization technologies in combination with different mass spectrometers became able to detect peptides at the femtomole level (10^{-15} M, or a few nanograms of a protein) under experimentally relevant settings. At the same time the accuracy of measurements at this high sensitivity reached a staggering 5–50 parts per million (ppm), and some modern mass spectrometers can achieve under 0.1 ppm for MS and MS/MS analyses under experimental conditions in the real world. Translated to measuring the weight of a person, this would mean we could deduce from the person's weight whether or not they were carrying a postage stamp. ESI remains today the more flexible (and slightly more difficult to handle) ionization method and can be seamlessly combined with LC analyses. MALDI is mostly used in the offline mode, that is, the measurement is not time critical and is independent of the analyte separation. ESI is mainly used in online mode, that is, the samples are ionized and measured as they elute from the final column, so the time one can spend on measuring a certain protein or peptide is limited to about 30 seconds, and this goes down to about 1 seconds in complex samples where different analytes will elute at the same time, all the time. We mentioned the advantages of combining ESI with LC methods in Section 2.4 – mainly purification and concentration of the analyte by several orders of magnitude – and these advantages are lost in offline ESI, where a (reasonably clean) sample can be sprayed at optimal flow rates for detection (i.e. at the highest sensitivity) for about 20–30 minutes allowing extensive measuring time, but only for a limited amount of individual analytes.

Next to superb sensitivity and versatility, the other important development in the field of mass spectrometry was of course that the whole measurement process (including ionization) became very robust, making it unnecessary to be an MS expert to be able to get good results for proteomic purposes. In fact, sample preparation and processing optimization are now much more of an issue than instrument sensitivity for the utilization of MS in proteomics.

Once the proteins or peptides are in the gas phase and ionized, there are (for the biochemist) an increasingly confusing variety of mass spectrometers that can be used successfully for proteomic analyses. These include time of flight (ToF), quadrupoles and a diverse set of ion trap instruments as well as combinations of these. Ion traps in particular have been developed into very versatile instruments, ranging from basic to high end devices, with corresponding performance. Although all of these mass spectrometers are immensely complex, they follow Antoine de Saint-Exupéry's notion that technology in perfection becomes invisible, in that for the end-user these very complicated machines are becoming easier and easier to use. They all have slightly different detection systems, tailored to the specific way in which ions are transmitted in the instruments. In most cases the ions are detected destructively, that is, they are accelerated towards a detector and the 'hit' is electronically monitored. Usually the time of the hit is correlated to the way the ions are manipulated inside the mass spectrometer and the correct mass of each ion is detected in this way. Another detection principle is used in Fourier transformation mass spectrometry (FTMS). Here ions are made to move along circular paths, sometimes perpendicular to a very strong magnetic field. Their movements can be measured without destroying the ions by measuring a tiny current induced by the motion of the ions; however, the complex summary of the data has to be interpreted by Fourier transformation (FT), to determine the mass and charge of every component. FTMS is one of the most accurate and sensitive, but also most complex, approaches, with applications in proteomics that go far beyond the scope of this book, such as peptide chemistry in the gas phase.

All these complications do not matter much to the typical beginner, and even an instrument with a very complex principle of operation can be easy to operate on a daily basis, as we will see in Sections 3.3.2 and 4.3. The aim of this book is to introduce the basic principles and how they influence the application of a certain mass spectrometer for a certain proteomic application.

At the end, all this technical wizardry can deliver the only information we will ever get from a mass spectrometer – the very accurate mass of a peptide or protein. To get even more specific information we can also apply MS/MS analyses; the mass of a certain compound (peptide or protein) is measured (first MS), and then one of the compounds is fragmented in a controlled fashion (second MS, MS^2, MS/MS or tandem MS) inside the machine and the fragment masses are measured. Since molecules tend to brake at predictable sites, we have the mass of the original (or parent or molecular) ion and the fragment (or daughter) ions. These as a set can even deliver the complete amino acid sequence of a peptide or protein, including PTM sites. Such a complete data set is not always needed in proteomics (see below) and can be difficult to get for several reasons, the most obvious being the amount and purity of the analyte, the time needed to measure it, and the time needed to interpret the data if you look at several thousand proteins and hundred thousands of spectra in a day-long experiment. The more material of a given analyte you have, the longer the time you can spend on analyzing a single peptide or protein, the better the results might be, but there is an element of 'luck'; some PTMs, proteins or even peptides are harder to detect than others, their sequence might not be conductive to proteolytic digestion, and they might not 'fly' in the mass spectrometer or be very unstable for a number of reasons. So getting very specific information can sometimes be very time-consuming or impossible.

To help with the classification of mass spectrometers you can sometimes see comparison tables, with values for accuracy, resolution or other performance specifications. All these specifications are met by the different instruments under slightly different conditions, so these specifications are not always directly comparable. It is a bit like comparing knives with forks or claiming you need four prongs on your fork – three are quite sufficient to do the job. Or, to put is plainly – whether your MALDI ToF mass spectrometer (according to its specifications) can achieve an accuracy of 50 or 10 ppm will not make a real difference if your sample is contaminated with salts, and you need to 'fire' at a LASER power setting higher than optimal; all signals will blur by about 80–150 ppm due to the LASER, which is too strong. Or as another example, comparing a resolution of a ToF instrument to an ion trap or quadrupole instrument is difficult, since you can trade off sensitivity and range in the ion trap/ quadrupole for a better resolution by changing the acquisition parameters of the instruments. In the following chapters I will therefore try to explain the utilization of the different MS technologies as they apply to proteomics. This should give a good indication of what will work best (or close enough to best not to matter) for the problem you might be facing.

It is commonly said that 'mass spectrometers don't lie', but all the information they deliver needs to be considered alongside the background of the sample, the instruments and the operator; there are 'ghost peaks' in ion traps when operated inappropriately, wrongly assigned charge states, circumstantial database hits, 'squashed up' ion distributions in ToFs at the border of detector saturation, and accuracies well beyond the expected limits, if the resolution is not as good as it should be, for whatever reason, on any mass spectrometer. This is why every experienced operator runs standards (or samples where they know what they should be seeing) before they start working or even during some measurements; the more one can trust the results of an MS measurement, the more can be achieved with the information in terms of answering biological questions.

A specific requirement of MS for proteomics is that a lot of time and effort must be put into interpreting the data after the measurement has been done. This step is sometimes quite confusing for the beginner, and one aim of this book is to communicate the basic principles that will allow you to 'dig into' your proteomic MS data set. Of course computers can do much of the interpretation, but if we do not understand what these machines do, we have no chance of understanding the implications of our data. Only when we have some form of understanding of these principles can we understand how far to trust our data. Mass spectrometers don't lie, but they don't tell the truth either; it is for us to tease out the truth (or the most likely version of the truth) from data we have gathered with impartial machines. From this point of view, the less material you have the longer the interpretation will take, and while very sensitive measurements can be done it is often easier just to start with more material, which

may also limit the different possibilities of interpretation. But no matter how much material there is, the less material there is at the beginning the more involved the whole process of experimental design, measurement and data interpretation becomes. The workload increases exponentially from trying to identify less and less of a given protein. From another angle, the workload also increases exponentially with our demands on the outcome: from the match of an MS data set to a group of database entries (homologues? fragments? splice variants?) to ensuring that a protein really represents one specific database entry, to ensuring that a protein is present in full length and full concordance with a single database entry up to detecting some PTMs, and finally to ensuring you can make quantitative measurements of all PTMs of a protein, which can turn out to be very difficult in a specific case. One should also never forget that there are hundreds of known (and potentially many more still unknown) chemical modifications, biological or artificial, that might affect your sample, and one is often hard pressed to explain more than 30% of all peaks present in a given spectrum. Even if we can only account for less than 70% of the peaks/ ions present in an experiment, we can still draw valid conclusions from the data set, such as the presence and amount of a certain protein, but not all present modifications.

Next to its very high sensitivity, MS has the advantage that you can apply it to any protein or peptide you are likely to encounter; due to the rapid development of RNA/DNA and protein databases we know (at least part of) the sequence of most proteins we are likely to encounter. Performing the same manipulations that we perform during an experiment with the theoretical sequences in a computer, we can predict the data from the mass spectrometric experiment. Now the mass spectrometer does not need to derive the sequence of a protein, or, to be more precise, it does not have to derive high enough quality fragment mass data that could be interpreted without doubt to be consistent with one specific amino acid sequence only. In most cases we merely need to match our experimental data to a data set of a particular protein in a database and we would say we have 'identified' a protein (or one its modifications in some cases). Other extremely sensitive methods, such as any of a number of immunodetection methods, need a specific component for every protein (e.g. the antibody). Thus massive parallel immunological methods can be prohibitively expensive to perform, and even impossible if we do not have enough knowledge (and raw protein) to create the specific antibodies in the first place. Moreover, MS is very flexible and can be used in a tailored fashion for a variety of approaches, be it the identification of proteins, their quantification or the analysis of chemical modifications such as PTMs.

Mass spectrometers are very expensive – for proteomic applications from about US$140 000 up to over $1 million for a single instrument. Their big advantage is that they allow an enormous amount of measurements over their lifetime, (say) 1 000 samples a day for a MALDI ToF, over a lifetime of 3 years, that is nearly a million samples, or some $0.20 per sample. The whole MS measurement, including sample preparation, can cost less than $1 per sample. These costs are similar to more classical analyses such as enzyme linked immunosorbent assays, without their typical limitations (see above).

Let us now look at some typical MS data. We have reflected upon some of the principles of MS data in Section 2.1.2. The mass spectrometers used in proteomics do not actually measure mass; they measure mass to charge ratios or m/z, where m is given in dalton (Da) and z is the number of charges (positive or negative). The derived unit for m/z is the Thomson (Th), and we will use this unit from now on, although it is not a recognized SI unit (yet?). However, as we will see, some ionization technologies (such as MALDI of typical tryptic peptides) deliver almost exclusively singly-charged ions, so measurements taken in Th (or m/z) are often laxly expressed as Da. Since most measurements are taken in the positive mode the MALDI ToF still does not measure the mass of our analyte, but the mass +1 Da for the charge carrying proton (Figure 3.2). But even MALDI sometimes produces double-charged ions of short peptides, and most certainly does so for proteins.

ESI is a very efficient ionization method and will always produce several charge states. For peptides one will find in typical proteomic experiments predominantly singly-, doubly- and triply-charged peptides. Proteins are always seen in multiple charge states; about 20 charges for a medium sized protein and about 50–60 charges for an IgG antibody (150 kDa) are typical, with a wide distribution of charge states observed in parallel, for example from 7 to 15, depending on the ESI conditions and the length and sequence of the protein (see Figure 4.11). In most proteomic measurements we will deal with

peptides, and typically we will find different charge states for each peptide (even in a complex mixture) at the same time (see Figure 2.35). Often one charge state is found to be predominant. Depending on the sequence, some peptides are more likely to be found in a certain state (e.g. tryptic peptides with multiple Arg or Lys residues are usually found in higher charge states) than others, but this can change according to experimental conditions. For example, we might find a peptide exclusively doubly-charged on one day and find a mixture of single and doubly-charged on the next day. The reasons for this might not always be obvious and can be anything from the position of the ESI needle, to contaminations with salts or changed flow rates or instrument settings. We will later see how we can still get a reproducible interpretation of the spectra (e.g. quantification with internal standards), but first let us discuss how we can determine the actual mass of our peptide.

To determine the correct mass from analytes with different charge states we need two elements: one is a beautiful quirk of nature (ultimately quantum physics) and the other is a high resolution mass spectrometer. We and all biological materials are made out of the ashes of exploding stars; all atoms up to the weight of iron are produced during the 'normal' life cycle and all heavier atoms during violent explosions. Due to these special circumstances during the creation of matter and the radioactive decomposition of elements since then, all atoms come in different isotopes. Some of these isotopes are unstable (i.e. radioactive), but many are perfectly stable over the lifetime of the universe. Take carbon as an example: 98.9% of all carbon we encounter in our samples is made up of the ^{12}C isotope, with 6 protons, 6 electrons and 6 neutrons (Figure 3.2); 1.1% of naturally found carbon is of the ^{13}C isotope, with 6 protons, 6 electrons and 7 neutrons. Carbon 13 is perfectly stable, although there are some differences between ^{12}C and ^{13}C, which have no real relevance to proteomic applications (e.g. the enzyme RuBisCO has a preference in reacting with ^{12}C over ^{13}C, so biological organic material has a very slightly higher $^{12}C/^{13}C$ ratio than abiotic organic material). In addition to carbon we encounter oxygen, nitrogen, hydrogen, phosphate and sulfate in proteomic samples. As we can see in the table in Figure 3.2, the natural occurring atoms are made up of nearly 100% of the most abundant isotopes, so except for ultra-high precision measurements in FTMS, we can concentrate here on the isotopic distribution patterns for

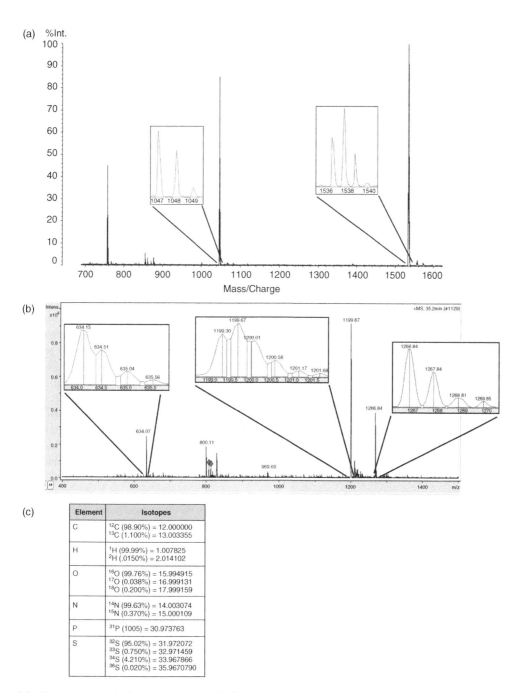

Figure 3.2   Mass spectrometric data for proteomics. (a) Shows a simple mixture of three peptides from MALDI MS. The insets show the isotopic peak distributions for two of the peptides. Notice the distance of the isotopic peaks is 1 Th, thus the ions are singly-charged. (b) Shows data from an ESI LC run using a Pauli ion trap. Insets show the isotopic peak distribution. Isotopic peaks are separated by 1 Th (single charged), 0.5 Th (doubly-charged) and about 0.3 Th (triply-charged). (b) The resolution of the ion trap under the experimental setting (optimized for sensitivity/scan speed) struggles to give a precise result for the triply-charged ion. (c) The table shows the isotopic distributions and precise isotopic masses for elements relevant in proteomics.

carbon exclusively (if we ignore the effect of ^{34}S, in peptides containing Met and Cys).

So how does this complication of the origin of matter affect our measurements and what is the influence of resolution? In Figure 3.4 we see how a singly-charged peptide of actual mass 1 046.01 Da appears at a resolution of about 2 500. The strongest peak appears 1 Th larger than the actual mass of the peptide. In proteomics most samples are measured in the positive mode. The positive charge is carried by a proton, which adds roughly 1 Th to the real weight. We should always make sure if a certain annotation of a peptide is made in terms of real mass, or as it appears in the MS (real mass + 1 for a single charge). Apart from the highest peak, we see a 'tail' of peaks 1 Th apart; these peaks are derived from molecules that contain increasing numbers of the heavier ^{13}C carbon. A peptide of mass 1 000 Da contains about 50 carbon atoms. Since 1% of these are of the ^{13}C isotope, we will get a statistical mixture of signals. Every mass spectrum is composed of the measurement of some 10^9 molecules and even the most sensitive measurements are still composed of 10^6 single molecules. For 50 carbon atoms with 1% ^{13}C and 99% ^{12}C, statistics predict that the strongest signal is the 'pure' ^{12}C peptide and this is what we measure, the strongest signal at 1 047.01 Da is composed exclusively of ^{12}C.

This 'king of the peaks' is called the monoisotopic peak (it only contains one isotope of carbon – or any other element, for that matter). It is the mass of the monoisotopic peak that we usually mean when we speak of the mass of a peptide or protein in mass spectrometry or proteomics. This is different from the 'nominal mass', expressing the mass of an atom as an integer only (e.g. the mass of nitrogen would be given as 14 or 15 and not 14.003074 and 15.000109; see table in Figure 3.2). The other isotopic peaks are very important as well; they can be used to determine the charge state of a peptide, as we will see in detail below. Since the isotopic peaks are made up of mixtures of ^{12}C and ^{13}C with increasing numbers of ^{13}C, they are 1 Da apart from each other in their absolute mass, but not always 1 Th apart from each other in the read-out of our instrument, as this depends on the peptide's charge state (see Figure 3.2b).

The monoisotopic peak is not always the strongest of the isotopic peaks, in fact, as the number of carbon atoms per molecule increases the chances are that at some point there is more molecules containing at least one ^{13}C than there are monoisotopic molecules. This happens for most tryptic peptides somewhere around 1 500 Da (Figures 3.2 and 3.3), or at above 80 carbon atoms. In fact, if the monoisotopic peak is strongest of all the isotopic peaks at 2 000 Da, you can assume that the peak is not derived from a peptide at all, but some contaminant containing less carbon than a peptide made out of standard amino acids. From about 2 000 Da onwards, the isotope containing two ^{13}C atoms becomes the strongest. From about 3 000 Da onwards it can become difficult to see the monoisotope clearly; it can become so weak as to disappear into the background and the third or even fourth isotopic peak becomes the strongest of the isotopic peaks (see also Figure 3.3). Here we encounter the problem that we need a good signal to noise ratio to detect our peaks and to determine accurately where they are. A peak starts to become visible at a signal to noise ratio of about 3. This means that the maximal signal intensity of the peak is 3 times higher than the average background signal intensity of the spectrum in the neigbouring m/z areas without any peaks. At low signal to noise ratios (3–10) the accuracy of the 'middle of peak' or 'top of the peak' determination (this is the mass we measure) is influenced by irregularities in the background signal (noise) and measurements at higher signal to noise ratios have the potential to be more accurate. Depending on instruments and sample, very intense peaks sometimes do not produce accurate measurements; the intensity of the 'tip' of the peak might be clipped, as the signal intensity is too high for the instrument's detector. This distorts the peak shape and diminishes accuracy (Figure 3.3).

The greater the mass of a peptide, the higher is the resolution of the measurement necessary to distinguish the peaks; in Figure 3.3 you can see how the isotopic peaks become harder and harder to distinguish with increasing peak mass. In Figure 3.4 the popular definition of resolution as full width at half maximum (FWHM) is shown; this definition is based on peak shape. One has to divide the apparent mass of the peak (here 1 246.63 m/z) by the peak width at half the maximum intensity of the peak (here 0.15 m/z). Since both values are expressed in the same unit (m/z or Th) the resulting resolution of about 8 300 is dimensionless or absolute. There are other definitions, such as the ability to resolve a peak from another mass at a given mass or the difference in intensity between the monoisotopic peaks and the

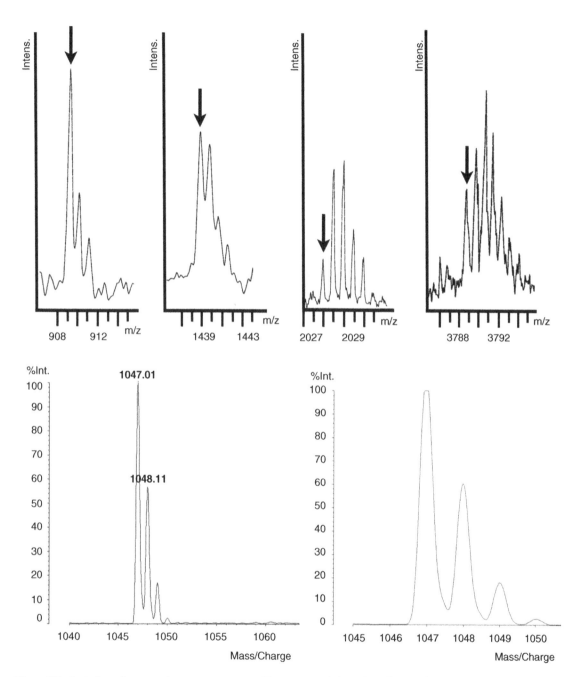

Figure 3.3 Isotopic peak patterns in mass spectrometry. The upper panel shows how the monoisotopic peak patterns of peptides change throughout the mass range. Arrows indicate the position of the monoisotopic peak. See text for more details. All measurements are from a reflectron MALDI ToF MS. The signal to noise ratio of the different peaks is 20, 15, 8 and 15, respectively. The lower panel shows two measurements of the same peak. The first has a signal to noise ratio of 360, the second over 500. Although the signal to noise ratio of the latter spectrum is higher, the measurement itself is not of good quality; the signal intensity was too strong for the detector of the mass analyzer, the tip of the monoisotopic peak is clipped, making an exact determination of the mass impossible. All measurements are from a reflectron MALDI ToF MS.

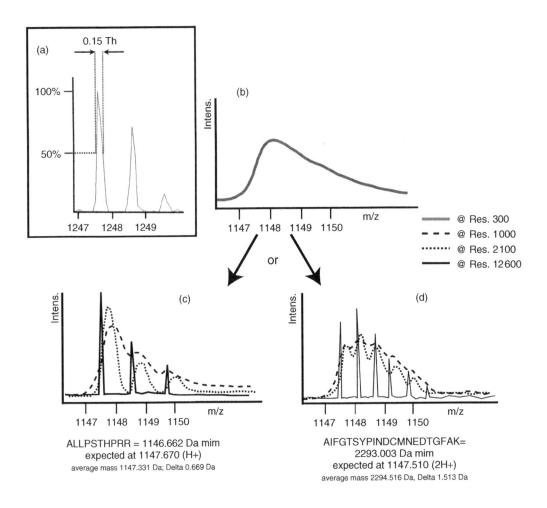

Figure 3.4 Definition of resolution in mass spectrometry. (a) Shows the definition of resolution as FWHM, see text. Note that resolution is peak dependent, not peptide dependent. The second and third isotopic peak will have different (lower) resolution, as their signal to noise ratio is also lower. The measurement of the peptide ATDFGAAIPAGTR was taken on a reflectron MALDI ToF MS. (b) Shows the measurement of a peptide at low resolution. At increasingly higher resolution the measurement could yield the results shown in (c) or (d). In (c) a singly-charged peptide is measured, in (d) a doubly-charged one. See text for further details. Mass analyzers used in proteomics have resolutions from 2 000 to 100 000 under standard conditions. Notice the difference between the different masses for the peptides indicated under the spectra; the monoisotopic mass differs by 0.699–1.513 Da from the average mass, which results from the integration of all isotopic peaks. All masses are measured in the protonated form, adding 1.008 Da to the theoretical masses, which then results in the experimental mass, sometimes also called the expected mass. Also note that the position of peaks changes with resolution; see Figure 3.6 and text for more details.

'valley' between them. We will stick with the FWHM definition of resolution throughout this book, because it is easy to understand. It is important to understand that all resolution values have to be related to a certain mass; in our example the measurement had a resolution of 8 300 at 1 246 Th (Figure 3.4). The resolution at 5 000 Th measured in the same mass spectrometer (and indeed the same spectrum) could be very different and would depend on the ionization method, sample quality and the type of mass analyzer used. So comparisons of resolution of different instruments (or for different applications) are worthless if they do not contain this information, as we will see later in this chapter, when we describe the different mass analyzers. It is important to know that if we wish to see the isotopes of a peptide as separate peaks, the FWHM resolution has to be at least 1.5 times (better twice) that of the mass in dalton that we are looking at. Let us work this out with an example. The spectrum in Figure 3.4 (b) could show either a measurement of a peptide of around 1 146.7 Da (singly-charged, minus 1 Da of the indicated mass for the H^+) or another peptide altogether, perhaps one of 2 293 Da (if it was doubly charged). At this resolution of 300 we cannot see the charge state. We actually have no useful measurement at all, as the mass could equally well be 'something like' 1 147 Da or 'something like' 2 293 Da. At a resolution of about 1 000 we start to see separate isotopic peaks (Fig. 3.4 (c) & (d)); at a resolution of 2 100 these transform into resolved peaks and for the first time we would be able to see a clear difference between the peptides of 1 246.7 (1+) and 2 293.0 (2+). If we take experimental effect such as statistical background (signal 'noise') into account, we can only tell the charge state and hence the mass of the peptides at the resolution of 2 100; for the singly-charged peptide we see clearly resolved peaks, while for the doubly-charged one we can imagine there is a peak between the two peaks at 1 147 Th and 1 148.5 Th, giving the isotopic peaks a distance of 0.5 Th, and thus showing the peptide to be doubly-charged. However, what if it was a triple-charged peptide? We can exclude a triple charge from the measurement in Figure 3.4 (d) clearly from a resolution of 2 100 and greater. From the data in Figure 3.4 (d) we would be able to distinguish a triple charge only just from a double charge at this resolution, since a triple charge would produce four peaks within the distance of 1 Th. If we needed to establish that a peptide was triply-charged we would wish to see all the isotopic peaks clearly separated, otherwise it could be quadruply-charged (4+). A triple charge becomes very clear from the date in Figure 3.4 (d) at the resolution for 12 600. Hence the notion that we need an FWHM resolution at least twice the mass of the peptide in dalton that we wish to analyze in order to clearly distinguish higher charge states.

The term 'resolution' is also used to describe the minimal difference in between two masses that can still be clearly separated in a measurement. A similar definition of resolution it is particularly useful to describe the capabilities of an instrument to select or isolate ions for further analyses, such as an MS/MS experiment (Figure 3.5). The resolution of selection for such a precursor ion for subsequent MS/MS is defined by $R = m/\delta m$, where δm is the distance between to peaks needed to clearly separate them and m is the mass of one of the peaks. There are several definitions of 'clear separation', but if you use the following definition you will not be far off the mark. If an ion of 1 000 Th is selected at resolution 200, then the ions outside a window of 5 Th centred around the selected mass are suppressed in intensity by 50%, outside a window of 10 Th by about 94% (assuming a Gaussian distribution). In other words, if one can select an ion at resolution 1 000, for practical purposes one has only ions of 1 000 ± 0.5 Th to work with.

Any mass spectrometrist will compare different instruments based on resolution, rather than accuracy. This is initially confusing for the novice; is an accurate measurement of masses not all we expect from a mass spectrometer? This is because absolute and relative accuracy need to be considered for precise measurements (see below) and because resolution and accuracy are linked; the level of accuracy is tightly linked to the resolution an instrument can achieve. Although we only spoke about resolution when we looked at Figure 3.4, you will have noticed that the peaks are not in the same position at all! Measurements at a resolution of 300 had their peak intensities about 0.7 Th away from those at a resolution of 12 600. This results in a difference of 0.7 Da for the singly-charged peptides and about 1.5 Da for the doubly-charged ones. This is explained by the nature of the measurements; at resolution 300 the measurement is like a blurred photo, the intensity of the peak is the sum of the intensities of all peaks present. These peaks are in our example of course isotopic peaks, and the measurement at low resolution measures the average weight of a peptide. The monoisotopic peak (in this example) contributes most to the average weight, but the peaks with 1, 2 and 3 carbon 13 atoms pull the average peak to the higher masses, till it nearly reaches the second isotopic peak. The average weight is not very useful for proteomic interpretation, and we are much better of with the monoisotopic mass of a peptide; remember that this is the mass of a molecule that contains exclusively ^{12}C atoms, and not a single ^{13}C atom. The monoisotopic peak is much 'sharper' and well defined, so it is possible to measure the monoisotopic mass at much higher accuracy compared to the average mass. Even when we start to see isotopic resolution, the results change with higher resolution; the numerical value of the mass of the

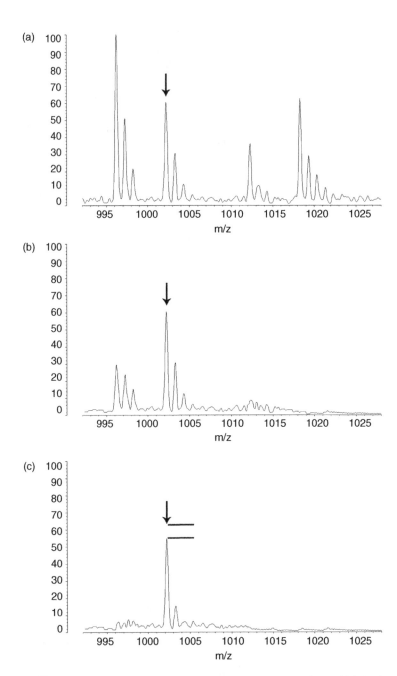

Figure 3.5   Resolution for the isolation of precursor ions. (a) Shows part of a spectrum from which the ion at 1 002 Th is to be selected for MS/MS fragmentation. (b) Shows the selection of this ion at the resolution of 200. (c) Shows the precursor ion selection at a resolution of 1 000. Note that selection at the resolution of 200 does not allow to select the ion 'cleanly'; the peptide at 996 Th still contributes about 30% of the total ion intensity towards the selection. In MS/MS experiments fragments from this ion would make interpretation difficult. At a resolution of 1 000 the ion is isolated sufficiently and even the isotopic peaks are suppressed. The signal of the chosen peak is also suppressed by a small amount. Most mass analyzers can isolate precursors at a resolution better than 500. All measurements taken with a reflectron MALDI ToF analyzer.

monoisotopic peaks is moving as the resolution increases. The higher the resolution, the sharper the monoisotopic peak becomes, and the centre of the peaks moves towards the left. This is an increase in accuracy, as the peaks as we see them result from adding up the signals for all the observed peaks; at low resolutions the signals of the isotopic peaks are very wide and stray to the left and right. Thus the centre position of the monoisotopic peak moves to the left, because stray signals from the other isotopic peaks contribute more to the left than to the right side of the peak. At resolution 1 000 the top of the monoisotopic peak is about 0.2 Th away from the top of the monoisotope at resolution 12 600. Here we come to the definition of accuracy often used in mass spectrometry, the accuracy expressed in parts per million (ppm). A fault in the measurement of 0.2 Th at 1 147 Th can be expressed as a fraction of the measured mass; 0.2 Th is about 0.17% of the measured mass, or 170 ppm. Expressing the accuracy in parts per million has the advantage of being independent of the size of the actual peptide; often if a particular measurement is accurate to 0.2 Da at 1 130 Da, it will be accurate to about 0.4 Da at 2 260 Da, or 160 ppm over the whole range of masses measured. Such accuracy is within the useful range for PMFs to match (or, as is often said, 'identify') proteins from databases. But of course, the more accurate the measurement, the more confidence one can have in the results.

So the higher the resolution the more accurate the position of the monoisotopic peak. But what is the actual position of the monoisotopic peak? Is it the highest point? Is it in the middle between the bases? Where exactly is the base of the peak? The centre of the peak when only the top 40% of the peak intensity are taken into account (see Figure 3.6)? Peaks are not symmetric, and repeat measurement result in slightly different positions and peak shapes, so any of the suggested methods could be used to come close to the real mass of the peptide. While it would go too far to discuss in this book all the implications that all the aspects of peak position determination have on the accuracy of a particular measurement, it is fair to say that the differences between all these different choices for peak position determination become very small at higher resolutions. At resolution 1 500 the peak width at 1 247 Th is about 0.75 Th (by definition, see above), at resolution 3 000 it is about 0.37 Th. At a resolution of 5 000 (typical for real measurements with a reflectron MALDI ToF MS of peptides from gels) it is

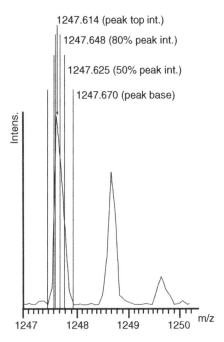

Figure 3.6 Accuracy of mass spectrometry data. Measurement of a peptide at around 1 247.6 Th at a resolution of 8 300. Depending on how the centre of the peak is defined (see text), the measured mass varies from 1247.614 Th to 1247.670 Th, a variation of 45 ppm, which is also the highest possible accuracy with this instrument. The peak width is 0.15 Th. At a resolution of 25 000, the peak width would be only 0.0499 Th, which would allow an accuracy of about 5 ppm. At resolution 60 000 the peak width would be only 0.021 Th, allowing an accuracy of about 2 ppm. These accuracy estimates assume perfect calibration of the instrument. The accuracy then becomes dependent on the resolution of the measurement.

about 0.23 Th. The difference in between the four possible choices of the 'real' peak position (Figure 3.6) is about 40 ppm or 0.045 Th. For several reasons that we will discuss below, this is typically the borderline accuracy in the real world for reflectron MALDIToF MS measurements of tryptic peptides from proteins separated on gels. This is also a very good accuracy at which to perform PMF analyses, as we will see later in Chapter 4.

At really high resolutions (e.g. 25 000; see Figure 3.4) the peak width in our example is only 0.045 Th or 40 ppm. We can see that the differences between the possible choices for the 'real' mass of the peak must be much lower, typically below 1 ppm or a staggering 0.001 Th. Nevertheless, the accuracy of a reflectron MALDI ToF mass spectrometer with this specification will not be 'below 1 ppm', but typically 10 ppm, as real samples pose different limitations which we will discuss later.

All the above statements about accuracy are made under the assumption that the mass spectrometer we use is perfectly calibrated, that is, that it 'knows' exactly where the absolute m/z values really are. As we will see in the mass analyzer descriptions, this is never really the case for any mass spectrometer, and different mass analyzer have different requirements for calibration. So, strictly speaking, the foregoing deals with precision and not accuracy; we measured a mass of 1 250.06 Da to perhaps ±0.03 Da (20 ppm), but it might be out by half a dalton (400 ppm) because we did not calibrate the instrument. If we measured the same peptide several times, on several days, we would always measure the mass in most cases as lying between 1 250.03 and 1 250.09 Da, an impressive precision. However, the real mass might be 1 250.56, and once calibrated correctly we would measure the mass between 1 250.53 and 1 250.59 Da. So our measurement was precise (on a relative base) but not accurate, as we did not know where on our m/z scale the 1 250.06 Da value really is. We can often calibrate an instrument after the measurement apply the new calibration to the data, and thus transform our precise measurement into an accurate measurement. Throughout the book we will not distinguish rigorously between accurate and precise measurements, as with proper calibration one can be transformed into the other.

In the case of MALDI based measurements there are certain aspects of sample preparation that will interfere with the accuracy and indeed resolution that can be achieved. ESI based measurements differ in this respect, in that the resolution that can be achieve is based on instrument performance and then beyond this often limited by signal intensity. Signal intensity in turn is dependent on sample amounts, sample delivery and spray quality. Furthermore, sample preparation influences how many contaminations are present, and some of these can in turn diminish sample ionization, similar to the situation in MALDI MS.

The limit of accuracy is determined by the resolution of the measurement, the calibration of the instrument and the amount of sample. The weaker the signal, the stronger the influence of background noise; this noise can be 'electric', which means it is derived by the instrument, or it can be chemical noise, coming through your sample and chemicals in the sample preparation process. We will also have to talk about contamination in your sample by peptides/proteins that do not belong there. While it is important to minimize the noise, it is impossible to exclude it totally. Depending on the achievable accuracy, from a signal to noise ratio of about 20 upwards the effects of noise on accuracy can be neglected (compare the rough shape of the peak at 3 789 Th with the smooth shape of the peak at 909 Th in the top panels of Figure 3.3).

In the real world accuracy is limited by a range of factors, and for different levels of accuracy the factors involved change. For example, at 500 ppm and resolution 1 500, it does not matter which of the following points of the monoisotopic peak you choose as your mass: the highest intensity of the peak based on raw data, after smoothing of the spectrum, or the centre of the peak based on the peak position at 50% or 80% of the maximal peak intensity (Figure 3.6). But all the above criteria affect measurements of relative accuracy. Every mass spectrometer needs to be calibrated to allow accurate measurement. Such calibration is performed with known compounds, ideally peptides and proteins that cover in size and charge state the analytes that we expect. How well the calibration can be applied is influenced by the type of mass spectrometer and by the way it is designed, which will be discussed later. The best achievable mass accuracies range from several hundred parts per million to below 1 ppm for instruments that can be used perfectly well for proteomics. Another major influence on the absolute accuracy is the type of calibration; external and internal calibrations are used. External calibration means that the calibrant is measured separately, ideally before the analyte. The instrument is then calibrated (i.e. it is told what peak is at what mass) and this calibration is kept for the measurement of the next sample, or couple of samples. Internal calibration means that the calibrant is actually measured together with the analyte in the same sample/measurement. A typical entry level reflectron MALDI ToF MS would deliver about 1 000 ppm accuracy without calibration, about 150 ppm with external calibration and about 50 ppm with internal calibration. Some instruments cannot easily be internally calibrated and a typical entry level Pauli ion trap would deliver accuracies in the range of several hundred parts per million. However, due to differences in the way both instruments are used and their different capabilities (e.g. MS/MS) the very 'inaccurate' ion trap can still deliver more reliable protein identifications (see below and Section 4.3). Also an instrument with an assumed accuracy of 200 ppm can deliver

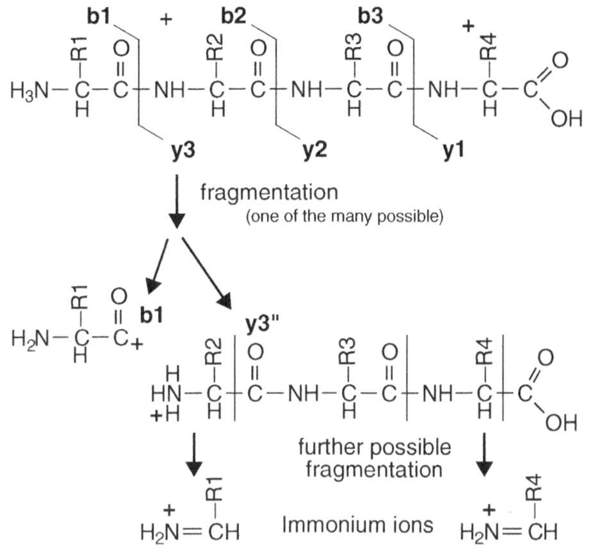

amino acid	Ala (A)	Arg (R)	Gln (Q)	Gly (G)	His (H)	Leu,Ile (L,I)	Lys (K)	Met (M)	Phe (F)	Pro (P)	Ser (S)	Thr (T)	Trp (W)	Tyr (Y)	Val (V)
immonium ions (related ions) Da	44	(100, 12)	101	30	**110** (166)	(44)	(112, 129)	61	**120**	**70**	60	74	**159** (117, **130**, 132, **170**, **171**)	**139**	72 (41, 55, 69)

Figure 3.7 Fundamentals of peptide fragmentation in mass spectrometry I. Collision induced dissociation (CID) most often leads to the generation of b and y type ions. The ions are named according to the example in the figure. Note that only fragments that retain a charge are visible. Molecules with additional protons get one hyphen for each proton; y ions with two additional protons are depicted as y″. Charge distributions for peptides in the gas phase are different from peptides in solution. Loss of water (−18) and ammonia (−17) is also often observed and ions are depicted as, for example, b^0 or b* (loss of water or ammonia, respectively). Immonium ions are often generated by CID, particularly at higher activation energies and gas pressures. Immonium ions and related ions are listed in the table. Bold figures indicate often observed ions. Amino acids producing only rarely/weak immonium ions are not listed. Note that not all immonium ions are specific for one amino acid only (e.g. 112 Da or 86 Da).

measurements that are within 50 ppm most of the time. The important consideration is the level of error at which we would assume a suggested peptide can be excluded due to it being measured outside the accuracy of the instrument. It is thus better to be conservative in the estimation of the accuracy of your mass spectrometer.

3.1.2 Introduction to tandem mass spectrometry and basic principles of peptide/protein fragmentation

Earlier in this section we recalled that for most proteomic application the mass determination of a protein or peptide is not enough; it is just the beginning of the analysis and we also need to determine the masses of fragments of our samples in order to obtain meaningful answers. We now look at how MS/MS measurements can be performed when using specific instruments, as every type of mass analyzer has a different way of achieving these measurements (often even more than one way).

However, the principle steps for MS/MS measurements are (most of the time) identical:

1. Take a measurement of all ions present.
2. Choose and isolate one ion for fragmentation (see Figure 3.5).
3. Transfer enough energy to this ion to fragment it (i.e. break its chemical bonds).

4. Measure the masses of the fragments that retained a charge (see Figure 3.7).

Depending on the type of instrument, the isolation of the ion to be fragmented (step 2) is achieved by ion separation in space, ion separation over time or in terms of energy transfer to only one ion. This target ion for the second MS is called the molecular ion or also the parent or precursor ion. The fragments may be called daughter ions or product ions. Instruments vary in their ability to isolate the parent ion; the resolution for the isolation of parent ions can be as low as 100, or as high as 1 000 or more. This means that at resolution 100 for a peptide of m/z 1 000 the ions from 995 to 1 005 are isolated. At a resolution of 1 000 the monoisotopic peak can be isolated from the other isotopic peaks for fragmentation (m/z 999.5–1 000.5) and thus de-clutter the fragment spectrum by removing about two-thirds of the peaks. The resolution for parent ion isolation is important to ensure that the fragments are only derived from one precursor ion. This, together with the sequence specific nature of the fragment pattern, ensures that either parts or the complete amino acid sequence of the parent ion can be derived from the fragments. Thus MS/MS data can identify the presence of a certain protein in a complex mixture of proteins based on one or several peptides. The presence and exact position of PTMs can also be analyzed in this way.

The fragmentation follows certain rules. Not all chemical bonds in the amino acids break with the same frequency and result in fragments that retain a charge. The fragmentation patterns are sequence dependent and can be partially predicted, in particular if trypsin is used to generate the peptides, as this guarantees a positive charge retaining amino acid on the N-terminus of each peptide (Arg or Lys). Figures 3.7 and 3.8 show the nomenclature of peptide fragments and the main fragments occurring in collision induced dissociation (CID; see below). Ideally the sequence of the peptide (or at least parts of it) can be derived from these fragments. The most important fragments for this task are often the b and y ion series with some of their derivatives (see the legend to Figure 3.7 and, for example, Figures 2.34 and 4.14–16). The extent to which different fragmentation patterns emerge is strictly sequence dependent, and some (but not all) fragmentation rules are known and can be used to interpret a spectrum. During this interpretation the aim is to match the

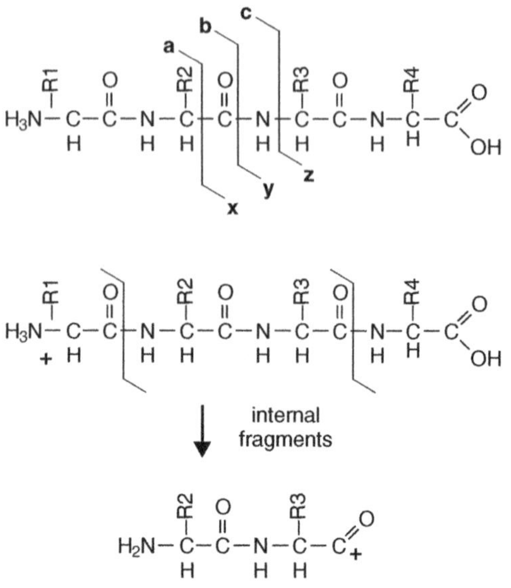

Isobaric amino acids	accuracy needed at 1000 Th
Leu/ Ile (113.084 identical)	– – –- (w/ d ions)
Gln/ Lys (128.059/128.095)	better than 30 ppm
Met-Ox/ Phe (147.035/147.068)	better than 20 ppm
Ala-Pro/ Trp (186.100/ 186.079)	better than 10 ppm
Ala-Gly/ Lys (128.058/ 128.095)	better than 30 ppm

Figure 3.8 Fundamentals of peptide fragmentation in mass spectrometry II. Fragmentations can occur at various bonds along the peptide backbone, and according to which fragment keeps the charge they result in a, b, c or x, y, z type ions. Next to b and y, a type ions are most common. If the charge-retaining fragment results from several backbone fragmentations, internal fragments result. Given the relatively large number of ions (hundreds of millions) used for MS/MS experiments, the resulting fragments follow statistical rules depending on experimental conditions and peptide sequences. Often more or less complete series of one to three ion types are observed, where the peptide backbone was broken between different residues (see text for details and Figures 2.34 and 4.16). Not all amino acids can be distinguished purely by mass in these series – those that cannot are called isobaric. Leu/Ile can be distinguished by fragmentation events involving the side chain, but only in high energy CID (resulting in w or d ions). For practical purposes nearly isobaric amino acids are also difficult to distinguish. It is also very difficult to decide in some cases whether a series has one fragment missing, or if it contains a large amino acid (see table for examples).

fragment pattern with the calculated, theoretical fragment pattern of one specific amino acid sequence. This interpretation can be performed manually or automatically and can easily take a lot longer than the entire data acquisition process (Chapter 4). The interpretation is complicated by many factors (e.g. data quality, preferred fragmentations leading to 'gaps' in fragment series of the peptide), but some of the problems are fundamental; several amino acids are absolutely or practically isobaric, that is, their mass is identical or the difference is so small that it cannot be measured with sufficient confidence (Figure 3.8). At increased amounts of energy transferred to a peptide, multiple fragmentations can become predominant. These produce smaller, internal fragments while larger fragments are missing, which can make the interpretation of a spectrum rather challenging. Internal fragments can be very helpful in defining chemical modifications of the peptide or distinguishing isobaric amino acids (see table in Figure 3.8 and Section 4.3). Immonium ions for instance, are very helpful as they can serve as indicators for the presence of certain amino acids (Figure 3.7).

Different mass spectrometers produce different fragment types, and often the intensity of different fragments varies dramatically between instrument types (e.g. quadrupole and ion trap). Sometimes there are even variations among mass spectrometers of the same type, from manufacturer to manufacturer. This is based on the specific design of the mass spectrometer. Design features that influence the nature of the observed fragments include the layout of the ESI or MALDI source, spatial arrangements, ion lenses and voltages applied, gas pressure and time spent in different sections of the mass analyzer, and of course the way the energy needed for fragmentation is transferred to the peptides/proteins.

The fragmentation type used for the vast majority of proteomic analyses is collision induced dissociation (CID), also known as collisionally activated dissociation (CAD). In this process ions are activated by collisions with neutral molecules, typically chemically inert gases such as argon, nitrogen or helium. To achieve enough energy, or 'speed', to induce fragmentation by collision, the ions are accelerated by voltages between 2 and 100 V for so-called low energy collisions and up to several thousand volts for high energy collisions (in MALDI ToF/ToF analyzers), against the stationary gas molecules (Figure 3.9).

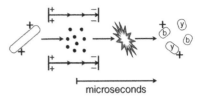

microseconds

Figure 3.9   Collision induced dissociation. For CID, ions are accelerated by electric fields and collide with inert gas molecules (e.g. Ar, N_2, He). In ESI based applications doubly-charged ions are usually chosen for fragmentation. In MALDI based applications similar fragmentation processes can occur with singly-charged ions. Here the energy for fragmentations is often delivered by the LASER energy. CID can also be added to MALDI based mass analyzers in the form of a dedicated gas chamber. The time between collision and fragmentation is relatively long, allowing dissipation of the energy and breakage along the 'weakest link', which often leads to series of a, b and y type ions. For fragment spectra see, for example, Figure 2.34.

CID is performed in a small part of the mass analyzer and the gas is pumped in and out of this collision cell, because ion transmission and performance of mass analyzers depend heavily on a good to very good vacuum. Only a part of the collisional energy between ion and gas is translated into internal energy of the ion. The energy that can be transferred is correlated to the mass of the gas ion used for collision, so heavy gases are preferred. Pauli ion traps are limited to helium, and thus also is the energy limited they can transfer into a peptide for CID (Section '3D quadrupole or Pauli ion traps'). The larger an ion, the more energy is needed for fragmentation. The basis for this is the fragmentation mechanism in CID; the fragmentation does not occur in the instant of the collision, but some time (usually up to microseconds) thereafter. In this time the energy is spread along vibrational nodes, which are basically the atomic bonds open to rotation. This type of delayed reaction with energy spread throughout the molecule is called ergodic. The larger the ion, the more vibrational nodes are present among which the energy can be distributed, which means more energy is needed to allow for enough energy to be present at a single bond for fragmentation, when it eventually occurs at "the weakest link". It also means that larger ions take longer to fragment, which can be a limiting factor for some mass analyzers or measurements.

It can be very difficult to assign all peptide fragments observed during MS/MS to a certain peptide. Often there are not enough fragments observed to be able to match them to one peptide only in a database. MS/MS is less sensitive than MS; some ions are lost during CID (scattered in the instrument, with loss of charge for all fragments) and the signal intensity of one parent ion is distributed among many fragment ions in MS/MS.

Fragmentation patterns also depend on the charge state of the parent ions, or more precisely on the charge location of the ion during fragmentation. In gas phase chemistry charges on peptides/proteins can be more or less localized, depending on the presence of charge-retaining groups in their sequence of amino acids. Tryptic peptides are predictable to work with; we know that they contain a positive charge retaining group in form of the free amino group at the amino terminus and another amino group at the carboxy terminus: based the sequence specificity of trypsin, the carboxy terminal amino acid is either Arg or Lys, both of these having an amino group in their side chain (see Table 2.3). Doubly-charged tryptic peptides usually fragment efficiently using low or high energy CID. 'Efficiently' in proteomics terms means there is an abundance of fragments, most of them of similar intensity, and there are fragmentations along the backbone of the peptide, covering most of the amino acid sequence, ideally with successive ions of the same fragment series (Figure 3.8). Most of the fragment ions derived from doubly-charged tryptic peptides by CID are singly-charged ions of the b and y type, which makes interpretation of the spectra a lot easier. Triply-charged peptides fragment less efficiency and often contain doubly-charged fragments. These are confusing, as it is not always possible to distinguish charge states of fragments (e.g. because the resolution of analyzer is too low, the signal to noise ratio is too high or only the monoisotope was used for fragmentation). However, in analyzers that can perform MS^3 (ion traps) the doubly-charged fragments can be analyzed further. Singly-charged peptides do not fragment very efficiently with low energy CID, but do so with high energy CID. This explains why fragmentation data from singly-charged peptides are poor on triple quadrupole or quadrupole–ToF analyzers but very informative on ToF/ToF analyzers with high energy CID (see Section 'ToF/ToF mass spectrometers'). The acceleration voltage for optimal informative fragmentation of doubly-charged ions is only half that of singly-charged ions, as the actual acceleration depends on the charge and the applied field. So singly-charged ions appear to be hard to fragment in low energy CID, because the instruments

are set for doubly-charged ions. Doubly-charged ions in low energy CID do provide more balanced series of b and y ions which are better suited to identifying peptides from databases or confirming sequences. All the above is true for peptides up to about 20–30 amino acids; above this length it is very difficult to achieve fragmentation at all or even most of the peptide bonds, and CID MS/MS spectra are very difficult to interpret. Below about 6 amino acids the information from CID MS/MS also starts to be less valuable for proteomic purposes; it is always difficult to obtain fragments from the last two amino acids on both termini and, even if the full sequence of a peptide can be derived, short peptides are not as specific in searches as larger ones (Chapter 4).

If too much energy is transferred to the ion in (low or high energy) CID then internal fragmentations become more prevalent and the resulting fragments become smaller and smaller, losing the ability to provide useful structural information. Multiple collisions in particular will destroy peptides in this way. Therefore, it is important to keep the gas pressure and acceleration voltage at a level which avoids these excessive multiple fragmentations. However, some peptides might not yield good fragmentation series at low energy levels; it thus becomes desirable to combine different CID energy settings, starting with low energies to identify possible peptide sequences, and use increasing energies to substantiate the findings with more details, such as PTM or side-chain fragments. Very useful small internal fragments are the immonium ions, which can give important hints as to which amino acids are present in the peptide (Chapter 4).

Often only certain fragmentation reactions are occurring, in accordance with sequence-based preferences of the fragmentation patterns (see also Figure 3.7 and Chapter 4). This can be limiting when trying to assign one or several PTMs on a certain peptide to specific amino acids in bottom-up approaches or when complete proteins are measured by MS/MS in top-down proteomics. With complete proteins it can also be difficult to transfer enough energy for fragmentation. Here two alternatives to CID have recently emerged. In electron capture dissociation (ECD) the parent ions capture single electrons and fragment almost instantaneously. The fragmentations thus occur also in positions that are not energetically favoured, providing more complete ion series and fragmentations in long peptides/ proteins where in CID the energy would be spread to a point where no fragmentation can occur

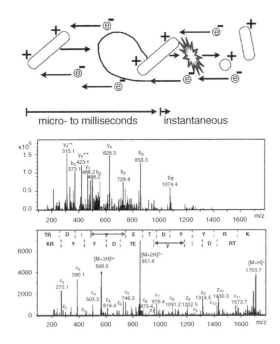

Figure 3.10 Principles of ECD and ETD. In ECD/ETD electrons or radical anions need to be brought together with the analyte for prolonged time for the electron capture/electron transfer to occur. Once this happened, the fragmentations reaction is very fast, so that no energy spread along the molecule occurs and not only the 'weakest' bonds break; a multitude of fragments is observed. The spectra show a comparison of CID and ETD results for the phosphopeptide TRDIyETDYYRK, with Y being the phosphorylated amino acid (a Tyr/Y). While the b and y ions detected in the normal ion trap CID MS/MS spectrum could have been sufficient to identify the peptide (if the phosphorylation was assumed!) the c and z series in ETD cover a lot more of the sequence and show clearly which of the three tyrosines in the peptide is phosphorylated. Spectra from Tang *et al.* (2007). © Agilent Technologies, Inc. 2007. Reproduced with permission, courtesy of Agilent Technologies, Inc.

(Figure 3.10). As there is no time for energy spread, these fragmentation reactions are called non-ergodic. In ECD it can take some milliseconds for the electrons to be captured. ECD is only possible in special mass spectrometers (some ion traps and ion cyclotron resonance (ICR) cells) as it difficult to keep the oppositely charged species in close proximity for long enough (see Chapter 3). A source for low energy electrons also needs to be integrated into the mass analyzer design. In electron transfer dissociation (ETD) a radical anion attacks the positively charged peptide/protein, resulting in the transfer of an electron, and similar fragmentation results as

in ECD. For ETD an anion generator (radical anions are short-lived) needs to be integrated into the mass analyzer design. Again the anion and the analyte have opposite flight paths and for ETD the measurement needs to be performed in a discontinuous mode; several milliseconds (10–100 ms) of ETD reaction/anion delivery without MS or MS/MS signals interrupt the measurement, and thus inhibit very high throughput measurements ETD has the advantage that it can be implemented with a wide range of (affordable) mass analyzers. ECD and ETD can deliver a wealth of internal fragments from nearly all peptides, which are observed less often with CID (see Figure 3.10). While this information is not ideal for matching the data to a peptide in a database, it can be very useful for assigning complete PTM information to a peptide previously identified (Chapter 5). In top-down proteomics on complete proteins electron induced ergodic fragmentations allow breaks to occur along significant proportions of the protein backbone, thus allowing characterization of PTMs from a whole protein, which has significant advantages over peptide based analysis of PTMs (see Figure 4.23). However, bottom-up proteomics can be applied to small amounts of complex mixtures of unknown proteins (if the genes are known) while top-down approaches need larger amounts of highly enriched/purified proteins.

3.2 IONIZATION METHODS FOR SMALL AMOUNTS OF BIOMOLECULES

We now turn to the ionization methods and mass spectrometers commonly used in proteomics. We describe their characteristics regarding handling, resolution, accuracy and sensitivity and also touch on their use in different proteomic strategies.

There are many different methods by which ions can be generated. For biomolecules these methods need to be efficient, sensitive and 'mild' enough not to destroy the analyte. Today there are only two ionization principles for biomolecules that fulfil these requirements, MALDI and ESI. Of these MALDI does not even produce stable ions; it actually produces metastable ions which fragment in milliseconds. However, in most mass analyzers this time is significantly longer than the analysis time, and in some specialized mass analyzers the advantages of MALDI outweigh the loss of a considerable portion of ions to metastable decay.

3.2.1 Matrix assisted LASER desorption ionization

The ionization of biomolecules by MALDI is the stuff of science fiction movies; we fire short LASER pulses at a sample prepared on a steel target plate. If the LASER energy is just right – not too strong, not too weak – the sample explodes (usually in a very high vacuum), the peptides/proteins get ionized in the expanding plume of the explosion (Figure 3.11) and are then accelerated away by usually very high electric fields. If the sample is abundant and well prepared (about a picomole of peptide on the target) a single LASER shot can deliver enough ions for a satisfactory measurement. However, for real samples several shots have to be accumulated, that is, the readout of the mass spectrometer is combined/added up from the single shots. For most proteomic applications between 50 and 200 shots are accumulated and up to 1 000 shots might be needed for very small sample amounts.

Several theories have been put forward in an effort to explain how peptides and proteins are ionized in MALDI. It would go far beyond the scope of this book to examine these theories; suffice it to say that most proteins and peptides do not actually get ionized and that the 'few' resulting positively charged analytes are more than enough to make MALDI a very sensitive ionization method. MALDI even produces negatively charged ions

(somewhat less efficiently), although the whole environment is kept at a low pH so there is an abundance of positive charges. These protons are transferred during the expansion of the plume onto the analytes. It is also worth mentioning that there are several ionization suppressing effects in MALDI; too many different analytes will result in less ionization of each, too much matrix can suppress analyte ionization (while not enough is also not helpful), and an ideal molar ratio of matrix to analyte is around 10 000 : 1. The ions resulting from MALDI are metastable; most of them will fragment. However, it takes several milliseconds for most of them to fragment, longer than most analyses. Thus, for the time being, we will ignore the metastable nature of the MALDI ions.

It may appear surprising that the chemically quite sensitive proteins and peptides survive all this exploding and heating relatively unharmed; in fact, MALDI can be used under conditions so mild that even complete viruses survive this ordeal intact (some even remaining infectious). This miracle is achieved by an ingenious trick: the LASER is normally not absorbed by the peptides/proteins but mainly by a so-called matrix. The matrix has to be in very close contact with the analyte, and they are crystallized together. The matrices are chosen so as to allow this co-crystallization. To allow the peptides/proteins to

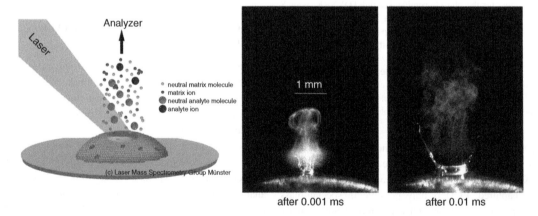

Figure 3.11 Principles of MALDI. (a) A LASER is fired at a preparation of mixed matrix/analyte crystals. The ions of the resulting plume are accelerated towards the analyzer by an electrical field. It is interesting to note that most analyte molecules (peptides/proteins) actually do not get ionized at all, and despite this MALDI is still one of the most sensitive ionization technologies. (b, c) Show pictures of a real gas plume, using a liquid matrix, not usually used in proteomics but for nucleic acids, for example. The violent explosion after absorption of the LASER energy by the matrix is clearly visible. For the influence of LASER intensity on results, see Figure 3.13. Cartoon: http://impb2.klinikum.uni-muenster.de/research/hillenkamp/index.html. Picture: http://chemistry.lsu.edu/site/People/Faculty/Kermit%20Murray/item1096.html. Reprinted with permission from Fan & Murray (2010), *J. Phys. Chem. A*; 114(3): 1492–7. © 2009 American Chemical Society

be charged positively the whole process is performed in rather acidic conditions, and the matrices are also chosen to be weak acids, that is, good proton donors. Because there has to be a much larger quantity of matrix molecules than of analyte, it is also important that the matrices have a much smaller molecular weight than the analyte. This is one of the reasons why masses below about 700 Da cannot be measured very well in typical MALDI based proteomic experiments; the matrix molecules are about 200 Da in size and also form dimers/trimers that do not dissociate in the MALDI process. So trying to measure masses below 600 Da is like staring at the sun while trying to see a firefly. MALDI for biomolecules can be so gentle that not only matrix molecules can form dimers, even protein/protein interactions can survive the ordeal. However, the interactions have to be stable in the gas phase, which can impose requirements different from those for interactions in solution.

MALDI is usually used in high vacuum, typically below 10^{-6} mbar. As the analytes literally explode at high speed this is optimal for guaranteeing the highest possible transmission. Transmission is a term we will come across several times with mass spectrometers; ideally and most of the time we will only consider the behaviour of ions that 'make it' to the detectors, which are said to be transmitted. So high transmission means minimal loss of ions and high sensitivity. MALDI can also be performed at atmospheric pressure (AP MALDI) using the same principles and reagents. AP MALDI is rarely used for mainstream proteomics, as its sensitivity is orders of magnitude lower and not all ions are transmitted; MALDI in high vacuum has no size limitation, while in AP MALDI larger ions are lost due to collisions with air molecules. While MALDI is usually combined with ToF instruments, AP MALDI can be combined with any type of mass spectrometer. It is generally an even 'softer' ionization than traditional MALDI, but has much lower sensitivity.

Typical well established matrices are shown in Figure 3.12. The matrices have different properties in terms of sensitivity, LASER intensity that is necessary for the MALDI effect, and crystallization. Some new matrices are being developed to enhance not only the sensitivity but also the selectivity of matrices; there are always some peptides (in all MS applications) that do not 'fly' very well. It is very difficult to find out in every case why that might be; the notion that a peptide/protein does not fly very well usually just means that it is not detected at the end of the measurement. There may be several reasons for this lack of detection (the peptide may be chemically unstable, not generated in the digest efficiently, sticky to plastics, contain negative charges, break up in source, not ionize well, etc.), but here we are interested in detecting a peptide better by changing the matrix.

Selectivity in MALDI is a big problem and any improvement is welcome. Different matrices have different properties as to how they transfer energy to the analyte; most notably both 2,5-dihydrobenzoic acid (DHB) and cyano-4-hydroxycinnamic acid (CCA) work nicely with peptides, with CCA perhaps delivering a slightly higher sensitivity. When it comes to larger peptides (2 000–4 000 Da), both work equally well. But when it comes to proteins, DHB has a clear advantage. This advantage is even more pronounced the larger the proteins get, particularly above 10 kDa, when CCA can not be used any longer. On the other hand, if the aim is to fragment the analyte with increased LASER power in order to perform post source decay (PSD) experiments, CCA is much more efficient. So it seems that DHB is 'milder' to the analyte and the loss of sensitivity with CCA for larger polypeptides can be explained by the fact that they are destroyed more easily.

There are several ways to prepare samples for MALDI. Dried droplet methods are the most popular choices. For this, basically a saturated solution of the matrix is prepared. Since the matrices are hydrophobic, a much higher concentration can be brought into solution by mixtures of acetonitrile (ACN) with water (40–70% ACN) or ethanol. This mixture is made acidic by the addition of a low concentration of formic acid. Formic acid is volatile and does not influence the crystallization. This is quite important as any substance influencing the formation of the mixed crystals of matrix and analyte will negatively influence the result of your MALDI MS. If salts are present (typical leftovers from buffers) they can be integrated in the matrix crystals and negatively influence the absorption of the LASER energy. The optimal ratio of matrix to analyte for the most sensitive detection of peptides/proteins is about 10 000 molecules of matrix to one molecule of analyte. When salts are integrated into the crystals, the transfer of energy between matrix and analyte can be influenced and higher LASER power settings are needed. MALDI is actually less sensitive than ESI to salt contamination which, depending on the type of salt, can be tolerated up to about 2 mM. Salts form a layer or crust on

Matrix	Structure	Wavelength	Major applications
Nicotinic acid	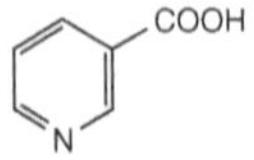	UV 266 nm	Proteins, peptides, adduct formation
2,5-Dihydroxybenzoic acid (plus 10% 2-hydroxy-5-methoxybenzoic acid)	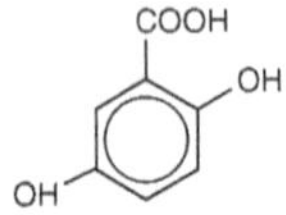	UV 337 nm, 353 nm	Proteins, peptides, carbohydrates, synthetic polymers
Sinapinic acid	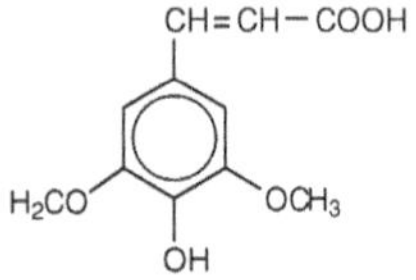	UV 337 nm, 353 nm	Proteins, peptides
α-Cyano-4-hydroxycinnamic acid	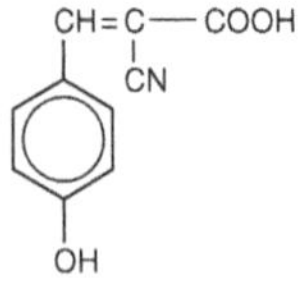	UV 337 nm, 353 nm	Peptides, fragmentation
3-Hydroxy-picolinic acid	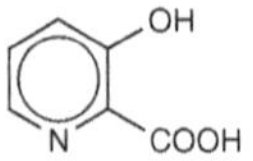	UV 337 nm, 353 nm	Best for nucleic acids
6-Aza-2-thiothymine	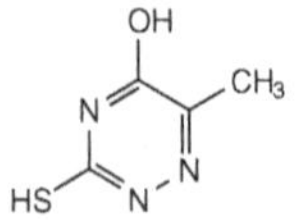	UV 337 nm, 353 nm	Proteins, peptides, non-covalent complexes; near-neutral pH
k,m,n-Di(tri)hydroxy-acetophenone	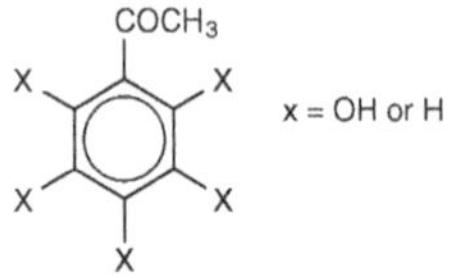	UV 337 nm, 353 nm	Protein, peptides, non-covalent complexes; near-neutral pH
Succinic acid	HOOC-CH_2-CH_2-COOH	IR 2.94 μm, 2.79 μm	Proteins, peptides
Glycerol	H_2C—CH—CH_2 / OH OH OH	IR 2.94 μm, 2.79 μm	Proteins, peptides, liquid matrix

IR = infrared; UV = ultraviolet.

Figure 3.12 Matrices commonly used in MALDI. α-Cyano-hydroxycinnamic acid (CCA) is by far the most often used matrix for peptides in proteomics. Reproduced with permission from Hillenkamp & Peter-Katalinic (2007). © 2007 Wiley-VCH Verlag GmbH & Co. KGaA.

top of the mixed matrix/analyte crystals, which influences the way we have to generate our MALDI MS data (see below). In a typical preparation 1 μl of the acidified, saturated solution of matrix in a hydrophobic solvent is 'spotted' onto the MALDI steel target plate, followed by the addition of another microlitre of sample, prepared in a slightly acidified water ACN solution (typically 20:1 or 5% ACN). The two solutions are quickly mixed and then left to dry in a quiet place (no vibrations) at room temperature. Within about 5–10 minutes (depending on humidity and actual room temperature) the sample will have dried completely and all you can see are crystals where your drop once was. These crystals are tiny, and this is what you want. The target plates are usually brushed and not polished, to give some starting points for the crystalliza-tion, thus favouring the formation of many, small crystals. On the other hand, you do not want to touch the target plate when you mix the solutions by pipetting up and down two or three times; this might create starting points for crystals and you get fewer, larger crystals and your measurements will not be as sensitive as they could have been. The ACN concentration of the mixed samples is typically about 25% (aceotrope) and ACN has the higher gas pressure; it vaporizes first, taking some water with it, so the peptides start precipitating as well, right into the growing crystals of matrix. Finally, the salt contained in the solution precipitates on top of our 'good' mixed crystals. In practice this step is very easy, you just need to get it right – some labs always use 50% ACN, oth-ers get no results at all unless they use 70%, so this is a matter of trial and error. As the crystals do not form uniformly in the dried droplet methods, one has to find an area with either a good crystal or several crystals to make sure signals are obtained. These 'sweet spots' are rela-tively easy to find in proteomic application (much harder for DNA or RNA) and most mass spectrometers will have an automated routine to get good signals from a variety of spots. If the sample preparation is bad, or the pep-tides/proteins are very low in abundance, there might not be many sweet spots in one sample. It is always a good idea to spot a sample twice, just to make sure, and usu-ally only 0.5–1 μl of the sample solution is needed for this. If the sample preparation leads to relatively large crystals this can affect the accuracy of the measurement. The crystals have a certain elevation from the target plate, and in ToF instruments this can be enough to influence the accuracy by up to 100 ppm, depending on the instrument geometry. Even if the crystals are smaller, the elevation at each one is slightly different. As a complete measurement is the sum of many different points, small differences make the peaks broader, resulting in apparent loss of res-olution. The 'thin layer' preparation methods avoid these problems very neatly: a thin layer of very small crys-tals of matrix is prepared on the target, and the sample is spotted on this hydrophobic surface in a small amount (0.3–0.5 μl) of an acidified solution containing matrix and analyte. When this droplet dries, it starts dissolving the very surface of the tiny matrix crystals and the analyte integrates into the top layers of the crystals, which grow due to the matrix present in the solution. If the sam-ple contains very low amounts of analyte this procedure

can be repeated. If the matrix layer was prepared with a small amount of nitrocellulose, the spot can be washed with water, and salts can be removed quite efficiently. As a result we get a measurement that is very uniform, since the whole surface is made up of a uniform 'car-pet' of crystals. Salt has been removed, so the smallest possible LASER power (see below) will guarantee best resolution, sensitivity and accuracy. As a drawback the sample preparation takes slightly longer, needs more solu-tions and these need to be made up precisely (e.g. if the sample contains too much ACN the layer will dissolve).

Another approach to optimizing MALDI is to use so-called anchor targets, basically a steel target plate coated with a hydrophobic substance except for tiny holes in the coating. The sample is prepared by the dry droplet method and spotted on top of one of the holes; as the solvents evaporate the droplets shrink around the hole, and crystals form only exactly in the hole. For low abundance samples the method can be more than 10 times as sensitive as the standard dried droplet method. The sample is also uniform, and there is no need to find a sweet spot, which is ideal for ultra-fast and sensitive automated measurements with very high throughput (thousands a day). However, all contaminants are also concentrated in one spot, so the samples need to be really clean. A standard way to clean MALDI samples up before spotting is to use reversed phase beads (about 0.5–1 μl) stuffed in the very end of pipette tips. These mini-columns can be purchased from commercial sources or easily self-made.

After preparation on the target plate the sample is placed in the mass spectrometer, where it will take a couple of minutes to reach a vacuum sufficient for mea-surement. Typically we need to evacuate to a reasonably high vacuum of 5×10^{-7} millibar before we are ready for the measurement. A field is applied between the target plate and the interior part of the mass spectrometer (see below for more details) to ensure that any ions generated will move towards the interior of the mass spectrometer. Target plates are usually made of steel, to guarantee a uniform surface, good electric conductivity and sufficient ruggedness for daily use. Now we fire our LASER at the sample. Most MALDI sources in proteomics use a nitrogen LASER, which delivers a pulsed (a few nanoseconds long) ray of light of 371 nm wavelength. Nitrogen LASERs have proven to be affordable, reliable and durable as well as being appropriate for a range of different matrices/analytes. The matrices in proteomics are chosen so that they have an absorption maximum around 370 nm. This (together with the prevalence of matrix in the sample spot) results in most of the energy of the LASER pulse being absorbed by the matrix and not the analyte. The LASER is focused on the target plate, so that the area of highest intensity is about 5–10 μm in diameter (Figure 3.11). The intensity of the LASER can be regulated. At the right LASER intensity the miracle happens: a small part of the surface of the crystals is ablated, and the gas plume contains intact peptides/proteins, around 10% of which are charged. How the ionization occurs in MALDI is still a matter of some debate; charge transfer in the 'cloud' of extremely hot molecules of the plume is one explanation, but the proton transfer might occur in the crystal before the plume in the gas phase is even formed. If the LASER power is too low no ionization/signal is observed. If the LASER power is increased above a certain threshold level, we get a weak signal and this signal becomes a little stronger and more stable if the LASER power level is further increased by a small value (typically below 5% of the range of your instrument). It is exactly this range of LASER power that is ideal for best measurements. Further increasing the power leads to broader peaks (lower resolution), peaks with 'chopped' tops (signal too strong) and deterioration in instrument calibration (see below). If the intensity of the LASER is increased even further (typically about 15–20% of the total range of the instrument above threshold), the signal is lost completely, as the analyte is destroyed in the source due to the huge amount of energy transferred; the sample is 'fried' (Figure 3.13).

The optimal LASER intensity settings are slightly different for each component in a mixture of analytes; usually the smaller (larger) the peptide/protein, the

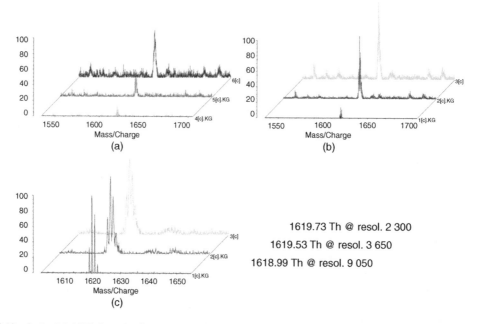

1619.73 Th @ resol. 2 300
1619.53 Th @ resol. 3 650
1618.99 Th @ resol. 9 050

Figure 3.13 Optimal LASER intensity for MALDI. The spectra in (a) are from a proteomic sample measured on a reflectron MALDI ToF MS. A small part of the total spectrum is shown. The spectra are summations of 20 LASER shots, at LASER intensities of 82, 89 and 90 (front to back spectrum, the units are arbitrary instrument units ranging from 0 to 160). At a power of 80 there was no MALDI signal at all. The intensity is adjusted so that the strongest spectrum reaches 100%. The spectra in (b) are summations of 100 LASER shots under the same conditions. Note how much lower the intensity of the background noise is, compared to the intensity of the peptide ion after 100 shots and how much background the 20 shot spectra show. The spectra in (c) are close-ups of the identical data set from (b); this time the intensity is set so that each spectrum reaches 100%. Note how the resolution decreases with higher LASER power. Note also that the measured mass is increasing significantly with the LASER power. See Section 3.3.1 for an explanation of this phenomenon. The weakest spectrum at the lowest LASER intensity setting provides the best results, in terms of signal to noise ratio, resolution and accuracy.

lower (higher) the LASER threshold for the MALDI phenomenon. In the real world one often ends up with a 'compromise' LASER power: strong enough to get signals from larger peptides/ions, but not too strong so as not to bring the most abundant peaks into saturation (see the 'chopped' tops on MALDI ToF peaks in Figure 3.3) and not to adversely affect resolution and accuracy for the smaller peptides/proteins too much (Figure 3.13). Generally, for the analysis of peptides MALDI detects ions in the range 750–1500 m/z at a higher intensity than larger ions, and in practical experiments it is not useful to try and detect peptides over a range of more than 750–4000 m/z. A similar rule applies for the measurement of proteins; the larger the protein, the lower the signal intensity, and in mixtures smaller proteins will always give much stronger signals (e.g. a 20 kDa protein will make it nearly impossible to measure a 150 kDa protein when present in comparable molar amounts). So in the measurement of protein mixtures by MALDI a range of 5–60 kDa is typical and 80% of the peaks will be below 30 kDa. This does not mean that the proteins in the larger range are present in lower amounts in the sample. Measured in isolation, it is quite possible to get sensitive measurements of even very large proteins (e.g. 200 kDa). This preference for certain analytes is also obvious for peptides. MALDI is susceptible to ionization suppression effects; the larger the number of different analytes contained in a spot, the less sensitively each will be detected. It is easily possible to detect less than a femtomole of a single peptide by dried droplet MALDI MS; in the presence of the same amount each of four other peptides, only the peptide with the best ionization characteristics will be detected. In proteomics on tryptic peptides each peptide contains either a carboxy-terminal Arg or a Lys. Arg-containing peptides are preferably detected by MALDI based mass analyzers, and typically 4–10 times more strongly detected than Lys-containing peptides (compare the search results in Figures 4.3 and 4.16). Peptides with multiple Arg will often totally suppress the signals of all other peptides. From these considerations it is clear that MALDI should only be used with great care for quantitative measurements.

If the sample (or the matrix) is contaminated with salts, the LASER power threshold for the MALDI effect is pushed further up. This can lead to a situation where signals cannot be generated at all and where further sample clean-up, or even a simple dilution, is required. It

is important to measure the external calibrants at the same or a similar LASER power setting as the sample; typically the external calibrants are very pure, and your sample is not, so different LASER powers are optimal for MALDI. If the LASER powers are too different, the calibration will suffer. Generally speaking, then, it is better to use slightly too high a power on the calibrant (and thus compromise resolution a little) in order to get a more accurate external calibration; the effects of different LASER power settings can be in excess of 300 ppm (see Figure 3.13). This is especially important when there is no way of applying an internal calibration. Why does the calibration change at higher LASER powers? This is very much a function of the design of the particular mass spectrometer; at this stage we just assume the plume expands further with higher LASER powers and the speed of the ions within the plume is larger (higher temperature: see more on this in Section 3.3.1). While higher temperatures would lead to a loss of resolution, changes in the plume expansion over time would change the position of the ions relative to the internal components of the mass spectrometer in time. For example, the ions would be closer towards the mass spectrometer within the first milliseconds after the LASER fired; this and their actual speed will have an influence on the later measurements. Many of the above described effects are more pronounced in axial MALDI sources (found in most MALDI ToF instruments; see Section 3.3.1) than in orthogonal MALDI sources (Figure 3.14). In orthogonal MALDI sources the ionization is uncoupled from the ToF stage and occurs far away in time and space from the ToF acceleration. Orthogonal MALDI sources operate at a moderate vacuum in the source and (as the name suggests) the ions are extracted orthogonally to their flight path into the ToF analyzer. Ions are cooled by collision with the residual inert gas and the spread of ions is a lot wider than the tight packing in axial MALDI sources, creating a near constant stream of ions. This ion stream is spatially and energetically very well defined, so that high resolution and accuracy can be achieved in ToF analyzers. However, in an axial MALDI, when combined with a ToF instrument, every single ion has a chance of actually being measured and there is no upper limit for the size of the ions analyzed. In orthogonal sources, the constant stream of ions has to be pulsed, so only part of the ions can actually be measured and the ions coming in at the time during which the first 'packet' is analyzed are

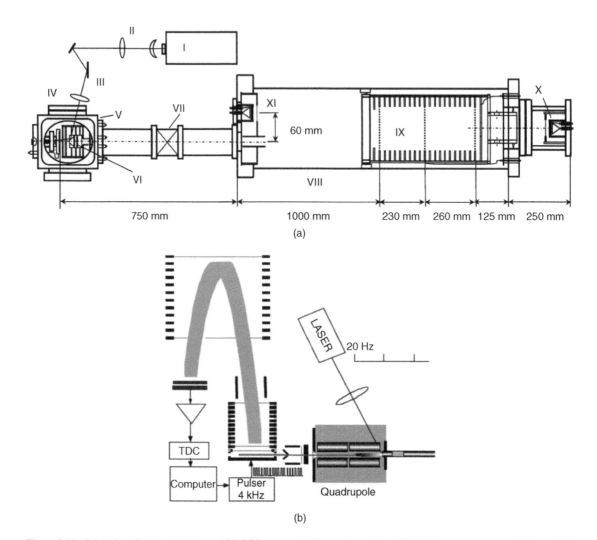

Figure 3.14   Principles of axial and orthogonal MALDI sources. (a) Shows a typical axial MALDI source arrangement. The LASER (I) is directed via lenses (II) and mirrors (III) to the ion source (IV) with ion extraction optics (VI). The ions are extracted/accelerated in a very high vacuum through the flight tube (VII and VIII) towards an ion mirror (IX) and onto multichannel plate detectors (X and XI). The ions are extracted along the axis of the ToF stage; changes in ion speed and energy distribution in the flight path are taken into the ToF stage, which needs to be adjusted for with several mechanisms. On the positive side nearly all ions that are produced get measured. (b) Shows a typical orthogonal MALDI source (path of ions is from right to left). The ions are extracted into a quadrupole stage for focusing and manipulation. Through the dampening by mild collisions with gas at this stage a nearly continuous stream of ions is produced. In the pusher region a pulsed field is applied to push the ions into the reflectron ToF stage, orthogonal to the ion extraction path. From here the ions go into a detector. The orthogonal source allows decoupling of ionization from ToF stage (ions have no energy distribution/differences in speed in direction of the ToF), which in turn allows easier calibration, uniform resolution over the mass range and surprisingly high sensitivity. (a) Reprinted with permission from Moskovets *et al.*, 2006, *Anal. Chem.*, 78, 912-19. © 2006 American Chemical Society. (b) For a detailed explanation see Krutchinsky *et al.* (1998). © Agilent Technologies, Inc. 1998. Reproduced with permission, courtesy of Agilent Technologies, Inc.

wasted. High sampling rates of several kilohertz help to keep these 'dead' times down. Furthermore, the ions are discriminated against based on their size and there is usually an upper mass limit around 20 kDa, as the pulse rate has to fall for longer flight times and thus the sensitivity suffers too much. Orthogonal MALDI sources are also used in ToF analyzers but mainly in hybrid instruments, ion traps or Fourier transform ion cyclotron resonance (FTICR) instruments. Orthogonal MALDI will be discussed in more detail in Section 'Quadrupole-ToF hybrid instruments'.

While all the above sounds very complicated, using MALDI in the daily routine is fairly easy. All modern mass spectrometers can run automated routines to find the LASER threshold and apply different patterns to target the LASER onto a certain sample, to ensure the software will find a 'sweet spot', test the data for resolution and intensity and then move on to the next sample. Typically hundreds of samples can be analyzed automatically in this way without human intervention.

A specialized application of MALDI is surface enhanced LASER desorption/ionization (SELDI). SELDI is no different than MALDI from a desorption/ionization point of view. It is, however, a sample preparation method; the target is coated with a surface that allows certain proteins/peptides to bind, but not others. The other analytes are washed off and the remaining peptides/ proteins on the target are analyzed, typically by MALDI ToF MS, but also by MALDI quadrupole ion trap (QIT) ToF or MALDI QIT orbitrap (Section 'Other hybrid mass spectrometers'). The currently commercially available surfaces are cationic, anionic, hydrophobic, hydrophilic or metal affinity-coated. There is also the possibility to have chemically reactive surfaces that can be loaded with proteins like antibodies or receptors. So basically, a lot of separation that is usually done in columns before an MS analysis can be performed under standardized conditions on a MALDI target, called a SELDI chip. The main application is biomarker discovery (see Chapter 5).

It would be not very practical to analyze a complex mixture of proteins or peptides (e.g. derived from tryptic digest of 20 proteins) directly by MALDI. Not only would the spectrum become 'crowded' with overlapping peaks, but also ion suppression would prevail. In simple terms, there are not enough ions available for ionization in MALDI to ionize every peptide/protein in a complex mixture. But there are even more reasons why MALDI is

a very "biased" ionization method. For reasons that are not completely understood, some peptides (e.g. containing several basic amino acids, such as Arg and Lys) are ionized much better than others. While in tryptic peptides Arg or Lys are always at the C-terminus, due to the specificity of trypsin, peptides with Arg in the C-terminal position are more easily detected in MALDI than those peptides with a C-terminal Lys. In a mixture of peptides the signal for most of the peptides is much weaker than when measured in isolation (see also Section 3.2.2). And finally, the larger the ion the more difficult it seems for it to be desorbed/ionized by MALDI and it cannot compete for detection with smaller ions. By removing most proteins/peptides from the analysis at any given surface, SELDI goes some way towards allowing the analysis of complex mixtures with MALDI. It is also possible to spot complete LC runs onto a MALDI target plate. Thus complimentary data from MALDI and ESI can be collected for complex samples, allowing a higher peptide coverage. However, the MALDI based analysis of a complete LC run takes about 6–10 times longer than the actual LC run, even with ultra-fast mass analyzers (Section 3.3.1).

3.2.2 Electrospray ionization

While MALDI is used to get biomolecules from a solid state into the gas phase, ESI does this from solutions. The trick is to spray the solutions in a very fine mist towards the mass spectrometer. Eventually the droplets get so small that we have a clean analyte, devoid of surrounding solvent; our peptide/protein is in the gas phase. Applying a high voltage between sample and mass spectrometer whilst at the same time ensuring a surplus of free charges in the solvents (usually protons) make for an extremely high efficiency of ionization in ESI (Figure 3.15).

Most of the ESI sources used in proteomics are so-called nano sources. These have extremely low flow rates ranging typically from 50 to 300 nl/min. The next size up are so-called micro sources, with flow rates of typically several microlitres per minute. The huge advantage of going 'nano' is the great gain in sensitivity, as already mentioned in Section 2.5. The smaller the flow rate, the less sample is consumed and the more efficient the ionization. To have a chance of equal signal strength in the mass analyzer, the sample in a micro source would have to be present at a higher concentration than the sample in the nano source. This explains why nano sources are

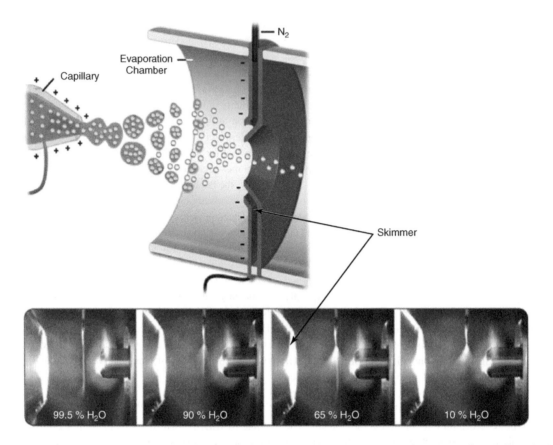

Figure 3.15 Principles of ESI. In ESI the buffer containing the peptides/proteins is sprayed from a very fine needle assisted by a high voltage and often by a drying gas sheet. This leads to very small droplets containing an abundance of charges. When the droplets get past a critical size limit the immense fields can lead to either a fission of the droplets or the direct desorption of analytes from the surface; the peptides/proteins are ionized and in the gas phase. Heated capillaries (see text) or gas streams (N_2 in evaporation chamber) can aid this process by drying the gas stream and providing smaller droplets. The 'Coulomb explosions' leading to ion desorption from the liquid ultimately depend on high field strength between the sample/spraying needle and the entrance of the mass analyzer, the skimmer cone (see above). Polarity switches are possible during a measurement (negative skimmer for positive ions as in the figure, and positive skimmer for negative ions). However, the choice of pH in the buffer will determine which one is more efficient for most analytes. The spray can change its shape considerably with flow rate and buffer composition (bottom panel, buffer changes from 0.5% ACN to 90% ACN). This so-called 'Taylor cone' can be critical for efficient and stable signals. As with MALDI, there are axial designs (top panel) and orthogonal designs (bottom panel, skimmer is on the left). If optical control is possible, one would try to adjust the Taylor cone in a fan (last picture, bottom panel) and bring it slightly off-axis in axial source designs, for optimal stability of the stream of ions entering the instrument. Note that the position of the needle giving the strongest signal is not necessarily the position allowing for best stability of the spray. (a) Reprinted from http://www.magnet.fsu.edu/education/tutorials/tools/ionization_esi.html with permission from National High Magnetic Field Laboratory. (b) http://www.chem.agilent.com/en-US/Products/Instruments/lc/1200serieshplc-chipmssystem/Pages/gp38427.aspx. © Agilent Technologies, Inc. Reproduced with permission, courtesy of Agilent Technologies, Inc.

the most sensitive ESI sources possible, as they allow the highest efficiency of ionization. Anything larger than a micro source (e.g. so-called turbo ESI) requires hundreds of picomoles of a protein for meaningful analysis, and these concentrations are not encountered in proteomics.

Basically, in a nano-ESI source all the analyte gets ionized and most of it finds its way into the mass analyzer, allowing sensitive analyses and long analysis times for low femtomole amounts of analytes. In a micro-ESI source a larger proportion off the analytes in the sample (increasing with increasing flow rates) does not ionize at all, the droplets of the spray never get small enough, and at higher flow rates most of the analyte is 'blown' in relatively huge droplets (micrometres) onto the orifice of the mass analyzer. To minimize this effect, micro sources, and the even larger turbo sources, use warm, chemically inert nebulizer gas (typically N_2) to make the droplets of the spray smaller and ESI more efficient.

ESI can be used as an online source, or as a so-called offline source. In the online application an ESI source is connected online to a separation column (typically RP, sometimes CE) and all analytes have to be ionized as they elute from the column. If things go wrong during the online measurement, the sample is lost. Each LC peak (peptide or protein) typically elutes in 30 seconds, so this is all the time you will ever have for a mass spectrometric analysis. There are ways to prolong the time available for analysis; using 'peak parking' at an MS data dependent signal, the flow over the column is virtually stopped, or slowed down to an absolute minimum, giving several minutes for measurement at this position in the LC run. When the required MS data are acquired, the LC run resumes, without any distortion. Another way to prolong the time available to measure a peak is the use of very shallow LC gradients (e.g. LC run times of 3 hours instead of 45 minutes), with peak elution times of 1 minute or longer. In practice, and especially with complex samples and repeated LC runs, these complications are not used very often, and a typical MS measurement cycle takes about 1–3 seconds. This cycle typically includes one MS measurement and the MS/MS analysis of several ions. This means that there is ample time for MS measurements during the elution time of a peak, even if it co-elutes with up to about five other peptides. However, if anything goes wrong during the measurement (spray stops, data recording PC crashes, etc.) all the sample is lost, and measurements cannot be postponed as in offline modes. On the positive side, online MS can be automated with many samples running one after another and the operator can get a good night's sleep while up to 100 000 automatic MS/MS measurements of an equal number of peptides are performed, a lot more than any offline mode (including ultra-fast MALDI ToF/ToF) can achieve.

In the offline application of ESI the measurement can be interrupted at any time (e.g. if the computer loose contact with the mass spectrometer), but usually the operator also has to be present to start the next run, which involves

positioning the needle. The sample (around 1 µl) is loaded into a capillary with a pulled spraying tip. Of course it will only become evident during the measurement whether this tip is perfect, while in online ESI the needle is tested before the first sample. The capillary can be filled with beaded RP material to clean up the sample and the sample is then eluted during the measurement straight into the MS (Figure 3.15). In effect the flow in offline ESI is often induced purely by applying a high voltage between the spraying needle (capillary) and the MS. This set-up is typical of nano offline ESI. The flow rates are about 20–50 nl/min. This means a sample of 1 µl can give about 20 minutes (sometimes up to 45 minutes) of constant, stable measurement time. This can be very helpful for the analysis of PTMs, as it allows different methods for analysis to be performed on one sample. For example, all peptide masses can be scanned and those that produce an immonium ion of 216 Da (because they are phosphorylated on Tyr) can be detected. In a second step very sensitive (i.e. long lasting) MS/MS experiments with different settings (e.g. collision energy ramping) can be performed to generate a different type of fragments in order to generate enough MS/MS data to identify the phosphorylated peptide in a database or to derive a de novo sequence. In addition, it is possible to perform a precursor ion scan to find more peptides (below the chemical noise level), as we will see in Sections 3.3.2 and 4.3. Offline ESI is also possible in slightly larger ion sources. In so-called micro-ESI sources the flow has to be be augmented by a little pump, for instance a syringe pump. However, flow rates are in the microlitre per minute range, and this is usually only feasible if relatively large amounts of protein (more than a couple of picomoles) are available.

Ideally, the solutions used to dissolve the analytes in ESI are acidic, so contain a lot of positive charges, and the proteins/peptides are highly positively charged in the solution and stay charged in the gas phase. The solvents are usually mixtures of water with some organics like ACN or methanol for best results (good evaporation and high ionization). ESI is so flexible that it can be used to spray the solvents used for RPLC directly; this helps to make the coupling of ESI with RPLC a very powerful online approach, as mentioned earlier. While the composition of the solutions during an RPLC run changes (from 5% ACN to about 50% in the part of the elution profile containing peptides/proteins), the ESI efficiency does vary, but usually remains good enough to provide excellent data from the mass analyzer (see also below). In offline ESI a solution of 50% methanol is used as a solvent for best results; methanol allows the highest signal intensity and lowest background noise as well as stability of the spray. While methanol can be used as polar solvent for RPLC, ACN usually delivers better separation profiles with higher resolutions and is preferred, but it may be worth testing it for certain applications. ESI is very sensitive to substances used to buffer the pH; even low concentrations of TRIS, for example, destroy any chance of getting a good signal from your analyte, as TRIS 'catches' most of the available ions. As a consequence, the buffers used to ensure high protonation of the analyte are volatile, like formic acid (FA).

Generally speaking, MALDI is more robust to many contaminants than ESI MS and samples that do not deliver any signals in offline ESI can still deliver good results in MALDI MS. However, if ESI is used in combination with LC most of the contaminations are separated from the peptides and ESI is even more sensitive than MALDI (see the discussion on LC effects on MS measurements in Section 2.5). Polymers are a class of contaminants that have a strong effect on ESI and can be difficult to remove by RPLC. They derive often from plastic-ware that was used at some stage in sample preparation (or LC buffer storage), so materials emitting polymers should be replaced by glass or high quality polymers.

While ESI can be more sensitive to contaminations than MALDI, it is by far the more efficient ionization technology of the two. A single peptide can easily be detected (in a 'clean' sample preparation) at a low attomole level in MALDI. As soon as several peptides are added, the same peptide might be invisible unless it is present in femtomole amounts. MALDI is very selective with peptides; some fly well, others do not (in mixtures), and in complex mixtures the effect of ion suppression is observed (see Section 3.2.1). ESI has no ion suppression effect, and even in very complex mixtures all peptides are ionized. Under these conditions (complex sample and much chemical noise) the signals for some ions can get lost in the noise (and can be isolated from the chemical noise by precursor ion scans, see Chapter 3.3.2.2 and Figure 3.27), but this not due to a decreased ionization as a result of increased sample complexity (ionization suppression), which is so typical for MALDI MS. In fact, ESI is such an effective process that most peptides measured in proteomic applications will carry multiple charges; typically up to three or four for peptides up to 20 amino acids. While under optimal conditions for ESI it is typical to observe a mixture of charge states for all peptides (e.g. single, double and higher charge states at the same time) it can happen that all or most observed peptides carry a single charge. This is either a characteristic of the specific ion source used or a sign that the parameters for ESI are not optimal; the flow rate can be too high, the sample is contaminated with, for example, salts, build-up of a blockage in the spray needle tip, the voltage between the tip and the skimmer is not optimal, the position of the needle is not optimal in relation to the micro-hole entrance into the mass spectrometer in the middle of the skimmer, or a combination of these problems.

It is also typical to obtain some fluctuations in the ion current and bubbling at the spray tip (Figure 3.15). In combination with changes in solvent composition during, for example, RPLC runs this can lead to a complete loss of signal or a significant reduction in sensitivity. It is therefore very important to establish conditions for a stable spray and ionization (which might not be conditions of highest signal) before experiments begin. ESI is used most of the time as an online technique; if things go wrong valuable samples might be lost. However, once stable conditions are established and the source is set up correctly it will run stably for long times (typically for a week). At regular intervals it will be necessary to change the spray tip. This is typically a gold coated glass capillary with a pulled and broken tip. These spray tips can be bought (at widely varying prices and of varying quality – these two factors are not always related to each other) or pulled in the lab. The tips need to be 'activated' before use, that is, a high voltage is applied and they can even be positioned to touch the skimmer, with voltage and flow applied. If your tip has been activated by the supplier it only needs to be installed and the spray head positioned according to the mass spectrometer manufacturer's instructions: all ESI sources allow the positioning of the spray tip in three dimensions in order to produce the most stable and efficient spray. To allow a reproducible alignment in all three dimensions with the instrument and optical control of the spray, most sources are equipped with one or two video cameras. While the best position of the needle depends on the design of the source, it will never be best to 'aim' the tip straight at the orifice of the instrument; a slight offside position usually delivers much more stable and sensitive signals.

To avoid infringing each other's patents each manufacturer has to come up with a unique source design, and some are not even spraying straight at the MS inlet; the ions have to follow a z- or s-shaped flight path. This can help to reduce chemical noise (by smaller ions, as these are following a different flight path) or even destroy the ions of interest if the voltage is not optimal. This so-called skimmer fragmentation is principally possible with every ESI source; it can be used as an analytical tool, producing in effect MS/MS-like data (see Chapter 5). Samples analyzed by different ESI MS instruments typically detect an overlapping but specific set of different peptides from complex mixtures; this is mainly due to differences in the design of the ESI source, and again can be used to increase the coverage of peptides during an analysis (see Chapter 5). Depending on the experimental set-up, the overlap of peptides is typically around 70%; that is, 30% of the detected peptides are more or less unique to a certain instrument, mainly based on the design of its ESI source. So it is a good idea to measure a sample with different instruments, ideally even ESI and MALDI instruments, to get the best possible coverage (Chapter 5).

As mentioned before, a strong electrical field is applied between the solution and the mass spectrometer entrance, the skimmer. This helps to make the droplets in the spray extremely small and is essential for the whole ESI process. In order to obtain small droplets, a jet of hot inert gas (N_2) is sometimes applied around the liquid spray. This stream of heated inert gas is usually employed in so-called micro-ESI sources, with flow rates in the microlitre per minute range. Nano-ESI sources (you guessed it! – in the nanolitre per minute flow range) can work perfectly well without them. Within the distance from the spraying needle to the skimmer, the droplets have to evaporate completely, which is quite a feature, given that the distance is a couple of millimetres. Once the droplets are very small, the surface of the droplets is charged to an extremely high density and they explode, leaving highly charged peptides/proteins with no water/solvent clusters around them. Again, as with MALDI, the exact mechanism of ionization and dehydration is still being researched. For the use of ESI in proteomics it is not necessary to go into all intricacies, rather we want to give an impression of what is involved and how it might influence the data that you can generate. The application of a high voltage between the ESI spray tip and the skimmer leads to the formation of the so-called Taylor cone at the very end

of the capillary. It is from this tiny cone that droplets are sprayed towards the mass analyzer in nano-ESI. It is important to note that once a high voltage is applied and the Taylor cone is formed, no additional flow is needed to expel the sample from the spray tip; it is 'sucked' out by the electric fields alone, generating a flow of some 20–50 nl/min. In typical nano-HPLC systems the flow is about 300 nl/min. This still allows a decent sensitivity and efficiency of ESI; however, the lower the flow the better the sensitivity, so offline ESI is more sensitive (on a clean sample) than online nano-ESI, but online ESI has the additional benefits of cleaning the analyte and concentrating it (see Chapter 2.4.1). The flow rate in nano-HPLC cannot be reduced easily any further for many reasons (e.g. it is very, very difficult to mix solvents at low flow rates or produce a stable flow) and the force of the electric field alone is not sufficient to 'suck' the sample through an HPLC column. Also the flow rate during a typical RP run is not stable – it can vary from 150 to 400 nl/min, depending on the viscosity of the solvents and the column back-pressure. So for online nano-ESI the flow through the column needs to be adjusted to allow some 'leeway' for the typical changes in flow-rate and sample composition during the RP run (see Figure 3.15). Should the spray be interrupted during a measurement, it may or may not start again without external intervention. If the flow stops due to the high back-pressure of the column (all flow directed to the flow splitter waste pipe . . .), the ESI may start on its own again once the flow is re-established. It might be necessary to ramp the voltage to establish a stable, non-pulsing ESI again. If the needle tip is blocked (salt build-up) it may be necessary to gently touch the skimmer with the needle tip (under full voltage) to make the needle spray again. After this procedure, stable settings might need to be re-established, and these might be different (position, flow rate in offline ESI, and voltage) from the situation before, since the needle tip might now have a different form or inner diameter. Again, depending on usage, it might be a good idea to change the needle about once a week for online ESI anyway. Once the flow is stopped for more than a couple of minutes, the liquid inside will evaporate, potentially blocking the needle with crystals, and then it is touch and go whether the needle will work again satisfactorily. For these reasons alone it is well worth keeping the ESI spraying at all times, even when there are no samples being measured (e.g. over the weekend), this will also keep the HPLC pump heads out of harms way. It is not only the cost of the needle, but also the need to disconnect the connectors in the hyphenated system that should be avoided; it might take minutes most of the times, but sometimes you introduce leaks or dead volumes, which can be time-consuming to detect and solve, due to the extremely low flow rates of such systems. A stable online ESI can sometimes run for weeks on the trot or you might have to optimize it several times a day, depending on the quality of your components (stable flow, good columns, trapping column to filter out debris) and the composition of your sample.

Let us now take a closer look at the ESI process. We mentioned earlier that the capillary in nano-ESI is at some point coated with gold, to allow a good electrical contact. The tip of the needle has a very small inner diameter, typically 2–10 µm, which will depend on how the needle was produced, and there will be needle to needle variations. At the top of the needle a Taylor cone is formed, due to the application of a considerable field (typically 1500 V over a distance of about 1 cm) and the properties of the conductive liquid. The tip of the Taylor cone is a lot smaller in diameter than the needle and it is here that at the end of a short jet, a very fine spray is formed. The liquid in the droplets that makes up the spray evaporates rapidly and there are two main concepts to explain the formation of ions from here; the charge residue mechanism (CRM) and the ion emission mechanism (IEM) sometimes also called the ion evaporation mechanism. The CRM implies that the droplets get smaller and smaller until they reach a certain limit. They carry excessive charges (in proteomics usually positive) so they cannot get smaller beyond a certain size (as a ball-park figure, imagine 10 nm droplets with 40 charges) and then undergo Coulomb explosions due to the immense forces of the repulsive charges in such a small space. After the first such fissions, the droplets have less charge, get smaller, explode again . . . until we end up with our analyte ions, devoid of solvents but now charged. According to the IEM theory, once the Coulomb explosions start the field strength becomes so large that single ions are desorbed from the droplets (hence the name 'ion emission' or 'ion evaporation'). There are indications that both mechanisms work together, depending on the analyte, charge state and external parameters. For now it does not really matter how the analyte ions are formed, but it might be important for future proteomic strategies, as controlled ionization can help to analyze

peptides or proteins more sensitively and may help to design better conditions for ESI, for example better solvents to get more robust ionization for automated applications. The existing ESI (and especially nano-ESI) applications are already very efficient; under standard proteomic application conditions one can expect to see on average doubly-charged peptides and one charge for every 3 kDa of a full length protein. The charge states are spread out; as several charge states for a peptide are detected at the same time (see Figure 2.35), even more charge states for a protein are detected simultaneously (e.g. +38 to +73 for IgG, with the strongest ion at about +55 or about 2 678 m/z; Figure 3.16). The size of the protein responsible for the detected ions has to be determined by deconvolution, that is, the automatic software based reconstruction of the theoretical protein mass (see Section 4.2.3) based on the multiple charged ion species observed (Figure 3.16). The charge states (and their relative intensities) that are detectable in any given experiment depend on many parameters, such as the quality of the spray and sample preparation. These considerations become very important for instance when trying to quantify MudPIT type experiments, with hundred thousands of ions that need to be considered. While it is relatively easy to find the charge state of each peptide and deduce its real mass (given an instrument that has a high enough resolution!), it is very difficult to do so during a run, and choose which peptide to use for MS/MS experiments. It is easy to end up analyzing different charge states of the same peptide while ignoring weaker ions in not strictly reproducible fashion, thus resulting in apparent differences between MudPIT runs when it comes to comparisons (Section 4.4). It is also interesting to note here that denatured proteins usually show higher charge states than native proteins.

The ionization voltage, or the voltage applied between spray needle and the mass analyzer orifice, is typically variable from 1.2 to 1.8 kV and depends on the positioning of the needle; the closer the needle to the orifice, the lower the optimal voltage, and the less time the solvents have to evaporate. If the needle is too far away from the orifice the voltage has to be increased to values so high that peptides may start to fragment before they have a chance to enter the mass analyzer. Also the flow rate will determine how high the voltage will have to be and where the best position for the needle is. Somewhere between those extremes, an optimal position for the needle can be

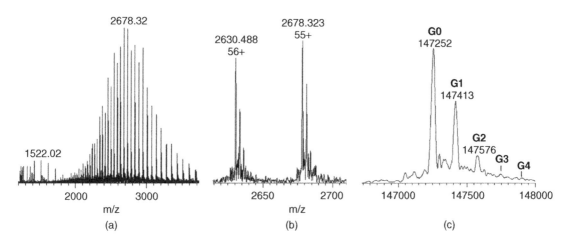

Figure 3.16 ESI spectrum deconvolution. ESI MS spectrum of a highly purified monoclonal IgG molecule generated using offline ESI on an orbitrap FT mass analyzer set to a resolution of 15 000. (a) Raw spectrum. (b) Detailed view of the most intense peaks, at charge states +56 and +55. The spectrum was then deconvoluted (see text and Figure 4.11). The resulting spectrum combines the data from all ions and recalculates it to the charge state of +1. The accuracy of the measurement was within 2 Da = 13 ppm. The different peaks represent glycosylation variants (+261 Da) of the molecule. G3 and G4 can be seen more clearly in online nano-HPLC ESI FT MS measurements. For an example of multiply charged peptides, see Figure 2.35. A very good MALDI analysis of intact IgG would deliver accuracies around several hundred dalton or hundreds of parts per million. The resolution of the measurement above was over 5 000 for G0, a MALDI measurement would have a resolution about 50, and all the peaks would be covered in the base of the only observable peak. Reprinted from Bondarenko *et al.*, (2009), *J Am Soc Mass Spectrom*, 20, 1415–24, courtesy of Elsevier. © 2009 Elsevier.

found, usually by starting at a not so optimal position and moving the needle slowly into the right place, slightly offset from the orifice for best results. However, due to the nature of the analyte, the flow rate and the solvents used, this position might change. ESI can be used to analyze protein/protein interactions; these will be observed optimally with lower voltage settings than small peptides. Proteins (if they are to fold and interact in a meaningful way) cannot be dissolved/sprayed in the same buffers as small denatured peptides, and their analysis in the native state might thus be less efficient (see Chapter 5).

While the optimization of the needle position for different analytes sounds very complicated, for most proteomic applications it is perfectly acceptable not be on the optimal settings, as nano-ESI is so impressively efficient and sensitive. Depending on the settings and mass analyzers used, it is easily possible to detect sub-attomole amounts of peptides. Furthermore, the ions are very slow moving and have low energy when they come out of the ESI spray, compared to MALDI generated ions. ESI is thus very versatile and offers excellent accuracy with a great many mass analyzers, from ToFs to ion traps and FTICR instruments. Basically every mass analyzer used

in proteomics can be efficiently connected to ESI and, in particular, nano-ESI sources.

From the ESI spray the sample (and any remaining unexploded or evaporated droplets) is literally 'sucked' into the mass analyzer; ESI is performed under atmospheric pressure (even if the source is encased sometimes), while inside the mass analyzer a reasonable vacuum is essential. Without a vacuum in mass analyzers ions would collide with gas molecules, which brings them off-course and can transfer enough energy into the analytes to fragment them. Both effects would diminish the yield of the measurement, also called the transmission of a particular instrument. Fragmentation by collision with gas molecules is an important tool for MS/MS analysis, but in order to achieve controlled collisions, the instruments generally have to provide a good vacuum inside (see Section 3.3). The vacuum gradient and the application of a high electric field in a particular shape (ion guide) work together to ensure that most of the sample actually ends up inside the mass analyzer, although the entrance into the analyzer is very small, often a hole of only 100 μm diameter.

Some designs of the ion source/mass spectrometer interface use a heated capillary right after the spray and before

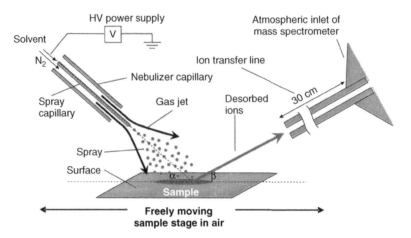

Figure 3.17 DESI from solid samples. DESI can be used on samples spotted on surfaces or, for example, on tissue sections. DESI currently achieves spatial resolutions of about 200 μm, while scan speeds across the surface are in the region of 300 μm/s. So to take an 'image' of a tissue slice of 3 × 2 cm takes about 100 minutes. This is faster than MALDI based imaging (Section 5.1), and DESI is fully compatible with MS/MS analyses. However, the MALDI matrix gives a better specificity towards peptides/proteins, and DESI is mainly used for metabolites and lipids from tissue slices. For proteomic applications, DESI can be used on samples spotted on a surface. Thus, several hundred samples can be measured in an hour, allowing high throughput 'ESI'. From Takats *et al.*, *Science*, 2004; 306: 47–73. Reprinted with permission from AAAS.

the skimmer, to help the solvents evaporate from the sample completely. This capillary is typically about 20 cm long and heated to over 100°C. After this capillary (or after the curtain gas, before the skimmer, in designs without such a capillary; see Figure 3.15) the ions (most of them now without any solvents attached) are in a vacuum which, through differential pumping, falls to below 0.1 mbar.

A relatively new development of ESI is desorption electrospray ionization (DESI), where a sample is basically extracted by a stream of hot solvent and nebulizing gas. The extracted peptides/proteins and other compounds are than collected into a mass analyzer (see Figure 3.17). This technology is likely to become more popular for imaging type mass spectrometers (see Chapter 5).

3.3 MASS ANALYZERS AND MASS SPECTROMETERS

The two main ionization technologies used in proteomics MALDI and ESI, as well as variations thereof such as SELDI and DESI, can be combined with a variety of mass analyzers to form a mass spectrometer. Although for various applications there are more or less suitable combinations, in proteomics MALDI is most often combined with ToF mass analyzers to form MALDI ToF MS or MALDI ToF/ToF instruments, while ESI is combined with all types of mass analyzers that are routinely used in proteomics and that will be introduced here; ToFs, quadrupoles, ion traps and hybrid instruments. Only recently have more instruments combining MALDI with other mass analyzers become commercially available, mainly ToFs and ion traps. There are several reasons for these preferences that will become clearer in due course.

Every mass spectrometer has the following components: an ion source, mass analyzer, detector, and a computer for control of instruments parameters, data acquisition, analysis, storage and retrieval of data. For MS/MS analyses, where certain ions are selected for fragmentation, some mass spectrometers might have two or even three mass analyzers, although all MS/MS or even further consecutive MS^n measurements can be achieved in principle by one mass analyzer alone (ion traps).

We will, as rule of thumb, not discuss the very important stages that allow transmission of ions in between the compartments. Although these are of the utmost importance for a practical instrument, this goes beyond the scope of this book. You will also fail to see a single mathematical formula, as the emphasis will be on understanding the basic principles of operation as they affect the usability of instruments for proteomic applications.

3.3.1 Mass spectrometers with time of flight mass analyzers

Linear time of flight mass analyzers

ToF analyzers are based on a very simple principle which is explained very clearly by an everyday example. If you had to push your broken-down vehicle home, it would take you a lot longer to do so if you had a heavy limousine than if your vehicle was a little moped. In the world of mass spectrometry, suppose we have two ions of different weights (say, ion A is 1 000 Da and ion B is 3 000 Da). If these two ions carry the same charge (e.g. +1) and are accelerated by the same electrical field (e.g. 14 000 V over a short distance in the centimetre range) they would be accelerated to different speeds, as both are accelerated by the same force (Figure 3.18). If we now measured the time from the beginning of the acceleration to when they hit a detector, we would see a difference between the times for ions A and B. We do not need recourse to the mathematics – your gut instinct will tell you that the lighter ion A will reach the detector first. The time difference immediately after the very short acceleration stage will be very small. If I just take the detector and move it 1 000 mm away from the end of the acceleration stage, the time difference will increase dramatically because the lighter ion A will have been accelerated to a higher speed than ion B. After the acceleration stage the ions will just drift, but at very different speeds. The longer the 'field-free' drift region is, the bigger the time difference will be, and hence the more accurately you can measure the time difference, and hence the difference in mass in between the two ions. This is the basic principle of every ToF mass analyzer and describes perfectly well the basics of a linear ToF analyzer. It also gives you an idea as to why bigger instruments will give you more accurate results; the flight tube is longer, so in principle measurements can be performed with better accuracy. However, ions will only drift in a field-free region if they do not collide with other molecules; ToF analyzers thus need a very high vacuum, not necessarily perfect, but good enough to guarantee that most ions do not collide with any residual gas in the flight tube. This also means that the longer the flight tube, the better the vacuum has to be (and the deeper your pockets need to be).

The principles behind ToF analysis mean that MALDI is very easily combined with ToF mass analyzers. ToF measurements have to be taken discontinuously. In combination with continuous ion sources such as ESI, a proportion of the ions have to go to waste. In combination with MALDI, basically a new measurement starts with every LASER pulse and thus every single arising ion can be measured. To increase sensitivity for proteomics the signals of several LASER shots are simply added up to give a mass spectrum (see Figure 3.13). ToF analyses need a clearly defined point in space from which to start the acceleration; the point at which the LASER hits the sample on a MALDI target and the resulting plume is much more clearly defined (micrometres) than the nanospray from an ESI source (ions derive from an area measured in several millimetres). As we will see later, ions from an ESI source can be focused so as to become very useful for ToF analyzers. Because all the components of the acceleration stage are very small and the ToF analyzer has to accelerate the ions to very high speeds anyway, ToF analyzers can cope quite well with the very fast ions arising from a MALDI plume. Since MALDI ToF mass spectrometers are such a winning combination, everything we say about ToF analyzers will be in the context of MALDI ToF mass spectrometers, unless stated otherwise.

The detector in ToF analyzers needs to be sensitive and fast; the ions of one LASER shot keep coming in fast succession, and if the detector needs a lot of time to measure them, heavier ones will arrive while lighter ones are being measured. The detectors in use today are multi-channel plate (MCP) detectors, where every ion that hits one of the walls of a microchannel creates an avalanche of electrons that are detected. Usually the whole detector is made up of thousands of such channels, and for MALDI ToF instruments the channels are V-shaped, so that the ions have a very good chance of hitting one of the walls (see Figure 3.18). The ions also need to hit with a certain speed and energy; if they arrive very slowly they cannot be measured very sensitively – all the more a good reason to accelerate the ions to high speeds towards the detectors. However, if the ions arrive too fast the differences in time between the ions get very small, and they can no longer be distinguished. That is why ToF detectors need a very high temporal resolution. They can also not be made too large in diameter; if the distances on the collecting surface exceed a couple of centimetres the temporal resolution is affected since the generation of the electron avalanches takes time, and if it comes from too large a surface the signals are not resolved in time. For all these reasons some instruments have special high mass detector modes, where temporal resolution (which is not crucial at high

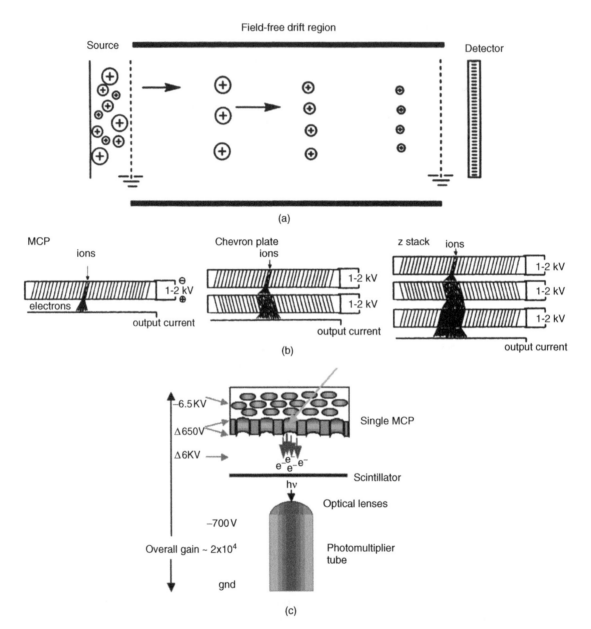

Figure 3.18 Principles of a ToF analyzer. (a) Typical discontinuous ToF analysis; at time point zero, ions are accelerated from the source into the field-free drift region. The acceleration stage is typically in the centimetre range, the drift region is around 1 m. Assuming all ions have the same charge (+1), the lighter ions travel fastest in the drift region and reach the detector first, followed in time by successively heavier ions. The time from starting the acceleration until the ions' impact on the detector is measured, from which is fairly easy to deduce the m/z of the ions. (b) At the detector ions usually hit a multichannel plate and induce secondary electrons in the walls, which are measured as current. Combining several MCPs enhances the sensitivity. (c) Highly sensitive combined detector – consisting of MCP, scintillating plates and photomultipliers – allowing signal amplifications up to a millionfold. In the example an ion from the ToF stage produces a handful of electrons at the MCP, which are then accelerated towards a scintillation plate where they induce light-flashes which are detected by a photomultiplier tube. For high performance ToFs high temporal resolution is essential. The detector needs to be able to measure ions only nanoseconds apart and needs to produce very short bursts of signal for every ion received (Gross, 2004). (c) © Agilent Technologies, Inc. Reproduced with permission, courtesy of Agilent Technologies, Inc.

masses as they take a lot of time to arrive) is sacrificed for high sensitivity. For example, the ions can get accelerated towards the detector from a field right in front of it. The most sensitive type of detector used today for ToF analyzers consists of an MCP in combination with a scintillator plate and a photomultiplier (see Figure 3.18). Such an ensemble can create huge gains in signal strength (more than 10^6-fold amplification) with reasonable linear range and low background signals. Although linear and reflectron ToFs have similar requirements, high performance detectors are especially important for instruments with a reflectron (see Figure 3.19). Most reflectron ToFs can be operated in linear mode as well, and these instruments have two detectors. The detector for the linear mode has more emphasis in detecting high mass ions. They might contain acceleration stages just in front of the detector, to give a higher sensitivity for low speed large ions, sacrificing some of the temporal resolution, which is not needed at higher masses as the differences in flight time are longer for larger ions.

Finally, the data generated from the detector need to be recorded at high speed. Currently a sampling rate of 1 GHz is typical, and high end instruments can sample data points at up to 4 GHz. This means that the instruments takes a record of the ion intensity at the detector 4 billion times per second. Each measurement is assigned a certain point in time, which is later translated into a certain mass. The higher the sampling rate, the more data points are taken to define one 'peak' and about 7–10 data points per peak will deliver very accurate measurements. Typically the digitizers measure the data in 8-bit format, which means that the analogue signal from the detector is divided into 256 signal levels, or in other words the difference between the lowest and highest signal is 256-fold. While this is enough for most applications, it is clear that it cannot be enough to quantify all components in proteomic experiments, and among mass analyzers ToFs are among those with smaller dynamic ranges. In real measurements, data from several LASER shots (from 50 up to 1 000) are accumulated and this can increase the dynamic range to a factor where chemical noise is limiting. To increase the dynamic range one has to compare a cumulative spectrum of, for example, 50 shots (for the most abundant peptides) and 1 000 shots (for the least abundant peptides), but this is of course fraught with difficulties, as, for reasons inherent to MALDI, quantification

is not straightforward (see Section 3.2.1 on ionization suppression).

Here are some ballpark figures to give you an idea of the time scales involved in ToF analyses in proteomics. The LASER pulse of a MALDI source lasts only about 1 ns, and typically instruments work with more than 10 LASER shots per second, high end instruments with up to 1 000. The speed of the ions in the resulting plume is several hundred metres per second (or the speed of a jet aeroplane). The ions are typically extracted after a delay of around 100 ns. The ions are accelerated (depending on their size) to speeds of the order of 50 000 m/s (a hundred times the speed of a bullet). Temporal resolution at the detector is around 1 ns. This resolution at the detector would result in a theoretical resolution of 7 000 with a flight tube of 1 m length at 1 000 Th and 35 000 at 3 000 Th. In proteomic applications (i.e. with real samples), high end linear instruments can reach resolutions around 5 000 and reflectron instruments more than 15 000 (see below). This example just shows how close to theoretical limits modern mass analyzers perform.

Pure linear MALDI ToF instruments are not used very often in proteomics (except for SELDI); linear ToF is mainly used in form of a linear mode in MALDI ToF instruments equipped with a reflectron. This is because reflectron (see Figure 3.19) MALDI ToF instruments deliver higher resolution and accuracy, as well as MS/MS capabilities. However, even small linear instruments can be very sensitive, as most manipulations of ions result in some sort of loss and the detrimental effects increase with instrument size. Also, when the molecules get large (in MALDI terms, for example, large peptides but mainly proteins) the metastable nature of ions produced during MALDI does not allow the use of a reflectron. MALDI with high LASER intensities generates even more fragmentations. The fragmentations are not instantaneous, but delayed by micro- to milliseconds (hence the name 'metastable') and occur mainly in the flight tube/field-free drift region. For linear measurements this has no influence on the signal at the detector; if you drop a ball out of the window of a moving car it will have the same speed (i.e. time of arrival) as the car. Hence, fragments of a peptide arrive at the same time as unfragmented ions of the same peptide/protein at the detector and all signals are combined. So for large peptides and proteins even the best reflectron MALDI will produce data similar (in its linear mode) to a cheap linear-only instrument.

The calibration of ToFs is a very important aspect of MALDI ToF MS. 'Out of the box' the standard calibration is usually only correct within 1 000–1 500 ppm, or more than 1–2 Da out on peptides! MALDI ToF accuracy depends very much on LASER intensity, delayed extraction (DE, sometimes called pulsed extraction) setting, sample topology, target position/topology, application of the correct voltages and the length of the flight tube. These parameters change on a daily and even hourly basis. LASER intensities will depend on the matrix and generally the purity of all components. If a sample is left for more than several hours in the machine, the chances are you will have to change the LASER settings. The sample topology depends on preparation methods and on the physical shape of the target plate; using different plates can change the mass measurement by 1 000 ppm, and the position of the sample on a large target plate can change the measured mass by as much as 300 ppm. Of course, the length of the flight tube is set by the factory; but it changes constantly due to changes in temperature. The changes in the length of steel are as much as several micrometres per degree Celsius. This can result easily in deviations of 10–20 ppm/°C of the measured mass. That is one of the reasons why mass spectrometers are placed in windowless and temperature controlled rooms. The power supplies for all the accelerating voltages also have a power drift, resulting in changes around 20 ppm/°C. These changes can compensate or add to the changes in the flight tube with temperature. The temperature inside the instrument also changes during use; from standby to measuring for several hours the drift can easily result in 200 ppm differences in measured mass. All these factors are best cancelled out by repeated calibrations before, during and after measurements. A calibrant mixture (similar in concentration/signal strength and preparation to the sample) is spotted near the samples, so that no sample is more than three to five positions or 2 cm away from the next calibrant on the target plate. The calibrants are than used to 'tell' the mass spectrometer what the actual measured mass really corresponds to, that is, the theoretical masses entered down to below 1 ppm accuracy. The instruments then tries to match the theoretical mass with the arrival times of the ions. Depending on the design of the instrument and the quality of the sample preparation as well as experimental conditions, this calibration will result in the instrument assigning the calibrants masses within 10–50 ppm of the theoretical value. The calibrants should be spread out evenly over the mass range, as the calibration below or above the range of the calibrants may be out by a considerable factor. Three to five reliable calibrants are used in practice, and often the calibration does not get better with more calibrants; the linearity of the instrument is reached. Using this calibration on a sample very similar in amount and purity to this calibrant should result in nearly the same accuracy. However, if the LASER intensity has to be increased or the crystals of the matrix are different or the sample is far away on a target plate with a surface uneven by more than several micrometres, or the m/z of the ions measured is far away from any calibrant m/z, the sample will be measured accurately within about 200 ppm, with accuracies of individual ions ranging from 10 to 200 ppm (depending on intensity, signal to noise, size or position between calibrants). If there are known ions in the sample (e.g. trypsin or keratin derived peptides) the spectrum can be recalibrated after the measurement. This internal calibration will result in an even better accuracy, typically around 100 ppm, and limited by the resolution of the instrument, the linearity throughout the mass range and the sample preparation (topology and contaminations). Some groups have changed the acquisition software so that the data of every single LASER shot is internally calibrated, allowing for even better accuracies. The order of calibration should always be from external to internal, as it is possible to calibrate to a wrong peak, making all efforts futile. All the above considerations apply also to reflectron MALDI ToF analyzers, which have a much higher resolution and thus a higher potential accuracy. The effects of all the potential limiting factors are more important the higher the accuracy one wants to achieve in the end; MALDI ToF for proteomics currently has a limit of about 10 ppm internally calibrated, using reflectron instruments with a specification better than 1 ppm.

However, linear instruments (or the linear mode of reflectron instruments) are less affected by temperature drift and electric drift; if the flight tube expands, it will do the same in both instruments/measuring modes, but in the linear mode the ions only pass this length once, and there is no reflectron where the power supply could develop drifting voltages over time or temperatures increasing due to usage of the instrument or environmental changes. So overall linear instruments and the linear mode in reflectron instruments are more robust in everyday use.

Calibration is important for all mass spectrometers. ToF analyzers suffer the most from changes due to their size. Axial MALDI instruments suffer more due to the direct coupling of ionization and analyzer. However, all instruments need calibration, even if only once a year (e.g. quadrupoles, which have a relatively low level of accuracy). This applies to all mass analyzers; for high accuracy, calibrations should be performed very carefully and repeatedly. Depending on the instrument and the accuracy achievable, calibrations (or at least the measurement of standards to control instrument performance) should be performed once a day or even directly before an important measurement, if accuracies below 1 ppm are to be achieved, which is possible with some instruments if internally calibrated. These instruments are 'racehorses' and beginners will be much better off with sturdy 'ponies' that will deliver limited but dependable data.

Reflectron ToF mass analyzers

Hitherto we have confined ourselves to linear instruments; that is, instruments whose flight tube is straight. You will not encounter a single instrument used in proteomics (except special instruments for SELDI) that is a purely

linear ToF. All instruments posses the ability to switch between linear mode and reflectron mode. As the name implies, in a reflectron ToF the ions are first accelerated and drift just as in a linear instrument; instead of hitting the detector at the end of the flight tube they are reflected by electrostatic 'lenses' (Figure 3.19).

The ions are usually (but not always) reflected by a little more than 180°, slightly off-axis. This ensures that the detector for the reflectron mode does not have to be placed in the way of the ions coming from the target. At the end of the reflectron tube there is another detector. Thus all ToF analyzers used in proteomics actually have two detectors, one for the linear and one for the reflectron mode. Using the reflectron has several advantages. Firstly, of course, you nearly double the flight path of the ions in a relatively compact instrument. Twice the flight tube length should double the resolution. From what we said above, the resolution more than doubles when switching from linear to reflectron mode. Why? The reflectron works like an elastic bouncing device (of course, the ions

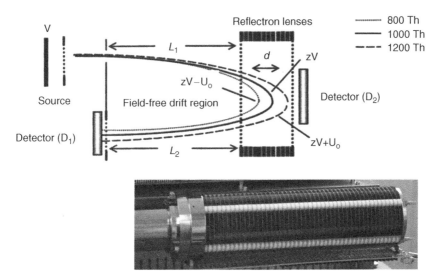

Figure 3.19 Basic principles of reflectron ToF mass analysis. The reflectron is a composite electrostatic lens with an increasing potential towards the back of it. Overall the potential is slightly higher than the acceleration potential that the ions experience after the source; this ensures that all ions will 'bounce back'. The picture of a (quadratic field) reflectron demonstrates that it is nearly as long as the field-free drift region. The flight path differences experienced by light and heavy ions are thus considerable. If an ion of 1 000 Th experiences a potential zV due to the reflectron, the heavier ions experience an increased potential $(zV + U_o)$ and the lighter ions a lower potential $(zV - U_o)$. It is important to focus the ions with these different potentials on one point, the detector. Curved field (or quadratic field) reflectrons can do this over a wide mass/potential range, linear reflectrons only over a narrow mass/energy range. In practice, these differences only become important for PSD applications (see text), and there are ways to ensure that all ions have similar energies and that they hit the detector when linear field reflectors are used. Photo courtesy of Jeff Dahl, http://www.flickr.com/photos/26142319@N07/.

never actually make any physical contact with the reflector); like balls flying attached to rubber bands the length of the linear flight tube, the ions slow down when they enter the reflector. Eventually they stop and are accelerated out of the reflector again. All the rubber bands in our analogy have the same strength: heavy ions stretch the rubber band further, hence fly deeper into the reflector before they stop, and then they are also accelerated for a longer time out of the reflector. Heavy ions (as compared to lighter ones) already fly more slowly, as in the initial acceleration stage at the MALDI source they were accelerated to a lower speed, in the reflectron ToF now they also have a longer flight path. Since a typical reflector is about 50 cm deep, the differential flight path can be quite substantial. This effect can be used to spread the time difference between the arrival of ions of different size (m/z really) as heavier ions travel more slowly and further. But the effect is also used to focus ions of the same weight (better, m/z) but with different initial speeds. The different initial speeds are derived from the different speeds ions achieve in the source, as an effect of the acceleration by the explosion caused by the LASER impact. The higher the LASER power, the higher is the speed of all ions and the wider is the speed distribution of ions of the same m/z. DE goes a long way to giving ions of the same m/z a more uniform speed distribution, but it can not eliminate all differences completely. The reflectron focuses these ions further (Figure 3.19). Ions of the same m/z but with different velocities also arrive at different times at the reflector; the faster ones arrive earlier than the slower ones. The faster ions penetrate the reflectron field more deeply, and thus come out of it later, much closer in time to the slower ions of the same m/z, which do not penetrate the field as deeply and thus travel a shorter distance. All these effects combine to give a roughly three- to five-fold increase in resolution when switching from linear to reflectron mode and allow top end MALDI ToF MS to achieve resolutions of 40 000 in reflectron mode. But there is even more to using the reflectron mode; it enables the analysis of fragments from metastable ions, similar to MS/MS experiments (see below). We will also see later that the fact that MALDI produces metastable ions is part of the reason why big reflectron instruments are very difficult (i.e. expensive) to build without a loss in sensitivity.

The reflector is an electrostatic mirror, made of a stack of metal rings with an increasing potential between them. The total potential difference over the reflector has to be slightly higher than the accelerating voltage if the ions are to be reflected back. If, for example, the acceleration voltage is set to 18 000 V, the total potential over the reflector is set to about 19 500 V. To make full use of the reflector, ions should be flying relatively rapidly towards the reflector and penetrate deeply into it, which is one of the reasons why the acceleration voltage in the reflectron mode is typically higher than that for the linear mode (e.g. 18 000 V compared to 15 000 V). In fact, in the linear mode the acceleration voltage should not be set too high so as not to wear out the detector with very fast ions. So for small ions (peptides) the voltage will be set lower than for very large ions (proteins), which otherwise might arrive at such a low speed that their detection might be impaired. To increase the sensitivity for proteins, some instruments have special high mass detectors, which are more sensitive but not as fast as the typical MCPs. A high temporal resolution is not needed for very large ions, as they fly very slowly and time intervals between the impacts of different sized ions in large mass (and m/z) ranges are accordingly much longer.

An increase in resolution in any mass spectrometer can be translated into an increase in accuracy if all other limitations are removed. We have not discussed the accuracy of ToF instruments yet because there are many factors that influence the accuracy, not least the way the ions are generated. In MALDI ToF instruments the ions result from an explosion plume. In other words, the ions are not generated at a fixed point and are constantly moving. To minimize the effect that this movement will have on the accuracy and resolution, a DE procedure was introduced by all manufacturers of MALDI ToF instruments. The idea is that heavier ions move more slowly in the expanding plume after a LASER impact than lighter ions. This statement is only quantitative, as even ions of the same mass show a distribution of speeds, according to the energy that was transferred to them during the LASER induced explosion. If ions are accelerated away (extracted from the source) as soon as the plume arises, the component of the movement vector that points in the same direction as the acceleration of the ToF will be added to the ultimate velocity that the ions reach at the end of the very short acceleration (see Figure 3.19). A typical value for the speed of ions during the plume expansion is 600 m/s. This speed is added to the 50 000 m/s the ions will get from the ToF acceleration potential. It may not sound like a big contribution, but it may result in an apparent decrease in mass of more than 1% or 10 000 ppm; that is, at 1 000 m/z the initial velocity from the LASER induced explosion translates to a decrease in flight time equivalent to a loss of mass of about 10 Da! But the news is not all that bad, since all the ions are accelerated in generally the same direction, so differences in speed resulting from the LASER induced explosion are less than 10 000 ppm, more like 2 000 ppm, which would translate to 2 Da at 1 000 m/z and one proton (typical of MALDI ions). What happens during DE is simply that the ions are extracted several hundred nanoseconds after the LASER shot, that is, the ions have 100 ns to move with their differential velocities from the LASER induced explosion. Lighter ions will move faster, that is, more towards the electrode used to apply the accelerating field. When the field is now applied after 100 ns of this movement, the slow ions get to be accelerated by the full potential and the fast ones by a fraction, that is, full potential minus the fraction of the distance within the field they have travelled before. If the design is right (and there are many variations) then the ions of different movement velocities are focused and the resolution is much improved. The use of DE makes it easier to create instruments that are easy to calibrate; most instruments need a multipoint calibration to achieve the highest possible accuracy, and one to two point calibrations cannot cover the whole m/z range. Before the use of DE, MALDI ToF MS delivered measurements of peptides with an accuracy of about 1 000 ppm. This is an error of up to 1 Da on a 1 000 Da peptide. With DE this error was reduced to 200 ppm and the resolution increased from about 1 000 to about 5 000. At 1 000 ppm accuracy it is not really possible to perform PMF analyses; there will too many potential hits and the real protein will be obscured by statistical hits. Modern MALDI ToF reflectron instruments deliver accuracies of about 100 ppm externally calibrated and better than 20 ppm internally calibrated in everyday use; the instrument specifications are much better than that, approaching 1 ppm internally calibrated in the top end models. The specifications for resolution are around 20 000 and in excess of 40 000 for the best instruments available; however, with real proteomic samples such excellent values can rarely be achieved. One of the reasons for this is the 3D shape of the sample on the MALDI target and the spatial spread it causes. Also a higher LASER output will be necessary for real (contaminated) samples than for ultra-pure calibrants. The method of sample

preparation is essential to achieve high performance and uniformity, hence the introduction of anchor targets (see Section 3.2.1). Orthogonal MALDI ToF (as opposed to axial MALDI ToF as described hitherto) offers many advantages in this respect, making it easier to achieve highly accurate measurements (5–10 ppm) at high resolutions (see Section 'Quadrupole-ToF hybrid instruments' in chapter 3.3.6).

We mentioned accuracies of mass spectrometer measurements earlier in Sections 3.1 and 3.2. With our more detailed knowledge of ToF analyzers we can now add to our considerations some more factors which influence accuracy. Ideally for ToF measurements all ions are generated at one sharply defined point in time and no differences in relative speed. While MALDI goes some way towards this goal, the ions have a certain spread of energy, and the higher the LASER energy used (the fluency) the wider this spread is. We have seen how DE can be used to counteract this effect to some extent. Most commercial MALDI ToF instruments have in fact a staged acceleration with two electrodes close to one another very near the source to the able to optimize the DE settings. Most instruments allow the DE settings to be optimized towards a certain m/z range, which will make these ions more sharply focused and the resulting peaks stronger, with higher resolution and better mass accuracy. It is advisable to set the DE optimum towards the larger end of the useful measurement window as the larger ions tend to be underrepresented in MALDI experiments (see Section 3.2.1). The initial energy spread from the MALDI process derived ions is a limiting factor for the resolution of MALDI ToF instruments. For small ions the energy spread of ions with the same m/z can be counteracted by using the reflector, as we have seen above. However, beyond a size of about 5 000 m/z the use of a reflector becomes less and less feasible in MALDI ToF instruments. There are several reasons for this, but the main one is that MALDI produces metastable ions. As we will see below, these ions fragment within micro- to milliseconds. The timescales in typical MALDI ToF measurements for proteomic applications are very much shorter than this; LASER pulse about 3 ns, plume expansion and DE about 500 ns, acceleration and flight time before reflectron 30 μs. The time for ions to reach the detector in the reflectron mode is typically longer than 60 μs. All these times are ballpark figures for peptides of 1 000 m/z and they vary between instruments and settings. However, the flight times increase with the square root of m/z. So for a 14 kDa protein we have four times longer flight times and for medium sized proteins (64 kDa) the flight times are eight times longer. This would mean about half a millisecond of flight time before the ions hit the detector in reflectron mode, by which time there are hardly any ions left, due to metastable fragmentations. This affects in particular large ions, as they need a much higher LASER power for ionization (see Section 3.2.1). Larger ions are not as fast as smaller ions in the developing plume, but they have more contact with matrix molecules and more energy per molecule is transferred, and we need more LASER power for their desorption, hence overall the energy spread is wider than for small ions. So there are several problems that combine to make the ultra-sharp sword of MALDI ToF MS (resolution) blunt for larger peptides and, in particular, proteins. We cannot use the reflectron mode for proteins, the energy (speed) spread from the LASER induced explosion cannot be corrected by the reflector, the energy spread is larger, and since large ions are not accelerated as fast by the ToF acceleration field the effect of the energy spread is more pronounced. So while a peptide can be measured at a resolution of about 20 000, proteins are measured at resolutions of 200–400, the accuracy slips from 20 to 20 000 ppm or 0.2%. In addition, it is very difficult to calibrate a MALDI ToF for proteins, because at these low resolutions we are looking at mixtures of molecules that might have been oxidized, or de-amidated, or even lost larger groups like phosphates or complete amino acids. All these isoforms merge into one peak of a hard to determine exact mass. A mass accuracy of 0.2% still means that we can detect the loss of a single amino acid (57–163 Da) in proteins around 30 kDa. This is much more accurate than SDS PAGE or density gradient centrifugation (5–20%), but if we need accurate masses of proteins we need to use ESI based mass spectrometers (see Section 4.2.2). ESI ToF instruments (not a mainstay in proteomics) can deliver resolutions in excess of 15 000, almost irrespective of protein size, so we can get monoisotopic resolution of proteins up to 15 kDa with undiminished accuracy in the region of 5 ppm. Monoclonal antibodies (about 160 kDa) can be measured with an accuracy of about 20–40 ppm (3–6 Da) in ESI ToF, whereas MALDI ToF delivers accuracies of about ±300 Da. Moreover, with ESI ToF MS and its high resolution we could look at different glycosylation variants (160 Da apart from each other), while with MALDI ToF MS all peaks would merge into one.

MALDI produces many metastable ions, and the longer the flight time, the more time these metastable ions have to fragment. This means that an instrument with a long flight tube has to be designed to be extremely accurate so as not to lose sensitivity over a much smaller, cheaper bench top instrument. Where the high end instruments come into their own is their high resolution and thus high accuracy, but also in observing the fragments from metastable ions in MS/MS approaches; there are simply more fragments produced in the time frame accessible for analysis in a larger instrument with longer flight tubes.

The metastable ions are a mixed blessing: they inhibit the accurate analysis of larger ions (as seen above) but enable very useful MS/MS analyses for proteomic studies. The process of fragmentation is (like the ionization itself) not completely understood. The LASER induced fragmentations (or LASER induced dissociations (LIDs)) are distinguished, disregarding any mechanistic aspects, by the place they occur in the mass analyzer; in-source decay (ISD) occurs before or during the ToF acceleration, post source decay (PSD) occurs after the acceleration, in the field-free drift region of the flight tube (Figure 3.20). The higher the acceleration voltage, the more energy can be transferred to the ions during the acceleration by hitting each other, thus enhancing PSD. This distinction in space between ISD and PSD is of course mainly a distinction in the time at which the ions are generated after the ionization, or better: after the LASER pulse. ISD is only rarely of practical use as the fragments cannot be recognized as such by the reflectron ToF analyzer; they behave like molecular ions and contribute to the background noise. PSD can be used to help the structure determination of peptides and thus proteomic peptide/protein identification as the parent ion for fragmentation can be chosen and the fragments measured as such in the reflectron ToF analyzer.

Let us look more closely at the consequences of PSD. The ions formed during MALDI get enough energy transferred into them to break a bond (or even more than one) within the molecule. This fragmentation occurs at some time after the energy transfer; by definition, the fragmentation can occur at any point during the drift in the tube (for most of the ions roughly 1–20 µs after the LASER pulse). In the linear mode this fragmentation goes unnoticed, as all fragments still fly at the same speed, regardless of size or whether they retained charge or not (see Figure 3.20), and all fragments would contribute to the signals as if there had never been any fragmentation.

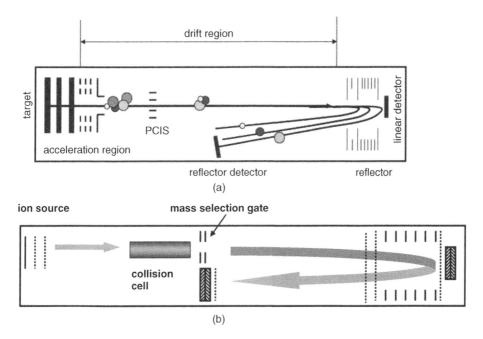

Figure 3.20 MS/MS with reflectron ToF: PSD. (a) PSD on a MALDI reflectron ToF mass spectrometer with a linear reflectron. The MALDI source produces metastable ions of different sizes, which are accelerated away into the flight tube. Here a single ion is selected at the precursor ion selection (PCIS) gate. Due to its metastable nature, the ion fragments while drifting in the flight tube. The fragments of different sizes stay together, and are only differentiated in flight path and time in the reflectron (see also Figure 3.19). The linear reflectron can focus some but not all fragments onto the reflector detector. To measure all fragments, repeated measurements have to be performed, in which the reflectron potential is lowered, so that smaller fragments are focused into the reflectron detector. (b) In a curved field reflectron instrument all fragment ions are focused into the reflectron detector, allowing all fragments to be detected in a single experiment. Note that the instrument is also equipped with a CID chamber (collision cell). The ions fly so fast that they do not fragment by CID until they are outside the collision cell. This allows the placement of the PCIS gate (here called a mass selection gate) after the collision cell. The further the mass selection gate is from the source, the better the resolution for precursors, as ions with different masses are more separated in time. (a) Reproduced with permission from Baumgaertel *et al.*, (2008), *Macromol Rap Commun*, 29, 1309–15. © 2008 Wiley-VCH Verlag GmbH & Co. KGaA. (b) Reprinted from Cotter *et al.*, 2007, *Jour. Chromat. B*, 855, 2–13, courtesy of Elsevier. © 2007 Elsevier.

If the instrument were in reflectron mode, the fragments without charge would not be reflected (without charge the electric field has no influence on their flight path) and only the charged fragments would be reflected. As their mass is lower than the 'parent' mass, they will penetrate less deep into the reflectron field, have a shorter flight path and arrive earlier than the parents at the reflectron detector. So while there will be some considerable loss in signal intensity, we should be able to analyze the fragments of peptides and proteins in a reflectron instrument. A major prerequisite for the meaningful detection of fragments is the selection of a source of fragments ions, the parent ion. We could still detect all the fragments without such an ion selection, but it would be impossible to

tell from which of the potential parent ions in a spectrum they would be derived. For this parent ion selection most reflectron MALDI ToF instruments have an ion gate, usually a so-called Bradbury–Nielsen ion gate. In this type of ion gate, basically a high voltage is applied perpendicularly to the flight path of the ions. While the field is applied, all ions are deflected. The voltage is switched on at the beginning of the measurement, switched of at the time ions of the selected mass are expected and then switched on again. Because the ions have to be selected before they fragment, the ion gate has to be placed before the field-free drift region. This results in very short time differences between ions with different m/z at the gate. Therefore, it is very difficult to gate ions with a high resolution, and typically the resolution is about 100–200, that is, using the definition $m/\delta m$ for an ion of 1 000 Th (Section 3.1.1, Figure 3.5) the window by which a precursor can be selected is 5–10 Th, for an ion of 3 000 Th it would be 15–30 Th. This is a serious limitation, as other ions might be present near the selected precursor ion and the analyses simply cannot be performed because the source for the fragment ions cannot be selected. Advanced instruments (see Section 'ToF/ToF mass spectrometers') can achieve a resolution up to 1 000 for the precursor selection without suppressing the signal of the precursor ion too much. While this would give a mass window of 1 or 3 Th at 1 000 and 3 000 Th respectively, other mass analyzers can select ions with even higher resolutions.

As a complication, traditional 'linear' reflectors cannot focus fragments of all sizes (or, better, of all energies) onto the detector in a single measurement. The fragments (or daughter ions) carry a fraction of the energy of the original or 'parent' ion. Their energy is considerably less than of an unfragmented ion of the same size (see Figure 3.20), since the parent ions were larger and hence slower. The speeds of the parent and daughter ion are identical, so the speed of a small daughter is much less than the speed and kinetic energy of an unfragmented ion of the same size (if you drop a ball out of the window of a moving car, the kinetic energy of the ball is less than that of the ball and car together). There are several ways out of this predicament: either one has to adjust the settings for the reflector, or the reflector is constructed with a so-called curved or quadratic field, or the instrument is equipped with an accelerator right in front of the reflector (LIFT technology; see Figure 3.22). In instruments with a linear reflector the settings for the reflector will focus ions of up about 80% of the parent ion's m/z correctly on the detector (limited in size, see above). The experiment is performed and daughter ions up to an m/z of 80% of the parent ion are measured. Then the reflector voltage is lowered to about 80% of the original settings; now ions with an m/z from 80% to about 65% (i.e. 80% m/z of the 80% of the parent's m/z) of the original mass are measured. If this is repeated about 12 times, the smallest masses that can be measured are those of immonium ions derived from the fragmentation of amino acids. All the 12 spectra are then 'stitched' together to deliver a continuous spectrum reaching from the original parent ion over all fragments down to breakdown products of amino acids, the immonium ions. If

the acquisition of a normal reflectron spectrum takes a couple of seconds and maybe a minute for a very small amounts of peptides/proteins, the manual acquisition of a PSD spectrum typically takes around 20–40 minutes. Furthermore, a complete PSD spectrum uses a lot more sample than an MS measurement, as the spectra need to be taken 12 times and are much weaker than normal MS measurements in the first place!

For PSD spectra the instruments will also need a special calibration that can be complex to maintain in an instrument with a linear reflectron. PSD analyses are much easier to perform in an analyzer with a curved or quadratic reflectron field. The special design of the reflectron allows all fragments to be collected in one measurement. One drawback is that some designs with a curved reflectron do not have the same ultra-high resolution capabilities of top end liner reflectron analyzers in standard reflectron operation.

With PSD analyses, in contrast to many other MS/MS approaches, it is not possible to manipulate the conditions under which fragmentation occurs since PSD is a by-product of the ionization process. In practice, the use of certain 'hot' matrices (Figure 3.12) is necessary to achieve a reasonable quantity of PSD fragments and the LASER intensity is usually set a little higher than for normal measurements in order to get a higher yield of fragmentation. This yield is critically dependent on the time the ions spent in the field-free drift region, and hence on the length of the field-free drift region (remember, most MALDI ions will eventually fragment – the longer the observable time frame, the more fragmentation we see). It is also possible to lower the acceleration voltage for PSD experiments to slow the ions down a little; the negative impact on resolution is not critical, as the resolution and accuracy in PSD mode are much lower than in standard reflectron mode, due to several reasons involving the relative high energy spread of the fragment ions. Usually mass resolution (e.g. resolution 800 at 800 Th) and 150 ppm accuracy are achieved. In some instruments a limited CID capability is given to the instrument to enhance the fragments generated by PSD. This is not really comparable to the sophisticated CID methods in other mass analyzers we will come across later in this chapter. A small cell is introduced in the ions' flight path after the ion gate and before the field-free drift region (Figure 3.20). Inert gas, argon or nitrogen, is introduced to locally enhance the pressure to about 10^{-5} mbar. This

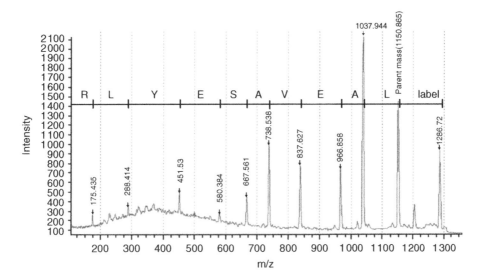

Figure 3.21 Using PSD in combination with chemically assisted fragmentation for peptide identification. A tryptic peptide from a 2D gel isolated *E. coli* protein was subjected to chemical modifications ultimately leading to a sulfonated N-terminus. This allows double protonation during MALDI with one delocalized proton, which in turn supports the fragmentation into b and y ions. Only y ions are detected, as b ions stay neutral due to the negatively-charged, sulfonated N-terminus. The peptide sequence is actually LAEVASEYLR (compare Figure 3.7 for nomenclature). Reprinted from Flensburg *et al.*, (2004), *J. Biochem. Biophys. Methods*, 60, 319–34, courtesy of Elsevier. © 2004 Elsevier.

is in contrast to the ultra-high vacuum in the flight tube (2×10^{-7} mbar) needed for sensitive ToF measurements, and after a CID measurement the instruments will need some time to reach a high internal vacuum to allow ultra sensitive reflectron measurements again. Again, the conditions of CID cannot be controlled very well, the gas pressure is more like on/off, and the speed of the ions is determined by demands other than CID efficiency. Any peptides that do not fragment within the initial 20 µs (in the field-free drift region on the way to the reflectron) cannot be measured, and again instruments with a long flight tube have sensitivity advantages as the time window in which fragmenting metastable ions can be measured is widened. So the benefits of CID are relatively small but measurable in MALDI ToF analysis of fragments, and under combined PSD/CID conditions immonium ions (Figure 3.7) become strong enough to predict the presence of many of the amino acids in a peptide.

Chemical modification of peptides can be used to enhance the information that can be made available via PSD. In this so-called chemically assisted fragmentation approach charged groups are introduced which allow different fragmentation behaviour of tryptic peptides.

Thus fragmentations are more efficient and directed towards a single ion series, making interpretation easier (Figure 3.21).

ToF/ToF mass spectrometers

To achieve much better MS/MS performances than from the above described PSD, a 'true' MS/MS approach is needed with two ToF stages, that is, using ToF/ToF instruments. In a 'true' MS/MS machine the precursor ions can be selected with high resolution, the fragmentation conditions are optimized to obtain a high fragmentation yield for different analytes, and the resulting fragments can be detected with high sensitivity, resolution and accuracy. There are designs of commercial ToF/ToF analyzers that fulfil these conditions using two quite different approaches. These ToF/ToF MS/MS machines allow the selection of a certain parent ion with reasonable resolution (about 400–1000) and can perform high yield fragmentation by either LID or 'true' CID, or a combination of the two. The LID or CID generated fragments can be detected with high efficiency/sensitivity (independent of acceleration parameters as opposed to the situation in PSD measurement), and the resolution is high enough to measure the monoisotopic masses of typical fragments with sufficient accuracy (several parts per million) to allow highly significant searches to identify proteins and peptides from huge databases. Comparing these capabilities of ToF/ToF analyzers with the MS/MS capabilities that PSD allows in reflectron ToF instruments, it becomes clear that the reflectron ToF MS is an instrument that can perform some limited MS/MS measurements, while ToF/ToF instruments are serious MS/MS instruments whose MS/MS capabilities can be exploited to good effect in routine analysis.

Nevertheless, the basic elements of a ToF/ToF machine are *in principle* similar to those of a standard reflectron ToF analyzer, with some 'extras' (see Figure 3.22). It is important to note that in reality every single component that is known from the reflectron ToF analyzer has to be modified to a great extent to make a successful ToF/ToF machine, as we will see after the main workings are covered.

Every measurement on a ToF/ToF machine starts like a reflectron MALDI ToF MS measurement would – by seeing which ions are present and available for MS/MS. As the MS/MS capability is not needed, the ToF/ToF instrument is run exactly like a reflectron instrument; the additional elements are quite literally switched off and a spectrum containing all the molecular ions in reflectron mode is acquired. Then the ions on which MS/MS

is to be performed are chosen one after another, as they would be in a typical MALDI reflectron ToF machine, by a timed ion selector (TIS). Here the differences between the implementations of commercial ToF/ToF analyzers start. We will first follow the operations of an instrument optimized for CID based ToF/ToF MS/MS analysis. The ions selected for MS/MS analysis enter a deceleration stage, followed by a CID chamber that is filled with inert gas. Here the relatively slow ions (for ToF flight conditions) collide with gas molecules and energy is transferred. These collisions within the collision chamber occur repeatedly, until eventually enough energy is transferred to 'heat' the molecular ion until it fragments. The time scales involved in the collisions and fragmentations are relatively long compared to ToF conditions in single-stage instruments. The times the ions spend in the collision chamber are very long since the ions have been slowed down, and waiting for fragmentations to occur after the collisions is not a major problem (compare with the paragraphs on PSD in Section 'Reflectron ToF mass analyzers'). The fragments (and some unfragmented molecular ions) are carried forward by their remaining speed after the deceleration stage and enter now a second source region, where they are focused and then enter a second acceleration stage that leads into the reflectron. This reflectron can focus all the fragments, with their

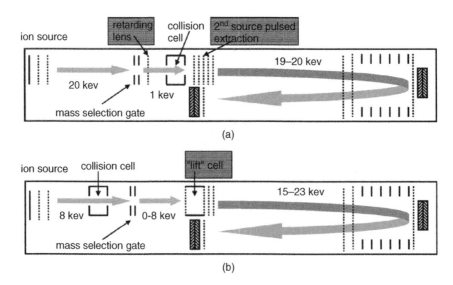

(a)

(b)

Figure 3.22 Commercial implementation of ToF/ToF analyzers. The combination of features used in the instrument (b) is called the LIFT technology. See text for detailed explanation. Reprinted from Cotter *et al.*, 2007, *Jour. Chromat. B*, 855, 2–13, courtesy of Elsevier. © 2007 Elsevier.

different energies and masses, in a single measurement. This is quite different from the typical linear or even curved field reflectron of a reflectron MALDI ToF in PSD, as the fragment masses can be measured with high accuracy because their energy spread is controlled very well by the design of the ion generation and the acceleration field in the second source region and there is no need to 'stitch' measurements of different m/z ranges together. The ions are reflected towards a detector, a typical multichannel device, much as in standard ToF instruments.

The above described ToF/ToF analyzer depends entirely on CID for the fragmentation and the generation of MS/MS data, and there are several reasons why this design is much more efficient than PSD in a standard reflectron ToF instrument. The first ion source is identical to a single-stage ToF, including DE and acceleration voltages of 20 000 V. The TIS (or the ion gate) is not placed right at the beginning of the field-free drift region as in a standard MALDI ToF instrument performing PSD type experiments, and thus the flight time of the ions is reasonable long and the precursor ions for MS/MS can be selected with resolutions typically around 400. This means that in a typical experiment at 1 000 Th (3 000 Th) the ions selected for CID have a bandwidth of 2.5 Th (7.5 Th). The example in Figure 3.23 shows how this improves the ability to perform MS/MS in a typical PMF experiment, compared to standard PSD. The ions are slowed down prior to collisions in the collision chamber until the remaining energy is in the region of 1–2 keV. The CID fragments retain the energy which represents their fraction of mass of the original ion. For example, if the parent ion was 1 000 m/z and the fragment retaining the charge was 100 m/z, then the parent ions' energy is 1 keV and the fragment retains 100 eV of the energy. Thus the energy spread of the fragments is 1 keV at the most. Note that a parent of m/z 2 000 also would have 1 keV, but a fragment of 100 m/z would only have 50 eV. This energy spread is comparable to the energy spread that DE has to cope with, and much smaller than the energy spread of PSD ions (around 19 keV). The ToF/ToF has an ion focusing stage right after the collision chamber with an electric field comparable to the DE to focus the fragments by DE before they are accelerated into the reflectron. Because the ions are well focused the acceleration voltage towards the reflectron (and the opposite voltage in the reflectron) can be in the region of 20 000 V, thus allowing high resolution measurement of the fragment ions. Typically fragmentation is performed on ions in the range of

1 000–4 000 Th and fragments are measured with a resolution in excess of 8 000. It is interesting to note, that (as in single-stage ToF instruments) the resolution at lower m/z can be limited by the sampling rate of the digitizer. It is also noteworthy that MS/MS measurements are less sensitive than MS measurement in any ToF/ToF. In the instrument in question not all molecular ions will fragment, as the gas pressure cannot be increased to a high enough level to ensure that every ion undergoes enough collisions. In addition, there are still PSD events from the first source; the metastable ions will fragment at various places inside the instrument and not contribute to interpretable fragments. And of course there is the principle problem in MS/MS sensitivity; the intensity of the original molecular ion is diminished according to the number of fragments it produces – we will return to this later. Since in this design all the fragmentation for MS/MS comes from CID, which in turn depends on gas in the collision chamber, through which all ions have to fly, constant changing from single-stage to ToF/ToF mode should be avoided as the gas will cause loss of sensitivity in normal reflectron ToF mode. Thus, a good strategy is first to measure all samples in reflectron mode, and then start taking all the measurements in ToF/ToF mode to avoid constant change from depending on gas to wanting to avoid it.

Another very successful implementation of ToF/ToF technology uses mainly LASER induced fragmentation (also called LID, see above). Without going into too much detail, the LID fragments used in ToF/ToF are in principle the same ions used in single-stage ToF instruments for PSD measurements (LIFT design in Figure 3.22). However, the design of the LIFT ToF/ToF allows for a much better generation, utilization and analysis of the PSD ions. The instrument can be used like a one-stage ToF instrument for MALDI ToF linear and reflectron measurements. When MS/MS data are acquired the instrument is run in a different mode. The LASER power setting has to be increased above threshold to produce more metastable ions, just as in a typical PSD experiment, with all the requirements for 'hot' matrices such as CCA. At the heart of the operation in MS/MS applications is a two-stage acceleration process; instead of accelerating the ions in the first (standard) source with a high voltage, they are only accelerated with a modest voltage into the field-free drift region. Since the acceleration voltage is low, the ions spend a long time in the field-free drift region before they reach a TIS (ion selection gate). As the ions

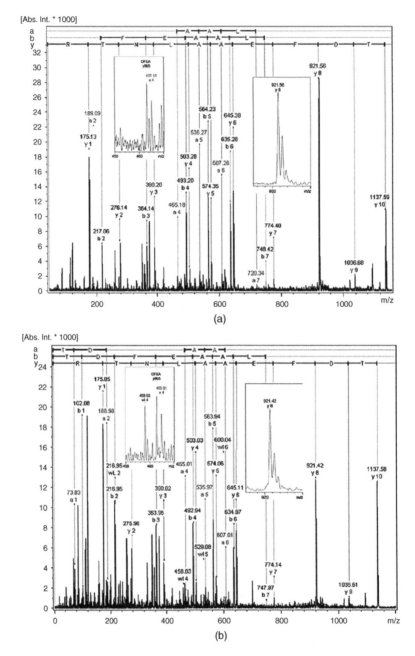

Figure 3.23 ToF/ToF performance. (a) Peptide fragmentation using the mass ToF/Tof analyzer described in Figure 3.22(b) under pure LID MALDI ToF/ToF conditions using the LIFT cell. (b) Here the collision chamber is used in addition to provide additional CID fragmentation. The additional CID fragmentation extends the "b" ion series, adds some new ions of the "a" type and results in the w_4 ion of 458.03 (see left inset in (b)). Without the w_4 ion the amino acid defined by y_4 and b_7 could only have been assigned as either one of the two isobaric amino acids Leu or Ile. w_4 at 458.03 defines Leu. Note that the CID process adds not only useful ions, but also ions that cannot be assigned to any specific ion series (48 unassigned in LID, 83 unassigned in LID + CID); these ions make an interpretation more difficult. Suckau *et al.* (2003) © Springer-Verlag, 2003.

are only accelerated to a modest speed, the differences in arrival time between ions of different m/z (and their fragments) are relatively large and an advanced TIS can operate with relatively high resolution of 500–1 000. The ions to be analyzed by MS/MS are selected and typically all the different isotopic peaks of a peptide are included in the analysis. The selected ions enter the so-called LIFT region, where they are accelerated by a large voltage, focused by a field with temporal variations (similar to DE in the ion source as labelled in Figure 3.22(b) – source 1) before they are sent into the reflectron. This whole LIFT process is essential in focusing the fragments of different kinetic energies and velocities into the reflectron, so that they can be measured with high resolution and sensitivity.

The source 1 region on this ToF/ToF is optimized for the generation of LID/PSD. Factors such as field strength right above the target plate and pressure (in the expanding plume) can have a large influence on the yield of LID derived ions. In addition to LID, the instrument can measure CID by including a gas chamber in the ion path. However, since the instrument is designed with LID/LIFT as main MS/MS mode, only high energy CID (8 keV) can be performed, which will not yield fragments usable for de novo sequencing of peptides or database searches. CID can, on the other hand, be used to generate immonium ions and internal fragments to differentiate isobaric amino acids (Leu/Ile) to supplement the LID/LIFT data. It is noteworthy that the CID chamber is placed before the TIS, to enable the TIS to be as far away from the source as possible (Figure 3.22). This allows a better resolution in the selection of ions for MS/MS. In the LIFT mode of MS/MS the total energy put into molecular ions is about 28 kV. This acceleration is divided into 8 keV after source 1 and 20 keV in source 2. The moderate acceleration after source 1 allows the ions to spend 10–20 µs in the field-free drift region, before the LIFT cell. Metastable ions have rate constants of decay in the microsecond range, so the low acceleration voltage enables a high yield of LID/PSD ions. When arriving at the LIFT region the kinetic energy for the fragments is roughly their fraction of the mass of the molecular ion, and thus ranges from 0 to 8 keV. For example, a 200 Th fragment of a 2 800 Th molecular ion has a kinetic energy of roughly $8\,keV \times 200/2\,800 = 0.57\,keV$. Due to the fragmentation (energies needed to break a C—C bond are a couple of electron-volts per bond), it comes to a slight energy spread of ions with the same mass and thus also to a

spatial spread, as ions with the same mass travelled in slightly different trajectories at different speeds, some of them 80 cm. It is this spread that causes a relatively low (500–1 000) resolution and accuracy in traditional PSD. This is where the LIFT approach comes into its own, by allowing a second focusing of these ions, just as DE does at source 1. LIFT also gives the ions an additional 20 keV in kinetic energy. This allows measurements at a high resolution using the reflectron. Even the smallest fragments have energies around 20 keV and thus fly with high kinetic energies towards the MCP detector and can be detected with high sensitivity. By comparison a normal PSD fragment of 86 Th (immonium ion of Leu/Ile) would only have around 1 keV if it came from a 2 000 Th peptide. Moreover, the energies of all fragments are relatively similar, ranging from 20 to 28 keV, and thus all ions can be focused on the MCP detector without changing the settings of the (linear) reflectron. Although the design of the instrument allows a high yield of LID, there will typically still be a significant portion of unfragmented, molecular ions. These molecular ions are excluded to a large extent by an additional ion gate after the LIFT process. This eliminates fragments that could form inside the reflectron, which would cause higher background or 'ghost' peaks appearing at positions not connected to their real m/z. Apart from using PSD and CID type fragmentations the LIFT ToF/ToF can also use LID that occours in the source, before the field-free drift region, to analyze whole proteins. This so-called ISD does not happen at a yield comparable to PSD, and, of course, because it happens so close to the target plate no TIS for ion selection can be used. Nevertheless for pure proteins and relatively high concentrations (picomolar) the LIFT ToF/ToF can be used to sequence large portions of complete proteins, but the details of this are beyond the scope of this book.

There are several issues that arise for the use of any design of ToF/ToF instrument which need to be addressed to make it a workhorse for proteomics. First of all, the sensitivity of any MS/MS analysis is lower than the corresponding MS analysis. The intensity (or number of ions) that is concentrated in one ion has spread among a variety of fragments in MS/MS analyses. And, of course, we can only see the fragments that retain the charge. It is also impossible to convert the molecular ion completely into fragments. If this were to be achieved in CID the only fragments observed would be very small, internal fragments and immonium ions; in LID this is not possible in any way, as not all ions will fragment. The signal intensities in MS/MS are reduced by a factor of 20–30 for many peptides. Since the noise in MS/MS spectra is also lower, lower signal intensities can be measured at a reasonable signal to noise ratio and with thus only slightly reduced accuracy due to the remaining noise. In MALDI MS/MS these sensitivity problems can be counteracted by collecting data from more 'shots' and thus increasing the intensity of weak ions above the noise level. A typical single-stage measurement in proteomics might have to accumulate data from 200 LASER shots, while a typical ToF/ToF measurement of a single molecular ion in proteomics will need about 1 000–2 000 shots. If there are about 10 molecular ions per sample this would result in up to 20 000 shots per sample! However, at 20 shots per second this can take quite some time (about 20 minutes per sample), particularly with large numbers of samples. Modern MALDI ToF/ToF instruments can have LASER repetition rates of 1 000 shots per second (1 kHz) to speed up the analysis. A complete LC MS run spotted on a MALDI target plate can be analyzed in several hours, including MS/MS analyses of thousands of peptides. This means that ToF/ToF instruments also need to have the electronics to deal with the fast calculation/storage of such a huge amount of fast accumulating data. This fast data acquisition rate, together with fast movement of the target plate and other improvements over standard ToF instruments, can allow the acquisition of a complete data set from a tryptic digest of a sample in under a minute, including automated MS/MS analysis of several peptides. These fast measurements are only possible with a high grade of automation. A typical acquisition routine could involve elements like this; the instrument is programmed to perform a measurement in reflectron mode of every sample, either accumulating 400 LASER shots at various positions within the sample while ramping the LASER power in a certain range, or stopping the acquisition when 14 peaks appear with a signal to noise ratio of at least 30 (whichever appears first). MS/MS measurements of the five most intense peaks are carried out, starting at the peak with the lowest intensity of the five. Data are collected from 2 000 shots with a fixed LASER intensity ratio dependent on the success in the first measurement (e.g. LASER intensity of MS measurement times 1.2), or the measurement is stopped when at least 10 peaks reach a signal to noise ratio of 10. An online database search is done with the acquired data while other samples are measured. We return to the initial sample if no significant match was observed to perform MS/MS of the next five most intense peaks.

The unique advantage of MALDI ToF/ToF instruments is that they deliver very good peptide fragmentation data in an offline mode. This means data dependent analysis can be performed within a reasonable time after the initial measurement (target plates with samples can be stored for some days or even weeks dark/dry at room temperature without major sample deterioration). This is very different from online ESI analyses. Moreover, MALDI ToF/ToF MS can be coupled to hyphenated separations and a complete 2D HPLC run can be analyzed within 8–16 hours on the instrument. MALDI ToF/ToF MS is also a very good choice for analyzing quantitative approaches of hyphenated separations, using SILAC or iTRAQ^{TM} approaches (see Chapter 5). The limitations of MALDI ToF analyzers for quantification mentioned earlier do not apply to these approaches, where the relative amounts of nearly identical analytes in the same sample are compared.

3.3.2 Quadrupole mass analyzers

The operation of a quadrupole

The quadrupole is an electromagnetic lens for ions, consisting of four parallel, finger-like rods (see Figure 3.24). These rods are also roughly the size of (thin) fingers and separated by only a couple of millimetres. The ion path is parallel to the rods and right through the middle axis in between them. The two opposite rods are connected, and changing voltages are applied to the rod assembly. The electrical fields applied to the rods consist of two components: a constant voltage (U) and a voltage changing with radiofrequency (RF) in which the voltage follows a harmonic change $V \cos(\omega t)$ with ω being frequency and t time. At any given setting of these voltages, ions, which are introduced along the long axis of the rods, will follow complicated flight path between the rods, similar to a corkscrew-like movement.

The quadrupole does not induce any movement on the ions through the rods, along the long axis parallel to the rods; the momentum for this movement comes from outside acceleration of the ions by static electrical fields. For ions of a specific m/z range (e.g. 500.0–500.2 Th) the corkscrew-like oscillating movement is stable; the ions pass between the rods and eventually hit the detector. All ions with other m/z values will have unstable flight paths

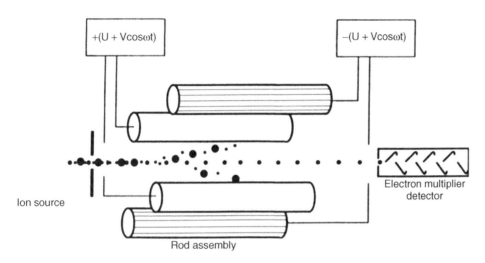

Figure 3.24 Quadrupole mass analyzer. The combination of constant voltage (U) and RF voltage ($V\cos(\omega t)$) makes this quadrupole a mass filter; only a certain m/z range can fly trough the centre of the quadrupole towards the detector at any given setting of the voltages, which are changed rapidly over time. Electron multiplier detectors are sensitive and have a large dynamic range. For the relatively 'slow' quadrupole there is no need for a high temporal resolution (compare ToF detectors, Figure 3.18). As all ions are nearly in the same place there is also no need for any spatial resolution of ions.

and will not hit the detector, but rather the rods. To make a quadrupole into a mass analyzer all that needs to be done is to change the voltage setting in a characteristic way over time. This would allow ions of a changing m/z range to pass through the quadrupole at any given time interval. All ions that pass will be led towards a detector and their presence measured over time. Since we know at what times the setting of the quadrupole allowed ions of what m/z to pass, all that needs to be done is correlate the intensity at the detector with the settings over time and we have a mass spectrometry readout. A whole scan of the entire mass range takes less than a second and is repeated over and over again so we generate a mass spectrum every half second or so and have a chance to generate many spectra over the elution of a single peak from, for example, a nano-HPLC column (peak width about 30 seconds). An important setting in this so called 'scanning' mode of quadrupole operation is the dwell time. This is the time the quadrupole spends at every specific m/z range to measure ions. The longer the dwell time, the higher the sensitivity, but the longer the scan takes! The dwell time and resolution are typically chosen such that each scan takes between 0.6 and 1 seconds; this allows good sensitivity, while guaranteeing that every peptide peak is covered by multiple scans. These multiple scans are usually taken together (summarized) to give the data

for one peptide. In modern ultra-high performance (or pressure) liquid chromatography (uHPLC) systems with columns of smaller particle size and higher back-pressure, the time for one run maybe only 10 minutes and a peak may elute in 2–5 seconds. Then the dwell time has to be lowered to allow more scans to cover one peptide. If the dwell time is too short, the quadrupole spends most of its time switching from one setting to another (e.g. from measuring 400–400.01 to 400.01–400.02 Th) and thus the efficiency and sensitivity fall rapidly. The switching time is very short compared to the dwell time, but there are also many switches to be made, typically thousands for one scan. Usually the manufacturer recommends certain settings, and for beginners it is worth sticking to these for standard applications.

Quadrupoles are very compact, merely several centimetres in length (compare this to ToF flight-tubes). The ions are moved along the axis between the rods by rather small potential differences by fields perpendicular to the quadrupole fields described above. The fields are just a few volts, which means the ions move slowly (compared to ToFs). The combination of compact dimensions and relatively slow ion movements means the demands on the vacuum inside the instrument are rather modest, which is not only good for keeping the instrument price low, but also allows the quadrupole to be coupled optimally to atmospheric pressure ionizations, like ESI. Of course quadrupoles work very well in high vacuums, they just do not depend on them. It is not straightforward to couple quadrupole analyzers to MALDI sources; MALDI ions are metastable and fly at high speed out of the source, have a relatively large energy distribution and are delivered in a burst of only a few nanoseconds. This does not give scanning quadrupoles enough time to do their magic, and most MALDI derived ions are prone to LID and would fragment while they are in the quadrupole and would thus be lost to the measurements. There are hybrid instruments combining MALDI with quadrupoles and ToF analyzers or quadrupoles with ion traps. In fact, most mass analyzers (except in MALDI ToF) contain more than one quadrupole, or their cousins, the hexapoles and octapoles. This is due to the huge versatility of quadrupoles/multipoles; they can be used to guide and focus ions and are instrumental in achieving high transmission rates of ions in mass spectrometers, that is, getting as many ions as possible from the entrance of the mass analyzer to the detector in such a way that they contribute to the specific signal. The more poles there are in a multipole, the lower the possible resolution and focusing strength, but the better the transmission and the mass range over which focusing and transmission can be achieved, so only quadrupoles are used for the business of analyzing masses in proteomics and other mult-poles are only used for ion transmission/focusing.

When we introduced quadrupoles earlier in this chapter we described the scanning mode of quadrupole operation – that is, one very small m/z range (about 0.02 Th) is transmitted at any given time, typically scanning an m/z region of 400–1 500 in less than a second. The highest resolution that a quadrupole can achieve for the isolation of this small mass range that passes though at any given time is defined by parameters such as the size of the quadrupole (length of rods and how close together they are), the accuracy of the surface of the rods and the accuracy/stability of the RF and the voltage amplitudes. Different parameters optimize a quadrupole in different aspects; the smaller the size (space between rods) the better the resolution, but (at some point) at the cost of transmission of ions (sensitivity), particularly for large ions. The longer the rods, the better the possible resolution and the higher the mass range at which the analyzer can scan or transmit ions, but the worse the transmission and low m/z range. Thus it is possible to

build well performing quadrupoles for ions from 100 to 2 000 Th or from 2 000 to 8 000 Th, but not for the complete range from 100 to 8 000 Th.

However, the quadrupole can be operated in various modes, which we cannot discuss in detail here. For a typical operation mode just imagine the setting in voltage amplitude and frequency of RF are set to let a whole range of m/z pass, for example 400–1 500. This is called the transmission mode (Figure 3.25). By manipulating the applied voltages this mode can be used to focus ions and separate them from chemical noise, that is ions that are very small or very large (e.g. salts or clusters smaller than 400 Th or larger than 1 500 Th) that could interfere with the analysis.

The quadrupole does not depend on a very high vacuum for sensitive operation, as we have already seen. Now imagine the quadrupole voltages are set to the transmission mode and a heavy chemically inert gas such as argon (or nitrogen) is let into the quadrupole (at the rather modest pressure of 10^{-3} mbar). This mode is called the collision mode, as the settings are chosen so that all ions collide with the gas molecules (Figure 3.25). The voltage used to accelerate the ions towards and through the quadrupole and the gas pressure is usually fine-tuned (or the voltage can be varied during an experiment) so that about 50–70% of the ions fragment. This usually allows the best structural information to be extracted from the fragments for peptides. This voltage used to induce the collisions is not related at all to the RF or DC voltages used to operate the mass filter function of the quadrupoles and acts perpendicular to these voltages. As the quadrupole is set to transmission mode and the fragment ions are generated halfway inside the devices, all fragments down to several Th are transmitted. Imagine an ion of 801 Th, being the doubly-charged ion of a 1 600 Da peptide. Some of the fragments will lose all charges. We cannot observe these fragments. Some fragments will remain doubly-charged; they will appear at sizes below 800 Th. Some fragments will only retain one charge. These ions can range in size from several Th to nearly 1 600 Th, so the fragments (or daughters) may appear to be 'bigger' than the parent ion! This is explained by the different charge states, which is why it is good to use Th or m/z units (see also Section 'Triple quadrupole mass spectrometers').

Another typical mode of operation for the quadrupole is the scanning mode, described earlier. In the scanning mode ions of a narrow m/z range are transmitted at any

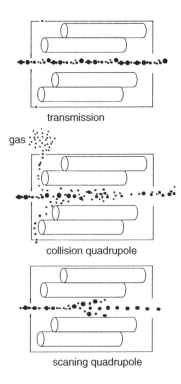

transmission

gas

collision quadrupole

scaning quadrupole

Figure 3.25 Operation modes of quadrupole mass analyzers. The quadrupoles are basically used in three different modes. In the transmission mode all ions (within the limitations of the quadrupole) are transmitted. This mode is used for ion focusing or transmission within mass analyzers. The collision mode is similar, except that gas is allowed inside the quadrupole and the ions are given a certain entrance energy by applying a field towards the quadrupole. In the scanning quadrupole only ions of a certain m/z and within a very narrow m/z range (e.g. 0.01 Th) are allowed through the quadrupole; all other ions have unstable flight paths. By changing the m/z range continuously (e.g. scanning in 0.001 Th steps from 500 to 1 200 Th within 1 second) the quadrupole becomes a scanning mass filter.

one time ('one ion at a time'). By scanning the entire accessible m/z range within less than a second this mode is used to determine the exact mass and intensity of all the ions present in a sample (Figure 3.25). Such a scanning mode measurement of masses is possible although the composition of the sample is changing all the time – the sample composition changes in the time frame of several seconds, while a complete scan over the entire mass range takes less than a second. In fact, for most proteomic samples the signal intensities from a single scan are not good enough for analysis. Typically one has to add up the results of (summarize) several

scans, for example 10. Suppose that each scan cycle takes 0.6 seconds, so 10 scans will take 6 seconds. If peptides are, for example, eluted from an RP HPLC directly into the quadrupole (online ESI) the typical time for the complete peptide peak to elute is about 30 seconds, or 50 scans. The instruments allow a 'sliding window' to summarize scans, so that we are sure to look at the peak when it is at its strongest and thus get the best signal to noise ratio, the best resolution, the highest intensity and, all together, the most accurate measurement.

In transmission and collision mode the sensitivity of the quadrupole is very high, while in scanning mode it drops significantly, as one ion is measured at a time. This means in turn that for the rest of the time (typically thousands of times longer than the individual ion measuring time) the ion is not measured, but thrown away. This explains the direct coupling of resolution to sensitivity in quadrupoles; the higher the one, the lower the other. Also, accuracy is coupled to resolution and thus inversely to sensitivity. Except for some applications (such as single-reaction monitoring (SRM)), measurement with quadrupoles always entails a compromise between these factors.

Triple quadrupole mass spectrometers

Having discussed how a quadrupole in scanning mode can be used as a mass analyzer, let us see how quadrupoles can be used as MS/MS analyzers. This is the field in which they are mainly used today and where they can really shine for proteomic applications. To use the full potential of the quadrupole for MS/MS they are usually connected in a series of three consecutive quadrupoles, the triple quadrupole or triple quad (Figure 3.26).

Quadrupoles are usually fed the ions via ESI. The skimmer voltage and flight path can be set so that many peptide ions will fragment. Such settings are not unusual, as some sources are designed to keep 'disturbing' ions (like salts) away from the mass analyzer. If such fragmentation occurs in samples containing only one ion, it is obvious that the fragments arise from this ion; thus, a single quadrupole can effectively be used as an MS/MS analyzer. A 'one single ion at a time' situation can occur when hyphenated methods of peptide separation are linked to quadrupoles, if the sample is not too complex. Some companies now offer a similar but improved version of this decade old approach even with complex peptide mixtures; even in complex mixtures peptides will rarely co-elute at exactly

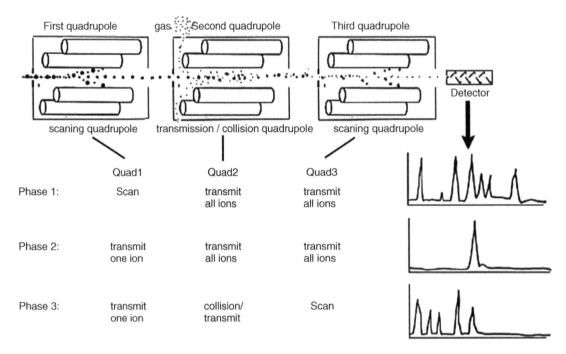

Figure 3.26 Tandem mass spectrometry using a triple quadrupole in product ion scan mode. The product ion scan mode is a popular application for triple quadrupoles. The top panel is for illustration purposes only; it does not show any of the phases of the precurser ion scan. In phase 1 of the product ion scan the first quadrupole is in scanning mode and the second and third quadrupoles are transmitting all ions. The panel on the right shows the total ion current at the detector. In phase 1 this shows an overview of all ions present in the sample. In phase 2 an ion for MS/MS fragmentation is chosen (the parent ion) by setting the first quadrupole to transmit just this one ion, while the second and third quadrupoles are set to transmit all ions. As a result the detector just shows this one ion. In phase 3 the first quadrupole is set to transmit the parent ion only, the second quadrupole is set for transmission of all ions under collision conditions and the third quadrupole is in scanning mode to measure the masses of all the fragments and the residual (unfragmented) parent ion. Collision conditions consist of an acceleration voltage between the first and second quadrupoles as well as the introduction of collision gas into the second qaudrupole. In practice (if MS/MS experiments are planned at all) the collision gas is introduced into the second quadrupole throughout all three phases, but the acceleration voltage is only applied in phase 3.

the same time from (say) a reversed phase column, so all daughter ions can be connected to their respective parent by taking the intensity change over time into account (see Section 4.3). This approach is very sensitive, as there is only one scanning step; switching between modes and 'concentrating' on the parent ions and thus ignoring all other signals is no longer necessary.

However, real MS/MS analyses (as opposed to skimmer fragmentation) are a lot more powerful, as all parameters related to fragmentation can be controlled and optimized, the parent ion can be clearly isolated and daughter ions can be assigned to a parent even if the sample contains a complex mixture of peptides during ESI. For MS/MS applications a triple configuration of

quadrupoles is used. As quadrupoles are relatively small this does not add at all to the size of an instrument. There are several configuration in which a triple quadrupole can be operated for MS/MS measurements. A typical configuration of a triple quad for MS/MS measurements is shown in Figure 3.26 for the so-called product ion scan.

In this set-up one first gets an overview of which ions are present (phase 1) and the same results as a single quad would deliver. Now a decision is made as to which ions will be analyzed by MS/MS. A certain ion is chosen and quad 1 is set to transmit this one ion only, at a reasonable resolution (e.g. the mass of the ion of interest plus or minus 0.5 Th). Often a quick test is done to see if we see the ion as expected (phase 2) and to establish the relative intensity and, importantly, establish the exact charge state (i.e. usually 1+, 2+, 3+ or even 4+). Best MS/MS results are achieved with charge states higher than 1+, and we will discuss this later in more detail. Now gas (N_2 or Ar) is allowed in quad 2, and quad 3 is set to scan all ions in the reasonably measurable range (phase 3). The voltage that accelerates the ions in between quads 1 and 2 is 'ramped', that is, a combination of gas pressure and voltage is found at which the original signal intensity of the parental peptide is diminished by about 50%. At these settings half of the peptides are fragmented; thus the signal for the parent ion gets lower and fragments keep appearing. These voltage and gas pressure settings are different for every peptide and strictly sequence dependent. The quality of the MS/MS data depends on these settings, as we will discuss later. The acceleration voltage between quad 1 and CID quad is anywhere from 20 to 200 V, and depending on the instrument 50–80 V is a standard range for many peptides. The MS/MS signals will take longer to record than the parent peptide. If, for example, the parent peptide's intensity gave a good signal (i.e. good signal to noise ratio of more than about 10, good resolution and intensity) after merging the data from five single scans, the MS/MS data of some 20–50 scans should deliver sufficient signal to noise ratio for most fragments. The MS/MS scan records the signal at a lower level than the MS scan, since the one parent ion can fragment into hundreds of daughters, most of which will loose their charge as well. The lack of chemical noise in MS/MS allows decent signal to noise ratios despite the severely lowered signal intensity, albeit at slightly longer measurement time (or more accumulated scans) than the MS measurement.

Note that the range at which the MS/MS ions are scanned should be slightly higher than the expected size of the singly-charged ion. If ions are detected at masses higher than the singly-charged parent ion, either the charge annotation was wrong (e.g. we assumed the charge was 2+ when it really was 3+) or another ion is co-isolated with the original parent (e.g. the parent is the doubly-charged ion at 601 Th from a 1 200 Da peptide, but there is a weak contamination of a triply-charged ion of a peptide of 1 800 Da). It is important to account for this problem, as either we have the wrong parent mass or some of the alleged fragment ions belong to another peptide altogether. Search algorithms can be robust enough to cope with both these situations in many cases (see Chapter 4).

In the offline mode of ESI one sample could be spraying for 30–50 minutes. This would leave more than enough time for a trained operator to have a quick look at all the ions present and earmark several for MS/MS analysis. A manual analysis would typically take several minutes, so 5–10 ions could be analyzed in 30 minutes, based on the signal strength and spray time (depending on the individual needle), even considerably more. In an online situation many peaks keep coming at the mass spectrometer relentlessly throughout an experiment. Only very rarely will the peaks for MS/MS analysis be chosen manually. If there is some prior knowledge of the sample it is possible to 'pause' an HPLC run at a certain time/peak and extend the elution time of that peak from a typical 30 seconds to perhaps 5 minutes, by stopping the flow from the pumps (peek parking). During peek parking the ESI can continue, due entirely to residual pressure in the system (the capillaries are pumped up, rather like balloons) and the high voltage at the tip of the spray needle/capillary, 'sucking' the positively charged droplets out of the capillary at a flow rate below 50 nl/min. So apart from the rarely used peek parking, the typical situation is that during an LC MS run peptides keep coming into the triple quadrupole. Each peptide has a typical elution time of 30 seconds and each peptide elution overlaps with at least another 4–15 peptides, depending on the complexity of the sample. There is no time to stop and think, this is the time to acquire as much meaningful data as possible. In other words, time for your computer and the software to help you out and measure in the manner you decided before the experiment.

Typical settings for a data dependent, automatic acquisition include the following parameters:

1. Keep acquiring data from the moment each RP run starts, until 5 minutes after the concentration of buffer B (ACN, the mobile hydrophobic phase used to elute the peptides) reaches its plateau.
2. Save files from consecutive runs belonging to a 2D HPLC run in a linked folder.
3. Start with the scanning mode over the chosen m/z range (typically 400–2 000) at medium resolution (e.g. 2 000) and medium dwell time.
4. As soon as an ion exceeds a certain threshold level, switch from scan to MS/MS mode measuring this ion for (say) five MS/MS scans each with a slightly increased acceleration voltage for CID.

5. Switch back to scanning mode.
6. If more than one peptide is over the threshold level, consider up to a predefined number of signals (e.g. the five strongest) and acquire successively (e.g. starting with the the weakest) four MS/MS scans for each parent ion.
7. Return to MS scans and ignore any signal that was chosen for MS/MS from consecutive MS/MS analyses for 30 seconds.
8. Add all MS and MS/MS files together for later analysis.
9. Create a file containing all peptides connected to their MS/MS data, extract all the peaks and create a peak list; this file is now ready to be searched against a database (Chapter 4).

Some of these settings seem trivial (but not always trivial to achieve), others seem more complicated at first, but it all follows a simple logic. In summary, these settings are described as data-dependent acquisition modes, as what and how the mass spectrometer measures is dependent on results of ongoing measurements.

Let us go through an example measurement. A scan range from 400 to 2 000 Th covers the peptide range from about 400 Da (for peptides with a single charge) to 6 000 Da (ion at 2 000 Th, triple charge) or from about 4 to 54 amino acids, which is impressive, and might be narrowed down for detecting the highest possible number of peptides per experiment (e.g. 600–1 500 Th). At the beginning of the individual LC run there are only minor contaminations coming down the column, their signals are not strong enough to start an MS/MS cycle, so they are measured in MS mode 'only'. The first peptide strong enough to trigger an MS/MS measurement elutes at e.g. $t = 12.5$ min or $t = 750$ s. The intensity settings are arbitrary instrument units. It is important to get these right: set too high and the instrument will waste time on ions where it has no chance of getting good MS/MS data in the time available; set too low and the instrument will only get MS/MS data from the most abundant peptides (if any), wasting a lot of time measuring the same peaks over and over again in the MS mode only. Setting the MS/MS threshold requires some experience and is sample dependent: an MS/MS threshold that is perfect for a total of 0.5 µg of a low complexity sample (e.g. isolated ribosomes) will be much too high for 2 µg of a complex sample (e.g. tissue homogenate), because in the complex mixture even peptides from abundant proteins will not be as abundant as the highly enriched components of a 'relatively simple' protein mixture from ribosomes. Complex tissue lysates may contain 10 000 different proteins, ribosomes hardly more than 200. However, if a certain threshold is very good in detecting most of the proteins in the ribosome, it might still be too high if you are interested in a minor component of a mixture (associated proteins in a co-immunoprecipitation) or in PTMs. At the end of the day we face a dilemma brought on by the resolution of 2D HPLC, which is too small to cope with the sheer amount of peptides or the limits in our methods to cope with the dynamic range of peptide (e.g. epidermal growth factor concentration in plasma is over 25 million times lower than that of albumin). Wrong settings can make sure you never get a good signal at all in MS/MS; if the intensity threshold setting is too low you do not gain sensitivity, but make sure you make MS/MS measurements of all peptides before or after they reach their maximum concentration.

Suppose that the setting for the MS/MS threshold was too low, the peptide is detected at a low signal intensity, the measurement has a low resolution and accuracy due to a low signal to noise ratio, and the triple quadrupole switches to MS/MS mode. It will take five scans, as the MS/MS signal is weaker, but also to give the machine a chance to ramp the collision voltage to achieve a region with good CID but not total dissociation of the peptide to small fragments like immonium ions and even smaller, completely uninformative breakdown products. As the optimal settings are sequence specific, a ramping of the collision voltage allows the best compromise. After 2–3 seconds (five scans) the instrument switches back to MS mode, but will exclude the peptide now (and for the next 30 seconds) from any MS/MS list, as it does not want to measure it twice. By the time this peptide can be considered again for MS/MS it has eluted completely and cannot be measured again. Had the MS/MS threshold been set higher, the MS/MS switch would have occurred several seconds later, at the highest peak intensity of the eluting peptide, giving the highest resolution, accuracy and – most importantly – the highest number of MS/MS fragment peaks with the best signal to noise ratio, which is important for automated search results (see also Chapters 4 and 5).

Concerning the switch back to scanning mode (point 5), it is important to determine the charge state of the peptide chosen for MS/MS and make sure that the m/z range for the MS/MS scan is chosen so that it covers the mass range in which we expect the singly-charged ions, with the limitations applying that we discussed above for the manual MS/MS measurement.

With regard to the number of peptides over the threshold level (6), if there are five peptides with sufficiently strong signals co-eluting at the same time than we have a good chance of obtaining decent MS/MS data for all of them. If there are even more sufficiently strong ions, it is important to make sure that they really are from five different peptides, and not just different charge states of the same peptides. A peptide of 2 000 Da can generate ions at 2 001, 1 001, 667.67 and 501 Th (charges from 1+ to 4+). One can expect to see at least two charge states for peptides of medium intensity and even more for peptides of very high intensity. Most software is unable to make sure that different charge states of the same peptide are fragmented, wasting the chance to analyze more peptides of the mixtures. Given the moderate resolution and accuracy of the quadrupole, it is also possible in principle to exclude ions from MS/MS measurements, based on the suspicion that they might just represent different charge states of the same peptides (e.g. is an ion measured at 667.5 Th the triple charge of a peptide of 1 995 Th or just an inaccurate measurement of the triply-charged original 2 000 Da peptide that was measured in the previous MS scan?). However, if there are five peaks strong enough for MS/MS analysis the next MS scan will be after 20 MS/MS scans, or perhaps 10–13 seconds, a considerable time, and if this was to be prolonged even further by analyzing more peptides, new peptides eluting from the columns might get 'lost', we have thus reached the capacity of the triple quadrupole to deal with co-eluting peaks.

Finally, with regards to point (9), to extract all the relevant data from thousands of MS and MS/MS spectra (even tens of thousands if several runs are combined for 2D HPLC) is quite a considerable task! There are also non-trivial tasks: is a peptide of the mass 2 000.2 Th eluting at $t = 1\,200\,s$ identical to a peptide of 1 999.8 Th eluting at $t = 1\,300\,s$? If so, did the peptide loose a labile modification (e.g. phosphorylation, which could change the elution time from RPLC slightly) at the skimmer, during ESI and entry into the instrument (so there was no chance to detect this modification by MS) or did the same peptide just 'smear' in the LC? These questions are answered only after further analysis of the results (e.g.

database search, Chapter 4), and sometimes cannot be answered. To address issues like this, it is very important to produce the right amount of redundancy in the results; if the same peptide is measured several times with slightly changed MS/MS values, it surely is present. If only one peptide of a certain protein is found this might be a chance hit, and most researchers set two or more peptides from one protein as identification criterion, even if the search engines suggest otherwise (Chapter 4).

The issue of data analysis from MS/MS approaches will be dealt with in Chapters 4 and 5.

Triple quadrupoles have been a mainstay of analytical mass spectrometry for several decades. They have been very useful for proteomics from the early beginnings. The appearance of hybrid instruments (e.g. q-ToF) with enhanced resolution, accuracy and sensitivity has led to the downfall of quadrupole usage for proteomics; however, there are several reasons why this is not entirely justified. The quadrupole offers a unique blend of MS/MS operation modes, see (Figure 3.27), not all of which can be reproduced by other analyzers. Some of these modes, like the precursor ion scan, are uniquely suited to analyze PTMs. While instruments like the q-ToF deliver superior resolution and accuracy, they cannot do this in all MS/MS modes, and the triple quad can have the same and even higher sensitivity in the precursor ion scan mode, or for multiple reaction monitoring (MRM ; see also Chapter 5) than a q-ToF instrument, for example. However, the main reason why the triple quadrupole should not be frowned upon its dynamic range, which is the most extensive dynamic range of all mass analyzers available. While most ToFs would struggle with a dynamic range of 10^4 and all but the most recent ion traps would be pushed to reach this level or even the next order of magnitude, the specification for dynamic range on modern quadrupoles is 10^6. While this may exceed the dynamic range in other parts of the analytical chain, it is obvious that proteomics struggles with the dynamic range of proteins and it is wrong to use mass analyzers because of their better performance in speed, resolution and accuracy and forget about the dynamic range. It is because of this high dynamic range and the 'install and forget' robustness of triple quadrupoles that they deserve their status in analytical chemistry.

The reasons why quadrupoles have been replaced by other mass analyzers for many proteomic applications are their modest performance in many fields. The

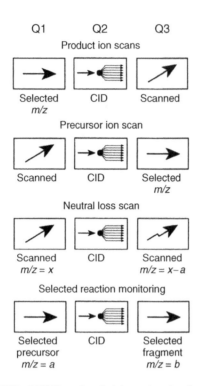

Q1   Q2   Q3

Product ion scans

Selected   CID   Scanned
m/z

Precursor ion scan

Scanned   CID   Selected
m/z

Neutral loss scan

Scanned   CID   Scanned
m/z = x   m/z = x–a

Selected reaction monitoring

Selected   CID   Selected
precursor   fragment
m/z = a   m/z = b

Figure 3.27 MS/MS modes of triple quadrupoles. See text for detailed explanation. Selected reaction monitoring is transformed into multiple reaction monitoring by switching the matched pair of selected precursor/selected fragments in quick succession. Reproduced with permission from de Hoffman, E. & Stroobant, V. (2007) *Mass Spectrometry*, 3rd edn. © 2007 John Wiley & Sons, Ltd.

resolution is usually in the range 2 000–4 000 (depending on the settings), the m/z range is 50–4 000, although special quadrupoles with better performance for large m/z values (up to 8 000) have been built, which in turn show limited performance for small ions. The resolution is normally set so that singly-charged ions can easily be differentiated from doubly-charged ions. Isotopic patterns of triply-charged ions are rarely resolved, and if the pattern looks even 'smoother' it must be a higher charge state, most likely 4+. With a limited resolution, accuracy is also limited to about 200 ppm. On the positive side, the resolution for the selection of parent ions is usually better than 1 000 and all performance values for MS are in principle retained for MS/MS, but low signal to noise ratios can cause problems here.

Triple quadrupoles have the unique capability of running in several MS/MS modes (see Figure 3.27). The first

and third quadrupole (Q1 and Q3) can be operated in scanning or fixed mode, while the quadrupole in the middle always performs the CID, thus it is often denoted by 'q'.

The MS/MS mode discussed hitherto for the quadrupole is basically available in many other mass analyzers, the so-called product ion scan. Q1 selects a certain parent ion, which is then fragmented by CID (in a well controlled manner) in Q2, and Q3 is set to scan the products (hence the name). Overall there is one scan stage with a reduced sensitivity as compared to a ToF analyzer; Q1 transmits 100% of the parent ions.

In the precursor ion scan MS/MS mode Q3 is fixed on one ion size only, the so-called product ion. Q1 is set to scan, while Q2 performs (as ever) the CID. This MS/MS mode is unique to the triple quadrupole. Uniquely to this mode, the quadrupole measures extremely efficiently (100% of the time) the production of one fragment ion, while it answers the question from which parent ion this fragment originates (scan Q1). This mode is very useful for detecting chemical groups that produce a certain fragment ion. This could be an immonium ion (e.g. to detect leucine or isoleucine, ideal to detect peptides of low level signal strength in chemical noise), a fragment specific to the phosphorylated group (to detect peptide phosphorylation or DNA/RNA attached to peptides), a sugar fragment (to analyze glycosylation) or any other prognostic fragment. In other words, in this mode we ask which peptide shows any of the above named modifications. Once a strong signal for the chosen product ion is detected, its precursor can be analyzed by a product ion scan to identify the peptide in question. This is very useful indeed, and the switch between the different modes can even be automated for online LC ESI.

In the neutral loss scan both Q1 and Q2 are scanning, with a fixed offset in mass range. If, for example, Q2 scans all the time at a mass 98 m/z smaller than Q1 scans, we would observe singly charged ions in our sample that possibly lost (and thus must have contained) a phosphate group = H_3PO_4 (mass = 97.977 m/z). Neutral loss scans are more common and useful in analytical chemistry than in proteomics; there is a higher chance of wrong interpretations than with a carefully chosen product ion in the precursor scan. For example, the apparent loss of 98 could not only result from the loss of a phosphate group, but also result from the loss of other groups, such as proline (97.053) or valine (99.068), or even from the loss of a larger group at a higher charge state. The neutral loss scan mode is also exclusive to triple quadrupoles. Neutral loss scans are quite sensitive; although both quadrupoles are scanning, the sensitivity is comparable to a single scanning quadrupole mode, because of the constant offset in the scanned mass.

Selected reaction monitoring (SRM) is an extremely sensitive MS/MS mode and is rather sought after in quantitative proteomics. In essence, this mode is perfect for quantifying a certain known component. The mass of a peptide alone is not specific enough to identify it in most circumstances with high confidence (in particular, with the limited resolution/accuracy of a quadrupole). However, fragmentation is sequence specific, and for every peptide there are some fragments that are very strong. If Q1 is set to the mass of the peptide in question and Q2 to the mass of a fragment known to be prominent, we can confidently ask how much of our peptide is present in the sample, as the quadrupole measures our analytical peptide with great sensitivity (100% of the time, no scan involved) and very specifically (as the chances of confusion with an unrelated peptide having the same mass and fragment mass are low). This method can take full advantage of the very good dynamic range of quadrupole linearity (more than 5 orders of magnitude).

MRM is a variation of SRM. Suppose that we wanted to measure the amount of 10 proteins over the course of an experiment. We would have to inject the different samples and set up the triple quadrupole to measure in 10 different SRM settings (one for each peptide) and change, for example, five times per second. Within 2 seconds we would have measured each single peptide with a sensitivity orders of magnitudes higher than in a typical MS scan, where the quadrupole only gets below 0.1% of its time for the mass range of 1 Th. In the example of MRM with 10 peptides the quadrupole would spend about 10% of its time on MS/MS for each of our peptides; 100 times more time and thus sensitivity. MRM is a powerful tool for quantitative proteomics of selected components, as with known elution times of peptides from LC runs the MRM settings can switch during a run and thus many peptides can be monitored. The MRM mode is not exclusive to the triple quad; MRM is also possible on ion traps, though not with anything like the sensitivity advantage of triple quadrupoles, and similar measurements can be emulated with hybrid q-ToF instruments (with some loss in sensitivity, speed and dynamic range, but improved resolution/accuracy). The Q-Trap hybrid instrument (tandem

quadrupole with linear ion trap (LIT)) can perform MRM at even higher sensitivities than the triple quadrupole instruments (Section 'Other hybrid mass spectrometers').

3.3.3 Ion traps

The mass analyzers hitherto discussed deal with ions flying more (ToF) or less (quadrupoles) rapidly through the instrument during analysis. Ion traps work on a very different principle: they capture or trap the ions for relatively short times (quadrupole ion traps, also called 3D ion traps or Pauli ion traps), intermediate times (linear traps, also called 2D quadrupole traps) or even extremely long times (FTICR or orbitrap) – up to several hours for some special applications. Of course, high throughput does not go well together with long trapping times, so we will focus for most proteomic applications on short to intermediate trapping times, which can be achieved by all the above mentioned analyzers.

Ion traps are based on a variety of principles, detailed discussion of which is beyond the scope of this book. All ion traps depend on catching the ions, and very fast moving ions are difficult to catch with high efficiency. For this reason MALDI is not usually the natural first choice of ionization mechanism to be combined with trapping instruments. Bearing this in mind, there are still successful combinations of MALDI with ion traps, either as dedicated instruments (Section 'Other hybrid mass spectrometers'), as 'add on' sources to an analyzer mainly using ESI (often seen with FTICR instruments) or in the form of AP MALDI (Section 3.2.1). The great advantage of ion traps is their potential of ultra-high sensitivity in instruments with a small footprint (except FTICR analyzers with huge magnets), comparable in size to quadrupole analyzers. The internal operations of traps are inherently much more complicated than those of, for example, ToF analyzers. However, in modern instruments used for proteomics, all this is kept well away from the user and the automation has reached a very high level to make ion traps very easy to use, versatile, reliable, low maintenance instruments (again not the FTICR analyzers with huge magnets). In the last decade Pauli ion traps have become arguably the most affordable, versatile and reliable analyzers for standard proteomic applications. Often higher costs, higher maintenance and more complex operation make the purchase of mass analyzers other than Pauli ion traps hard to justify. This is despite the fact that other mass analyzers may offer higher resolution and higher accuracy than Pauli traps and may also have the ability to carry out some specialist operations which may be difficult for Pauli traps. Recent years have also seen a change in the notion that ion traps are not well suited for quantification experiments; modern methods of internal operation have completely removed this shortcoming of most traps.

3D quadrupole or Pauli ion traps

The easiest way to introduce this versatile mass analyzer is to imagine that it is derived from a single quadrupole (as indeed the name also suggests). Imagine your thumb, index, middle and ring fingers are the rods of a quadrupole (Figure 3.28).

The thumb and middle finger are electrically connected, as are the index and ring fingers. Let us call the third finger *a*, the first *b* and the second *c*. Ions would fly in a corkscrew trajectory in between the rods, coming from in front of your hand. If you now bend the fingers towards the thumb, and imagine that the thumb shrinks, the corkscrew trajectory becomes bent. If you could bend your fingers till they are perfectly circular, the thumb would bend in on itself and disappear altogether (I can never do this with my thumb...). If your fingers are bent perfectly round, the corkscrew trajectory would fall on itself, like a snake biting its tail. Overall the movement would be very similar in principle to a Lissajous figure (Figure 3.29). The real flight path of ions in a Pauli

Figure 3.29 Ion trajectories in Pauli ion traps. A 3D Lissajous figure generated by charged iron and aluminium particles, simulating a stable ion path in a Pauli ion trap, omitting ion/gas interactions. Many different stable trajectories are possible. These will be different for ions of different m/z at any given setting of the voltages in the ion trap. Reprinted with permission from Wuerker *et al.*, *J App Phys*, 30 (3), 342–9. © 1959, American Institute of Physics.

ion trap depends on many parameters and involves some higher-order movements on top of the basic Lissajous pattern. Ions are usually trapped for around 50 cycles in a Pauli ion trap. It is this trajectory that gives Pauli ion traps the name '3D quadrupole traps', as opposed to 2D quadrupole traps (also called linear quadrupole traps, see Section 'Linear ion traps'), although of course strictly speaking ions in both traps have complex 3D movements.

We now have a basic understanding of how the ions in a Pauli trap fly in complex 3D trajectories in a 'cage', but how do they get into this enclosed space? Sticking with the hand analogy (Figure 3.28), roll your hand to the right, and now imagine that the ions come from the left towards the hole in the bend of your middle finger: the Pauli ion trap has a hole in electrode *a* to let the ions in. The entry cap electrode *a* on the left is followed by the middle ring electrode *c* and then by the exit or endcap electrode *b*, which also has a hole to let the ions out of the trap.

Obviously, the shape of the electrodes in a Pauli trap is very different from that of the rods used in quadrupoles. The electrode shape towards the centre of the trap is hyperbolic, and indeed there are also hyperbolically

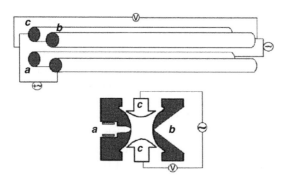

Figure 3.28 Pauli/quadrupole ion traps principles. See text for how best to relate the two mass analyzers to each other. Ions come into the analyzer from the left and leave on the right towards a detector, through a hole in electrode 'b'. Source: http://www.abrf.org/ABRFNews/ 1996/September1996/sep96iontrap.html.

shaped rods in quadrupoles (or LITs). The size and shape of the inner chamber formed by the electrodes in a Pauli trap is extremely important for efficient trapping, as is the way in which the phase of the RF at the endcaps and middle electrode and intensity of the applied voltages are changed. Pauli traps do not use a DC component, which means that they can trap negative and positive ions at the same time.

The ions have to be channelled and focused for the trapping to work efficiently, which is why there are usually octupole lenses in between the ESI/MALDI interface and the ion trap entry cap. The ion gates, which are absolutely essential for the trap function, are located between the lenses and the entry cap. Initial trapping of ions is the single least efficient step in Pauli trap MS, as we will see further below, and without precise focusing and speed control of the ions it would be even less efficient.

Commercial Pauli traps are available with ESI and MALDI sources. Operating the ion trap is a non-continuous process, and as such offers advantages when used with MALDI, also a non-continuous process. However, by far the majority of applications involve the combination of ESI with Pauli traps. MALDI ion traps became available around the time MALDI ToF/ToF began to be feasible, and were outcompeted by the latter approach with its superior speed, resolution and accuracy. However, MALDI ion traps are a valuable tool and have proven especially useful for the analysis of PTMs, such as phosphorylation and especially glycosylation.

As already stated, Pauli ion traps are arguably the most affordable, versatile and user friendly mass analyzers on the market. The aim of applying the complex voltages mentioned above is quite different from their use in quadrupoles; in the scanning mode of the quadrupole only a narrow range of m/z values will 'fly through' the quadrupole on stable paths. In the Pauli trap the initial aim is to trap ions with as wide a range of m/z values as possible; in a typical situation ions ranging from 500 to 2000 Th are trapped. All these ions have reasonably stable trajectories at certain voltage settings in a given trap design. The actual measurement of masses is achieved by ejecting the ions in a controlled manner over a relatively short period of time. Typically the flight path of the smallest ions is made unstable first and the entire m/z range of ions is ejected at a scan speed of between 6000 and 50000 mass units per second. These scan speeds allow all ions within the trap to be ejected from

anything approaching 0.1 s to 1 s. It is worth noting that while specific instruments might differ in the way the ions are ejected (using different settings of the voltages applied to the electrodes), the higher the scan speed, the lower the possible resolution of the measurement on any given instrument. Ejection scan speeds over about 6000 mass units per second will have negative effects on the maximal possible resolution in the real world. The ions are ejected from the exit 'hole' in the endcap electrode into a relatively simple detector (dynode converter), comparable to the ones found in quadrupoles. The whole process is discontinuous; the trap cycles between filling, trapping and ejection modes. Thus the faster the rate of ejection, the more cycles per unit time can be run and the more sensitive the instrument can be. The filling phase is also kept as short as possible. In practical terms it is very easy to 'overfill' the ion trap. Therefore many traps have a special feature: the trap is opened for a very short time to estimate how many ions are being delivered. The total ion current is measured (not the mass of any ions within the trap as in a 'real' scan) and, assuming a constant rate of ion delivery, the time in which the trap is filled for the 'real' measurement is determined. The more ions are coming per second the shorter the trapping time will be. Overfilling the trap leads to charge repulsion between the ions within the trap. In principle, negative and positive ions are trapped at the same time, but if the trap is 'overloaded', ions of the same charge start to influence one another (ion repulsion). This will influence the resolution at which these ions are measured as their flight path is influenced by the repulsion; this in turn influences the RF setting at which the ions have stable flight paths. In other words, the position/flight path of some of the ions of a certain m/z gets unstable 'sooner' during an ejection scan than that of the bulk of the ions with the same m/z; peaks get broader, resolution, trapping efficiency and accuracy suffer. This also means that quantifications are not very reliable. So there are many good reasons to avoid overfilling a trap and, thus, there are many good reasons to perform a short 'pre-scan' just to determine how many ions are likely to end up in the trap and to accept the time/sensitivity 'penalty' that comes with it. Also trapping times and efficiencies have been improved by changing the phase and amplitude of the RF voltage in recent years. Modern Pauli traps now have the capacity to store six to eight times more ions than a decade ago. Some Pauli ion traps can now store more than 20 000 ions at a time, approaching the number of ions that can be trapped in 2D or linear quadrupole traps. This higher storage capacity makes ion traps less dependent on accurate and correct pre-scans and, more importantly, allows for a larger dynamic range and better quantification using Pauli ion traps.

We will not go into quantitative details of Pauli ion trap operation but rather will try to give a qualitative understanding of its operation. For ion trapping, an RF voltage of constant frequency but varying voltage is applied to electrode c, the ring electrode (Figure 3.28). This voltage changes its amplitude and polarity with a radio frequency in the region of 1 MHz (e.g. going in a harmonic fashion from $-4\,000$ to $+4\,000$ V and back to $-4\,000$ V a million times per second). This RF between the endcaps and the ring electrodes is called the fundamental RF. It has a constantly changing voltage, but a fixed maximal voltage, which does not change during trapping; the amplitude of this maximal voltage determines the limit of the size of ions that can be trapped in a given trap. Only ions with a certain range of m/z values have stable trajectories in the trap. Increasing the maximal amplitude of the fundamental RF will make the trajectories of the ions on the small end of the stability range unstable. Decreasing the frequency of the fundamental RF will allow larger ions to have stable trajectories. By changing either of these two parameters it is possible to choose which ions are trapped in a given experiment (higher amplitude and lower frequency for larger ions and lower amplitude and higher frequency for smaller ions). If ions are ejected out of the trap by changes in the fundamental RF (usually the amplitude) they are called ejection ions at the stability voltage. It is one fundamental principle of the Pauli trap that only ions within a certain range of m/z values can be trapped at the same time. The quotient between the m/z of the largest and smallest ion with stable trajectories inside the trap is always about 0.3. There is no sharply defined cut-off point at the lower end of the m/z range, and, if there are enough ions, one can still detect 220 Th ions together with 1 000 Th ions (predicted cut-off with 1 000 Th would have been 300 Th), but the signals at the low end will be of lower intensity compared to a trap set up to have the higher cut of (say) 700 Th. The cut-off at the upper end of the stability range is a lot sharper than at the lower end.

Inside the trap the ions have complicated trajectories in the three dimensions (Figure 3.29). However, the movements of ions in the x and y direction are the same, as these dimensions point towards the symmetric ring electrode (the electrode c in Figure 3.28). Thus the ion trajectories in a Pauli trap can be dissected into movements along the z axis and the r axis of the trap, where r is any axis put through the plane formed by the x and y axis and the z axis is perpendicular to the r axis. Movements on the z axis shift ions from the centre of the ion trap towards one of the endcap electrodes (or vice versa), movements on the r axis shift ions from the centre of the trap towards the ring electrode (or vice versa). Returning to our model deriving the Pauli trap from a quadrupole, we can also distinguish different components of the movement; the movements along the diameter of the corkscrew and a circular movement resulting from bending the quadrupole rods into circular electrodes. The movement along the imaginary corkscrew results in the movement between the two endcap electrodes along the z axis. To eject ions the amplitude of this movement has to be increased, so that a fraction of the ions have a chance of getting to the detector, once they get too close to the holes in the endcap electrodes (part of the ions can not be measured as the 'eject' through the 'entrance' hole in electrode a, Figure 3.28). The frequency at which the ions oscillate on the z axis is very important for the operation of the ion trap and is called the secular sequence. This sequence is dependent on the size of the ion trap, on the radio frequency and, very interestingly, on the m/z of the ion in question. The secular sequence is much lower than the RF, usually in the region of hundreds of kilohertz. In figure 3.29 the z axis goes from left to right, and the r axis is in the vertical direction; the frequency of the oscillation in the r axis is about 30 times higher than that in the z axis.

The secular sequence is, in other words, the number of times per second that an ion of a given m/z value flies from its closest point to the exit endcap, to the closest point of its trajectory towards the entry endcap and back to the closest point to the exit endcap. Imagine now that you give this ion a gentle push towards the endcaps, but in no other direction. If this push is steady over several cycles, it will eventually push the ion too close to the entry or exit hole, and the ion will be ejected. All other ions stay in the trap, as they swing between the endcaps in another rhythm, with a different frequency, so the gentle movements will not accumulate into net movements in any direction; or in other words, they do not resonate as they have a different secular frequency.

To put it positively, applying an alternating current (AC) between the endcap electrodes at the secular frequency of a certain ion will lead to resonance with this ion and, if the AC is applied for long enough with a high enough maximal voltage amplitude, will lead to ejection of all ions with this particular m/z value only. This process is called resonance ejection, as opposed to ejection at the stability limit using changes in the frequency of maximal voltage amplitude of the fundamental RF. With these tools in hand an ion trap can be operated in different ways, by using one or the other ejection mode, or combination of both, or even exploiting overlapping phases in fundamental RF and AC. The user of an ion trap would never notice how the ion trap was run internally (which is a good thing for anybody who is not a specialist), as all the user has to do is to set the m/z range of ions in the trap, the type of experiment and the resolution she/he desires to achieve.

The limits of using the Pauli trap are not necessarily set by limits in trapping ions; ions also need to be ejected in a controlled way to allow the determination of their mass, as we will see later. One of the limiting factors in the operation of ion traps is the limiting ability to eject ions outside a certain m/z range in a controlled fashion. The higher the maximal amplitude of the fundamental RF voltage, the higher the m/z of ions that can be ejected at the stability limit. However, there is a limit of about 8 000 V in the amplitude of the highest voltage across the trap, given the size of about 8 mm across the trap at the narrowest point, as higher voltages would cause arcing in the trap (remember that the vacuum in the trap is far from perfect, and there is helium inside as we will see later). Ion ejection at the stability limit thus works for most commercial traps only up to 3 500 Th. However, if resonance ejection is added to working with the stability limit, the range of commercial ion traps can be extended up to about 6 000 Th.

Pauli traps work best for relatively small physical dimensions, measuring several millimetres to about a centimetre across. We cannot go into too much detail here, but with all these complex parameters it is still not possible to run ion traps at 100% efficiency (i.e. trap and measure every single ion flying towards the trap) and one additional complication is needed for normal operations: the Pauli trap is filled with helium at a low pressure (around 10^{-3} mbar) to provide relatively gentle collisions between the 'light' helium molecules and the 'heavy' peptides. These permanent collisions have the effect of 'cooling' the ions in the trap and allow an efficient trapping. If the ions in the trap were undisturbed by collisions with helium they would have more of a tendency to fly straight through the trap in the first place. In addition, once inside the trap, all ions of the same charge repel each other and the ions deeper inside the trap are shielded from the electrical fields by the ions further on the outside. This is called the space charge effect. The ions thus have the tendency to 'bounce' off each other and to push each other too close to the electrodes, outside the efficient trapping boundaries, or generally not to move according to their m/z. These effects lead to the loss of ions that could have been trapped based purely on their m/z. This detrimental behaviour occurs in particular at higher ion densities and becomes even more relevant during longer trapping times. Helium gas is very useful in minimizing the effects of ion repulsion, as any fast movements of the ions are slowed down by multiple 'gentle' collisions. This cooling allows a higher amount of ions in the trap, which in turn allows a larger dynamic range and better quantification of the ions in the trap. There has been a movement to increase the physical size of ion traps, to allow room for more ions. This enables a higher dynamic range but it is not easy to increase the size of Pauli traps without losing sensitivity and resolution, as well as the ability to trap ions with higher m/z and compromising speed and ease of operation (hence the move of some manufacturers towards 2D or linear traps; see Section 'Linear ion traps').

Although we have derived the basic workings of Pauli traps from a quadrupole, the way a Pauli trap is run for proteomic experiments is very different from a quadrupole; while quadrupoles can be run truly continuously, the Pauli ion trap is run in different phases (Figure 3.30). The bare minimum of phases is:

1. filling of the trap,
2. storage and cooling of ions,
3. ejection of ions.

During the filling phase ions are let into the trap. The RF and the voltages are preset to store ions with a certain m/z range. The principal range of Pauli traps goes from about 50 to about 6 000 Th, but it is not possible to trap the entire range; rather the lower limits of m/z is about 0.3 times the m/z of the largest ions that can be stored. Given the multiple charges introduced by ESI, this would

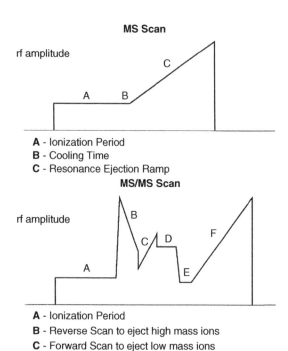

MS Scan

rf amplitude

A - Ionization Period
B - Cooling Time
C - Resonance Ejection Ramp

MS/MS Scan

rf amplitude

A - Ionization Period
B - Reverse Scan to eject high mass ions
C - Forward Scan to eject low mass ions
D - Resonance Excitation (Tickle) Period
E - Cooling Time
F - Resonance Ejection Ramp

Figure 3.30 Operation of Pauli traps. During the ionization period the ion trap is filled with ions from the outside. In the cooling time ions concentrate towards the centre of the trap due to collisions with He. This cooling period ensures that all ions are close together, important for the precise ejection allowing high resolution measurements. The ejection period completes an MS measurement. During the ejection ramp the masses are actually measured, based on the signal intensities at the detector and the ions expected to be ejected at this time point. If the cooling down period is too short, inaccuracies will occur. In a MS/MS cycle the ion for fragmentation is isolated by first ejecting all higher masses followed by ejecting all lower masses. During the excitation period the ion undergoes repeated collisions with He molecules, transferring enough energy for fragmentation. All fragments are kept in the trap and cooled down, to allow for an accurate ejection ramping. Reprinted from Jonscher, KR & Yates, JR, Anal Biochem, 1997; 244: 1–15, courtesy of Elsevier. © 1997 Elsevier.

still allow anything from small peptides to larger proteins to be analyzed by the Pauli trap, albeit not very efficiently in the same experiment. It is worth noting that, depending on the operations of the trap, negative as well as positive ions can be trapped at the same time, which is important for some fragmentations methods such as ETD (see

Chapter 5). The time it takes to fill the trap depends on how many ions are coming towards the trap per second. Filling times are usually in the range of 10–50 ms. If a peak containing 1 fmol of a peptide was eluting in 30 seconds, this peptide would produce about 2×10^5 ions in 10 ms, more than enough to fill the trap. Of course, there is no such thing as a 'pure' peptide and most of the ions entering the trap derive from anything but peptides, even after an LC run. Note also that the trapping efficiency is in the region of 5%. In any case, the filling time is the only time at which ions are actually taken into the trap, and one could look at this time as the only 'duty' time of the mass analyzer. If the filling time were 10 ms of a 1 second cycle, only 1% of the ions would have a chance of being measured. This may sound low, but in a scanning quadrupole the mass analyzer can only spend about 0.01% of its time on a given peptide (see Section 3.2.2), while in a MALDI ToF analyzer all ions have the chance of being measured, but MALDI is rather inefficient compared to ESI. At all other times ions are deflected form entering the trap by ion gates. How does the ion trap know how many ions are about to enter? In older instruments the filling time is set by the operator, with safety margins to avoid overfilling. This limits the small dynamic range of the Pauli trap even further and may lead to 'ghost peaks', low resolution and unreliable quantifications as the rate of ions per second varies during an experiment and at any setting there will be cycles when the trap is nearly empty or overcrowded. All modern instruments do a very short pre-scan with fast ejection that basically counts the total ion current and thus estimates how long the optimal filling time is, ensuring the same efficiency of ion detection at all times of the experiment, allowing best sensitivity, highest dynamic range, optimal resolution without space charge effects and robust quantification. The pre-scan only sacrifices several milliseconds in total, a low percentage of the total measurement cycle.

Once the ions are in the trap they are stored for a certain time. This storage period is also called cooling period. In this time the initially captured 'hot' ions collide with helium and are transferred into stable trajectories in the very heart of the trap. It is important for a controlled ejection of the ions to 'locate' them in stable trajectories first, a process which can also be described as cooling down. The longer this cooling down phase lasts the more ions will be lost due to unpredictable movements after some not so gentle collisions. The frequency at which the ions oscillate around the inside of the trap is the secular sequence and is dependent on the fundamental RF, but is a lot smaller than the fundamental RF, and the secular sequence is ultimately also dependent on the physical size of the trap and the m/z of the ions. Thirty to sixty trajectories per ion for the cooling down/trapping phase are typical. At a secular frequency of 800 Hz this would equate to a time of about 40–80 ms, which at a total cycle time of 1 second equates to only 4–8% of the time. In other words, it is a worthy investment in time to cool the ions down. During the cooling down phase ions are collected deep inside the trap, giving them also a sharp spatial definition, important for a good resolution in the ejection phase. So here the small size of the Pauli trap clearly works to its advantage, allowing high resolution scans in short times (see below).

The most variable phase in term of length is the ejection phase. This also happens to be the phase when the m/z of every ion in the trap is measured. Using a variety of possible ejection methods, the complete m/z range of ions is ejected over a short period of time while the current generated by impacting ions at the detector is measured. 'Knowing' which m/z is ejected at which time and comparing this to the ion current of the detector, the intensity/amount of ions at each m/z rate is measured and plotted. For ejection usually the amplitude of the fundamental RF voltage is changed or an AC with sufficient amplitude is applied and its frequency scanned, so that the trajectory of the ions becomes unstable, they get near the ejection endcap and are expelled out from there onto a detector. If the voltage are changed abruptly more or less all ions are expelled at the same time. This is useful after the pre-scan to determine how many ions are in the trap. However, the simultaneous ejection of all ions does not allow for an m/z measurement; this is achieved by gradually scanning the fundamental RF voltage or AC frequency, so that the trajectories of ions with different m/z values become unstable in a consecutive manner – 'one ion at a time', as it were. The scan mode and speed are important performance parameters for any particular ion trap. Most Pauli traps will be able to reach comparably high ejection scan speed, measured in atomic mass units (amu) per second. But depending on their design and the particular mode of ejection (RF amplitude voltage change or resonance ejection with AC voltages at secular frequencies; see Figure 3.30) used, they will have different resolution and mass accuracy at

a given scan speed. However, it is true for all Pauli traps that the higher the ejection scan speed the lower the best possible resolution. Suppose that an instrument was set to trap ions from 500 to 2 000 m/z. At an ejection scan rate of 1 500 amu/s the ejection phase would take 1 second. The instrument would most likely achieve its highest resolution of anything between 6 000 and 20 000, depending on the instrument. At a scan rate of 15 000 amu/s the ejection phase would only last 0.1 seconds. Depending on the instrument the resolution could be anything from 2 000 to about 8 000. Within the width of a peak from a reversed phase column (30 seconds) with the first setting some 15–25 scans could be taken, resulting in perhaps 1.5 seconds of total filling time over the elution of the peak. In fast scan mode some 250 scans could have been taken. While such a huge number of data points is surely not necessary for MS analyses (but might be needed for MS/MS measurements; see below), it would result in a total of over 12 seconds of filling time, if, as usual in mass spectrometry, the results of consecutive scans are summarized. So if sensitivity was needed, the fast scan had the chance to be eight times more sensitive! If many different MS/MS cycles were needed (see below) more consecutive MS/MS cycles of either the same parent ion (MS^n) or of various co-eluting parent ions could be performed. Of course, resolution (and with it also accuracy) will suffer with fast scan rates. Thus the resolution/ejection scan rate has to be adjusted to give the best compromise for the strategy behind the experiment in question. If ions are ejected too rapidly, they can undergo fragmentation during the ejection process. This can produce 'ghost peaks' at m/z values different form the real masses. Ghost peaks can appear under various operating conditions depending on gas pressure in the trap, the design of the trap, the RF and voltage amplitudes applied as well as the filling state of the trap. Modern designs and graphical user interfaces on the computers used to control ion traps make the appearance of ghost peaks impossible under correct operating conditions (e.g. stable, steady spray).

From the way the Pauli traps are operated to eject ions during a scan (making their trajectories unstable on the r axis by bringing their trajectories close to the endcap holes) it follows that only 50% of ions have a chance ending up at the detector; the other half loses its stability towards the entrance endcap and not the exit endcap.

This whole MS measurement cycle (see Figure 3.30) can take anything from 0.3 to 1.5 seconds, as two stages can vary during an experiment: the time it takes to fill the trap and the time it takes to eject the ions. As filling times are rather short, scan times have more influence on cycle length. Depending on the resolution setting, ejection scans may take anything from 200 ms to more than a second.

However, no one would prefer to use a Pauli ion trap based on its MS capability alone. It is their MS/MS and MS^n capabilities that make Pauli ion traps a mainstay in proteomics. Ion traps are operated in measurement cycles that are repeated perpetually. Each cycle consists of various steps, modified according to the type of measurement intended in the actual cycle. Some typical measurement cycles with their characteristic steps are described in the following (see also Figure 3.30). The first measurement cycle (pure MS) intended, to get an overview of all the ions present and to choose ions for further MS/MS analysis, involves:

1. filling of the trap;
2. storage and cooling of ions;
3. ejection of ions.

The second measurement cycle (MS/MS for one specific ion) consists of:

1. filling of the trap;
2. storage and cooling of ions;
3. ejection of all ions *except* for the chosen parent ion for MS/MS;
4. transferring of energy to the parent ion for fragmentation by CID;
5. storage and cooling of all fragment ions and the remaining unfragmented parent ion in the trap;
6. ejection of all ions for m/z determination.

From here the instrument might switch back to an MS (overview) measurement cycle, to an MS/MS measurement cycle with a parent ion of another mass/charge ratio, or alternatively the instrument can switch to an MS^3 experiment with the following steps:

1. filling of the trap;
2. storage and cooling of ions;
3. ejection of all ions *except* for the chosen parent ion for MS/MS;
4. transferring of energy to the parent ion for fragmentation by CID;
5. storage and cooling of all fragment ions and the parent ion in the trap;

6. ejection of all ions except for the chosen parent ion for MS^3;
7. energy transfer to the ion chosen for fragmentation by CID in MS^3;
8. storage and cooling of all fragments and the parent ion for MS^3;
9. ejection of all ions for m/z determination.

You can imagine the sequence of events for an MS^4 or an MS^9 experiment. Indeed, it is the very high efficiency with which ions (once trapped) are kept in the Pauli ion in trap and transmitted from one MS/MS experiment to next that makes the unique MS^n feature possible. If this efficiency were not close to 100%, the intensity of the fragments would soon be so low as to make any further analysis impossible, since every parent fragments into various daughters and thus every single fragment is *per se* a lot weaker than the parent. So MS^n is only possible given the extreme sensitivity of the Pauli ion trap and the high transmission efficiency.

For the isolation of the parent ion in MS/MS (point 3) the other ions in the trap can be ejected in any number of ways. For example, by scanning the frequency of the AC to match the secular frequencies of all other ions in the trap, only omitting the secular frequency of the parent ion. In another approach all ions larger than the parent ion can be ejected by decreasing the amplitude of the fundamental RF voltage in a first ejection phase. At the end of this phase, the parent ion has the largest m/z of all ions in the trap. Then an AC matching the secular frequency of a particular m/z range (smaller than all ions in the trap) can be added. By now increasing the amplitude of the fundamental RF voltage the secular frequencies of all ions change; the amplitude can be set so that all ions smaller than the parent ion are ejected in this second phase. At the end only the ion we wanted to be the parent ion for the next MS/MS is left in the trap. The resolution of the parent ion selection is comparable to that of the measurements, and thus high enough for precise parent ion selection.

Now let us see how energy is transmitted to the parent ion for fragmentation (point 4). We have already seen that fragmentation might occur (unwanted) during ejection if it is tried too fast, and indeed similar mechanisms for ejection and induction of fragmentation are used. If the amplitude of the AC voltage used for resonance ejection is set too low for an actual ejection, the ions will still be accelerated towards the endcaps. Such a voltage is called tickle voltage or excitation voltage. As the secular frequency stays the same during excitation, but the ions now oscillate from one wall of the trap to the other and not just in the centre, they fly a longer distance in the same time; that is, they fly a lot faster and all those 'soft' collisions with the helium get harder and harder until the ion fragments. The excitation period is in the range of 30–50 ms, giving the ions several tens of oscillations between the endcaps, depending on their secular frequency, which of course depends on their m/z value. As soon as the ions are fragmented, they are no longer in resonance with the tickle voltage, as the fragments are smaller than the parent and thus have a different m/z and thus a different secular frequency. This is very different from the situation in a quadrupole, where the initial fragments will still hit gas molecules and can fragment further (secondary fragmentations). Thus optimally in a quadrupole one chooses the acceleration voltage as to be able to see the unfragmented peptide at about 50% of the intensity that it had before fragmentation (see Section 3.3.2). For an optimal fragmentation in the Pauli trap the intensity in the parent ions should be reduced by the fragmentation to about 10–15% of its initial value. This is much lower than the value aimed for in quadrupoles, but in Pauli ion traps fragmentations are singular events, and in quadrupoles most measured fragments result from multiple consecutive fragmentation events. The fragmentation takes a relatively long time (30–50 ms; see above), as opposed to CID in other instruments, and the fragments in ion traps are cooled by multiple collisions with helium as they emerge. In other words, there is ample time for the peptides to dissipate the energy and brake at the weakest point, often the backbone, which is very good for the interpretation of spectra and identification of peptides from ion trap data. Another added advantage of the ion trap is that if a certain breakpoint is not observed, one could always perform an MS^3 experiment on one of the fragments encompassing this particular fragment. On the negative side, the energy that can be transferred to a parent ion for fragmentation is limited. For once the 'hammers' used for CID are very light, 4 Da (Helium) instead of the 40 Da (Argon) or 28 Da (N_2) 'hammers' used in quadrupoles, and the amount of energy one can apply to the ion in the trap is limited by some complex factors, the most obvious being the fact that the fragments need to be trapped during CID collisions. Ultimately, it is possible but very rare to encounter a peptide where the

ion trap cannot transfer enough energy for an efficient fragmentation. Usually only fragments up to 20% of the m/z value of the parent can be trapped efficiently, making ion traps not well suited for iTRAQ experiments, nor for experiments where one wants to see immonium ions.

While the resolution for choosing the ions that should be fragmented can be set as high as the resolution of the trap, it is also possible to fragment different regions of the m/z range or indeed all ions in the trap. This adds to the versatility of the Pauli trap and can be used also in hybrid instruments, containing ion traps as one part of the analyzer combined with ToF or FTICR (see later chapters).

To repeat, the user does not 'see' or 'feel' any of the complications of the inner working of the trap; choose the mode and resolution, and everything is done automatically.

Keeping 'fast' cycle times is important for sensitivity of the ion trap, as trapping and scanning are 'dead' times; any ions arriving at these times are rejected from the ion trap by the ion gates. However, sensitivity is not the biggest issue with the Pauli ion trap, as we will see later, and very fast ejection scan rates will result in poor resolution. So typically the scan rate (and resolution) are set to achieve just enough resolution to distinguish the charge states. Resolution of Pauli traps in daily use is typically 4 000–6 000, with some instruments achieving 20 000 for full range scans in cycles that take about a second. Most instruments can achieve higher resolution at the cost of prolonged analysis times and lower sensitivities. Because this would take too long over the whole m/z range, a small region is often chosen for scanning at ultra-high resolution (e.g. to define the exact mass and the charge status of one or two selected multiply charged ions).

We now turn briefly to the number of ions in the trap. If an experiment were set to allow 5 000 ions in the trap and this number were made up of equal numbers of eight different peptides, we would have 600 ions of a certain species. In other words, the Pauli ion trap would have no problems generating useful spectra using 0.001 attomole per scan! In reality we need more than one scan, there are many contaminants in each of our samples and the ion trap can work with far fewer than 600 ions. But this small example shows how extremely sensitive ion traps can be. They can even produce several consecutive MS/MS experiments with this small number of peptide ions. The maximum number of ions in Pauli traps just a couple of years ago was about 5 000, and you can expect to see good spectra in today's traps with even less than 1 000 ions. So for most proteomic applications the sensitivity of the Pauli ion trap is not likely to be the most severely limiting factor.

The Pauli trap has several very desirable traits for use in proteomics; it is extremely sensitive, able to generate data from only several thousand ions, it is perfectly suited to generating high quality MS/MS spectra of peptides (and even MS^n, as we will see later), it has a very small footprint, moderate vacuum requirements it is very easily combined with ESI. Pauli traps are arguably the cheapest mass analyzers to purchase and run. They are relatively easy to maintain and operate, although the way they work inside, hidden from the operator, is anything but simple. They have a modest resolution (around 4 000) and a sufficient accuracy for MS/MS analysis in proteomic settings. The accuracy (externally calibrated from 100 to 300 ppm depending on instrument, slightly better for fragments) is not sufficient to compete in PMF analyses with MALDI ToF instruments, but the in-built MS/MS capabilities more than make up for this. The huge sensitivity comes with some problems: the ion traps are very sensitive to contaminations (salts, polymers, etc.) as they have a relatively low dynamic range; if there is salt or polymers in your sample you often just see the contaminations and not the analytes! Therefore, they are not a good choice in offline experiments, triple quadrupoles and q-ToF hybrid instruments are much more robust for this application. However, combined with LC (1D or 2D), which delivers clean samples to the ESI, Pauli traps are very sensitive and can play out their advantage to the maximum. Pauli traps can be operated over a range from 50 Th to about 4 000 Th, some instruments achieving up to 6 000 Th. However, for reasons we will discuss further below, Pauli traps cannot measure in one measurement the complete range of ions; rather the smallest ions you can trap are a fixed fraction of the size of the largest ions, usually close to 0.3. So a typical range of ions in one measurement would be 500–1 500 Th or 1 000–3 000 Th. This also holds true for MS/MS experiments: if an ion of 1 800 Th is fragmented, the smallest ion observable is about 600 Th. Although these limitations can be partially circumvented by new excitation methods, they make Pauli traps unlikely candidates for iTRAQ type experiments (see Chapter 5). In iTRAQ, fragments of around 115 Th are used for quantification, allowing effectively only ions up to about 300–500 Th to be quantified 'in

one go' without having to resort to time- (and sample-) consuming MS^n experiments. However, quantifications using SILAC and isotope coded protein labelling can be used with no such limitations (Chapter 5). The way round this fragment size limitation for many other proteomic applications is to isolate a fragment ion from the first MS/MS (MS^2) experiment and fragment this again (Figure 3.31). When fragmenting an ion of 1 800 Th a fragment of 600 Th could be chosen for MS^3. This would then allow fragments down to 200 Th to be analyzed in the third MS. To reach even 'further down' one could isolate a fragment of the second MS/MS (MS^3) experiment and fragment this again (MS^4). This is about the upper limit of consecutive MS/MS experiments used in proteomic applications, but the Pauli traps are efficient enough to perform up to MS^{12}.

While the MS^n capability is invaluable for structure determination (as 'preferred' fragments often arise in the first MS/MS that preclude further structural analysis, for example around proline, which is 'stiff' in the peptide backbone), it is not often needed in proteomic applications. One of the biggest positive traits of the Pauli traps for proteomics is the 'smooth' way in which fragmentation is achieved, by a very gentle CID process (see earlier in this chapter). The CID process in Pauli traps takes a relative long time, thus there is ample time to spread the energy throughout the molecule before it fragments. Thus, mostly the (doubly-charged tryptic) peptides will break at their weakest points and produce relatively complete series of b and y ions (fragmentations along the backbone of the peptide; see Figure 3.8). These series of often complimentary b and y ions allow an easy interpretation of the MS/MS spectra and thus a relatively easy identification of peptides from databases, despite the low resolution and accuracy of the Pauli traps.

Another major drawback of Pauli ion traps used to be their rather limited dynamic range in terms of the manageable number of ions in a trap. This made Pauli traps a poor choice for quantification experiments, and indeed triple quadrupoles have set the standard for quantification for many years in the analysis of biochemicals. However, developments over the last couple of years have eliminated this limitation completely and ion traps can now be used for quantification experiments in proteomics as well.

With all its versatility, there are some modes in which Pauli ion traps cannot be run, such as precursor ion scan and neutral loss scan, which are so successfully used

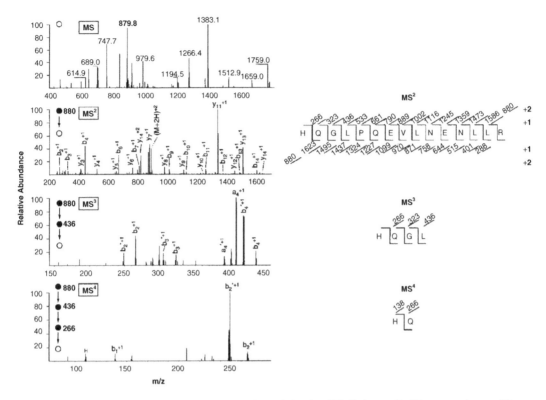

Figure 3.31 MS^n on Pauli ion traps. A tryptic casein peptide was infused at 200 nl/min on a Pauli ion trap, using an offline spray, at a sample consumption of 100 fmol/min. Tandem MS up to MS^4 data were generated. Note how most of the amino acids could be confirmed by MS/MS and only the smaller ions become more intense with MS^3 and MS^4. Characteristic of ion trap MS/MS, the fragmentation results in good series of b and y type ions. Reprinted from Jonscher *et al.* (1997), courtesy of Elsevier. © 1997 Elsevier.

by quadrupoles in proteomics especially for the analysis of PTMs. However, with their extended mass range capabilities, generally higher resolution than quadrupoles (even more so in zoom or ultra-high resolution scans), superior CID spectra and the option for additional fragmentation mechanisms such as ETD, for proteomic applications they are great workhorses providing very good value for money.

Linear ion traps

Commercial Linear Ion Traps (LITs) only became available after 2003. They are marketed as being able to overcome some of the limitations of ion traps, mainly in terms of dynamic range and resolution. While they surely improve in these fields on the Pauli ion traps that were available at the time of introduction of LITs, modern Pauli traps achieve similar performance to LITs in terms

of resolution, sensitivity and accuracy. Today, LITs are a varied mixture of mass analyzers with fundamentally different internal operations and different performance characteristics often disguised by the needs of marketing strategies. They are often used in hybrid instruments in combination with different mass analyzers or in the configuration of two consecutive linear traps of different configuration, which takes full advantage of their strength and enables higher sensitivity, higher capacity, better quantification and improved accuracy than Pauli traps, without sacrificing their unique ability to produce peptide fragments that are very useful for identification of peptides and their PTMs in proteomic applications.

LITs are sometimes also called 2D traps, to distinguish them from the Pauli traps, which are accordingly called 3D traps. This distinction can be somewhat misleading, as the ions in both traps fly in complex 3D trajectories.

Pauli traps can be regarded as derived from quadrupoles, with some major changes in geometry and application of voltages, and indeed LITs are also very similar at heart to quadrupoles. Imagine a quadrupole, with the corkscrew movement of ions and identical application of voltages, but with a bouncy 'lid' for ions on both sides; instead of flying right through the space between the quadrupole rods, the ions bounce back at the ends of the quadrupole and fly backwards through the space in between the rods to bounce back on the other side, and so on and so forth (Figure 3.32). As the ions do not fly in superimposed circular motions as in the Pauli trap, this is sometimes described as a 2D movement. This follows the notion that ions in Pauli traps fly around a point in space, while ions in a linear trap fly along a line. Compared to a Pauli trap the space inside linear traps is massively enlarged, hence the increase in the amount of ions that can stay in the trap. However, space charge effects still apply and the number of ions when such effects become limiting is around 20 000–100 000 for linear traps.

There are two principal ways in which the ions are ejected in linear traps: axial and radial (Figure 3.32). Without going into details of the ejection mechanism, the

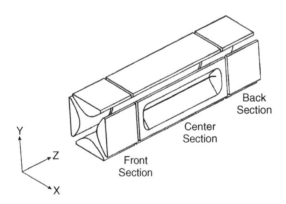

Figure 3.32 Linear ion trap with radial ejection. This design uses a quadrupole split in three sections to contain the ions in the space between the centre section, rather than two simple endcap electrodes. Also the 'rods' are formed hyperbolically. Note the slits in the centre section electrodes through which the ions are ejected. Radial ejection is sensitive and fast, but in hybrid instruments where ions need to be transferred from the LIT to another part of the analyzer, axial ejection (through the endcap electrodes) is often preferred. Both ejection types depend on different ejection principles. See text for details. Reprinted from Schwartz & Senko (2002), courtesy of Elsevier. © 2002 Elsevier.

differences in physical dimensions alone lead to different characteristics for the two types of LITs when it comes to ejection. LITs with axial ejection have a relatively low ejection scan speed over the range of m/z inside the trap. The ions move along a relatively long axis (several centimetres) and are thus less well defined in this dimension than ions in a Pauli trap or ions along the radius of the LIT. While this adversely affects the scan speed, it is very good for relatively simple instruments (with high transmission rates to other parts of the instrument) and inherently high resolution. However, the efficiency of ejection is relatively low at about 20%. Nevertheless, as ions are ejected along the axis of the quadrupole it is relatively straightforward to combine the axial ejection LIT with other mass analyzers. For example, most beneficial characteristics of triple quadrupoles and ion traps can be combined in instruments in which the third quadrupole of a triple quad is replaced by an LIT with axial ejection (Section 'Other hybrid mass spectrometers'). Axial ejection is achieved by applying an AC voltage to the endcap lenses of the LIT.

Radial ejection can make an LIT more sensitive if it is used as the final mass analyzer, as the path along which the ions are ejected is relatively short. Ejection is highly efficient, and is performed with similar methods to Pauli traps. Ejection scan rates are also similar to Pauli traps, as is the resolution. The ions are ejected through slits or cutouts in the quadrupole rods. Special detectors are made to collect most of the ions, and this can be highly efficient (Figure 3.32), in particular if both opposing rods contain ejection cut outs and the instruments have 2 detectors. However, this is not the ejection method of choice for transmission of the ions to other parts of a hybrid mass analyzer.

The trapping efficiency in LITs is much higher than in Pauli traps (60% in hybrid LIT instruments compared to perhaps 5% overall for Pauli traps), which is good for increasing sensitivity. However, axial ejection is not optimal for transmission of ions into other parts of a hybrid mass analyzer.

LITs still need helium to improve the trapping efficiency and to induce CID by excitation. With their increased space for ion travel in the direction of trapping, far less gas is needed, or more precisely a lower gas pressure. The helium pressure in commercial LIT is about 300 times lower than in Pauli traps. Helium in the trap is a double-edged sword; having a 'high' pressure is good for improved trapping efficiency and fast and complete CID. Fast ejection scanning (without fragmentation during ejection) is better done in a 'low' pressure trap. Indeed there are commercial hybrid instruments combining both environments with two consecutive traps: the first (high pressure) trap performs high efficiency trapping and resonance excitation, after which the ions are ejected (all sizes 'at once' without mass scan) into a low pressure LIT from where the ions are radially ejected with high resolution and scan speed. Overall resolution is about 4 000, with narrow scan range resolutions of more than 25 000. These instruments are very sensitive (trapping efficiency further increased by about 50% as compared to a single LIT), with a high dynamic range and very fast cycle times in the region of 0.1 s per cycle (which also aids the sensitivity). MS^n performance is similar to Pauli traps, sensitive and efficient, but again with faster cycle times. Accuracy is comparable to the best Pauli traps, at below 100 ppm, with the accuracy for fragments being slightly improved compared to those of the parent ions in the MS scan, again also typical for Pauli traps. The overall performance for proteomics experiments is very good, and, thanks to the 'smooth' generation of b and y ion series (typical of all quadrupole ion traps) and high sensitivity, linear ion traps can outperform instruments with numerically better specifications in terms of resolution and accuracy (see for example Elias *et al.*, 2005).

The performance of modern stand-alone Pauli and linear ion traps is very similar in most respects, allowing very fast measurement cycles with 10–20 measurements per second. This becomes more important when the new uHPLC systems with very short run times and short peak durations (typically around 1–3 seconds) are used.

We will see more examples of LITs when we consider hybrid instruments in Section 3.3.6. These instruments try to combine the versatility and valuable CID characteristics of quadrupole traps with the enhanced resolution and accuracy of other mass analyzers. It is particularly useful to use the trap to collect ions over time and inject them into another mass analyzers in a sudden motion. The second analyzer can then get to work while the LIT is collecting at the same time more ions for the next cycle. In this way high duty times in the cycle are achieved, and the good sensitivity, CID characteristics and storage capacity of the LIT can be combined with superior resolution and accuracy of other mass analyzers.

3.3.4 Fourier transform ion cyclotron resonance mass analyzers

Fourier transform ion cyclotron resonance (FTICR) mass spectrometers are somewhat exotic, particularly for beginners, but they might be of interest for some special proteomic approaches, and might with new technical developments become more available in the near future. FTICR mass analyzers easily cost three to five times more than entry level mass spectrometers and nearly twice as much as other types of high performance mass spectrometers. For this reason alone one is not likely to find FTICR mass spectrometers outside dedicated facilities. They are technically demanding, as they need very strong magnets to perform to the level they can. However, it is also fair to say that no other mass analyzer inspires the imagination like this one. This might perhaps be because most people will never have access to one, but the possible performance is indeed staggering; imagine a resolution of about, 1 000 000 or an accuracy better than 1 ppm paired with a sensitivity in the attomolar range – the stuff of dreams! However, filling up cryogenic tanks with helium and the long waits for the 15 Tesla (T) magnet to cool down (and warm up) when things go wrong as well as the huge capital commitment – these are more the stuff of nightmares. It is also important to know that FTICRs are not high speed instruments; to reach anything like their full performance potential the time per measurement (cycle time) should be 1 second and longer, which is not always compatible with high throughput analyses. ICR also needs a very high vacuum and this can cause a delay of seconds from sample entry until a measurement is possible, which can be complicated to adapt to LC separations. At least with the advent of (nearly) unlimited computational power at low price, FTICR MS is now much easier to run than ever.

The ultra-high accuracy that FTICR instruments can deliver opens up many new possibilities, which will be explored in more detail in Chapter 5. Here we will take a brief look at the implications of measurements at different accuracy levels. When a given mass is used to search for a peptide with a 'matching' or comparable mass, at an accuracy of 100 ppm you can expect several thousand peptides to be present in a representative database within the mass window (measured mass ± 100 ppm, e.g. 1 210.694 ± 0.12 Da). At 20 ppm accuracy you might still find around a thousand peptides falling within the expected mass window (±0.024 Da). If you manage to achieve an accuracy of 5 ppm only about a hundred peptides might fall into your error window (±0.006 Da). For 1 ppm accuracy (or even below) only a handful of peptides (and sometimes even only one) will fall into the window of the measured mass (±0.001 Da). For PMF one always has to use several independent peptides. If you consider eight peptides in the PMF, all these masses with their potential errors at 100 ppm create a total pool of several tens of thousands of peptides among which the 'right' hit (i.e. a proteins containing all the measured peptides!) is contained. At 5 ppm accuracy this pool will shrink down to below 1 000 peptides, and perhaps 50 for 1 ppm. Of these 50 entries in the database only one hit with overlap for several peptides of the PMF will emerge, allowing for very confident scoring and peptide/protein identification of one protein or a mixture of proteins contained in your sample (if your peptide with the modification is present at all in the database). Of course these numbers vary according to peptide size (the bigger the peptide, the more candidates) and the exact measured mass, as there are distinct distributions of the number of peptides for certain mass ranges, based on the average composition of peptides and the exact atomic mass of elements (see Chapter 4 for more details). From this simple example it becomes clear that there are cases where no MS/MS analysis might be needed, as the very accurate (below 1 ppm) mass determination of a small peptide can identify the presence of a certain protein on its own. This is a good example of a paradigm shift based on a significant change in instrument performance, in a technology driven area like proteomics.

FTICR MS combines several basic principles to measure the mass of ions. The first fundamental principle is the Lorentz force. All ions moving in a magnetic field experience the Lorentz force (Figure 3.33), which acts perpendicular to the lines of the magnetic field and perpendicular to the motion of the ion. This is the force that is exploited in, for example, any electric motor to convert electric energy into movement. As a result of the Lorentz force ions flying through a homogeneous magnetic field will be deflected to fly the segment of a circular path. The radius of this circular path depends on the charge, mass and speed of the ions, and the strength of the magnetic field. If the field is strong enough, the ions can get trapped into flying in circles (see Figure 3.33). Trapping in the plane perpendicular to the field lines of the magnetic field is achieved by the Lorentz force resulting from the strong static magnet and trapping in the

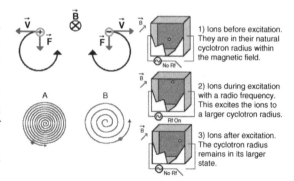

Figure 3.33 Principles of FTICR MS. All charged particles in a magnetic field experience the Lorentz force, leading to a circular movement (cyclotron motion), with positively and negatively charged ion circulating in opposing directions. If the magnetic field lines are directed into the plane of the paper, the ions circulate perpendicular to them. The radius depends on the field strength and the energy of the ion. The frequency of the cyclotron motion for a given field depends on the m/z value. If ions are excited by a radio frequency applied in the plane of the motion, their radius increases, with a frequency dependent on their m/z. Electrodes for measurements (see Figure 3.34) are perpendicular to the excitation electrodes. To trap the ions, changing voltages on endplates are used. The endplates are at the front (not shown) and back of the trap. Ions would travel in the same diameter circles forever if they did not collide with each other in a practical instrument, so ion density in the trap is an important consideration, as is strength of magnetic field. Source: http://ncrr.pnl.gov/training/tutorials/ft_icr_tutorial.stm.

dimension along the magnetic field lines by a small AC trapping voltage. The frequency at which the ions perform their circular motion depends on the ions m/z and the strength of the magnetic field. In other words, in a strong constant magnetic field the ions with their various m/z values each have a specific frequency, at which they travel one complete circle. Importantly, the speed at which the different ions travel has no influence on this frequency; rather, it increases the radius of the circle in which the ions travel. If the speed of the ions becomes too high, the radius increases beyond the physical size of the trap and the ions are expelled. The physical size of these so-called penning traps is rather small, measuring several centimetres along the magnetic field and about a centimetre perpendicular to it. The ions can be excited in their circular movement; applying an RF of the same frequency as that of the circular movement will pull (and push) the ions each time they fly towards the electrode of opposing (same) charge (Figure 3.33). The ions get faster as this happens. However, the frequency of their circular movement stays the same, thus the radius of their movement has to increase. The instrument is now accelerating particles in a circular movement, thus becoming a cyclotron, more precisely a cyclotron at the resonance frequency of the trapped ions, exploiting the ICR effect, which gives the mass analyzer part of its name. If not disturbed the ions would fly at the higher speed/larger radius forever after a short pulse of ICR; however, if the pulse were too long, or the voltage of the RF too high, they would be ejected out of the trap. This gives us a mechanism for expelling ions of a particular m/z, isolating ions of a particular m/z or performing CID, if we put some gas in the ICR cell (the trapping compartment).

Hitherto we have not discussed how ions are detected, and this mechanism is very different from all the other mass analyzers discussed so far – ions are detected without destruction. Every charged particle in motion can induce a current. After ICR excitation the ions fly close to the walls of the ICR cell. Moreover, the pulse also synchronizes all single ions of a particular m/z, so that they fly in discrete packages (see Figure 3.34) and the current induced by these discrete packages of ions of a specific m/z in electrodes placed in the walls of the ICR cells is measured. Technically a short ICR pulse, ramped through all resonance frequencies of possible ions in the ICR cell, is applied in a short (i.e. almost instantaneous) pulse, and the ions are allowed to fly

without further disturbance. Electromagnetic interaction, like induction for currents but also collisions between ions, slows the ions down over time, so the radius of their movement (but not the frequency) is again reduced, and thus the induced current gets smaller in amplitude, but not frequency. The induced or mirror image current, also called the transient signal, is amplified and recorded over time. If there was a single ion species in the trap, the mass could be simply deduced by the frequency of the induced current, the amount of ions by the intensity of the signal (see Figure 3.34). However, for a practical mass analyzer this is not feasible, as there are always a multitude of ions present in a sample. Indeed, with its ultra-high resolution, ICR MS is the best mass analyzer for extremely complex samples. To solve this problem Fourier transformation (FT) is applied. We can use the analogy of music to explain the process: each instrument in an orchestra plays single notes, which are composed of harmonic tones, each again composed of harmonic sonic waves giving each instrument its specific sound signature. So we are used in everyday life to situations where simple harmonic oscillations (or, to keep the analogy, ICR frequencies of ions with a single m/z) combine to give a rich tapestry of sound (i.e. the complex transient signal of induced currents in the ICR). The reverse process can be achieved by the mathematical process of FT: each curve (be it the pressure signal of a sound or the transient signal of an FTICR) can be imagined

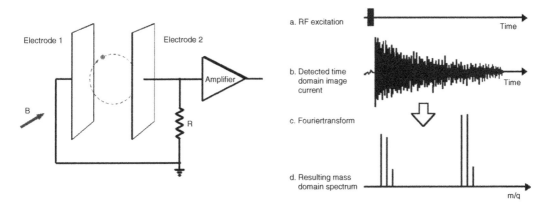

Figure 3.34 Transient excitation in FTICR. Ions flying next to the cell walls of an ICR cell induce a minuscule current, which can be used for non-destructive measurement of ion packets. A measurement cycle for an FTICR MS starts with an RF excitation after filling the panning trap (panels on the right). The resulting current is measured over time. The longer the measurement time, the higher the resolution and potential accuracy of the results. The transient records (change of current in the time domain) have to be subjected to Fourier transformation, resulting in harmonic frequencies which are translated into the m/z values and intensities of the ions present in the ICR cell.

and calculated as being composed of a finite number of harmonic oscillations. If this mathematical process is applied to the transient signal of the ICR, it will produce a harmonic oscillation at a specific ICR frequency for each ion and its specific m/z, and the intensity will reflect the amount of ions in the FTICR cell.

We now have all the components of a working FTICR mass analyzer. To summarize, ions are injected into a small cell located in a very strong magnetic field, where they oscillate at specific frequencies according to their m/z value. ICR excitation of each ion brings them close to the walls of the cells where each ion package of a certain m/z induces a current of a specific frequency and intensity. Only the latter changes over time and on the combined electrical signal of all ions in the trap (the transient) an FT is performed to give the intensity and m/z of ions in the trap.

When we said that the FTICR mass spectrometer requires a 'strong magnet', we really did mean a strong magnet. Usually magnets in the range of 9.5–15 T, some 100 000–150 000 times the strength of the earth's magnetic field, are used in proteomic applications. These magnets are very large (about 1.5–2 m in every dimension) and produce an extremely homogeneous strong magnetic field extending over several cubic centimetres. The size partially results from active shielding (so that, for example, pens do not become lethal 'spears') and the requirement to cool the magnet down, as such a strong magnetic field can only be produced by using superconductors, which at the present time need cooling by liquid helium. There is a good reason for the use of such strong magnetic fields: higher magnetic fields offer better performance. Resolution and accuracy increase linearly with the field strength, the upper mass limit, the maximum number of ions in the trap and the maximal trapping duration increase with the square root of the magnetic field strength. It is commonplace to enhance the performance of an 'ageing' FTICR instrument by upgrading to a stronger magnet.

In commercial instruments about 10 ions can produce a measurable signal and the dynamic range is just shy of 10^5. Resolutions up to several million in relevant m/z ranges have been achieved and internally calibrated accuracies of up to 50 parts per billion (0.05 ppm) can be achieved. It is important to realize that the resolution in the FTICR actually decreases with mass. In fact the resolution in FTICR MS is inversely proportional to m/z,

hence the need for stronger magnets if one wants to work on proteins rather than small molecules and peptides. Since the detection of ions is non-destructive and ions can be kept for hours in the ICR cell, gas phase chemistry experiments can be performed. Given the high resolution these experiments can, for example, be performed on full length proteins (top-down proteomics) to analyze PTM or even protein 3D structure via the exchange of heavy protons (deuterium). However, all these superlatives in performance require time; measuring times are of the order of several seconds; faster measurement (down to 1 second) will reduce performance and sensitivity to levels more comparable to (but still outperforming) other mass analyzers, at about 800 000 resolution and 1 ppm accuracy. MS/MS experiments of the MS^n variety are possible with high transmission rates, comparable to the situation in quadrupole ion traps. The ions are excited by RF application and fragment after collision with gas. However, the gas pumped into the ICR cell for CID needs to be pumped out again for measurements (to below 10^{-9} mbar, or about 1 000 times less than in a ToF stage), easily leading to a delay of 15 seconds before measurements are possible again. Other fragmentation modes (not depending on collision gas) can be used (e.g. infrared multiphoton dissociation, ECD, ETD) and ICR cells are particularly suited for ECD, as they can trap low energy electrons and positively charged analytes at the same time.

Overall the outstanding performance of FTICR can lead to a shift in proteomic methods; at high accuracies and sensitivities it is not necessary to perform MS/MS to identify a peptide from a huge database, and, in combination with other highly sophisticated technologies (e.g. LASER capture microdissection, UPLC with retention time database), one can analyze the proteome of several thousand highly enriched cancer cells to the same level as with standard LC MS approaches which need hundreds of thousands to millions of cells (usually of mixed origin).

Instruments with FTICR as the only mass analyzers are rarely used in proteomics nowadays, and we will discuss hybrid instruments in Section 3.3.6.

3.3.5 Orbitrap FT MS

The orbitrap is a new mass analyzer that only became commercially available in 2005. The ion detection is very similar to that in FTICR instruments: non-destructive

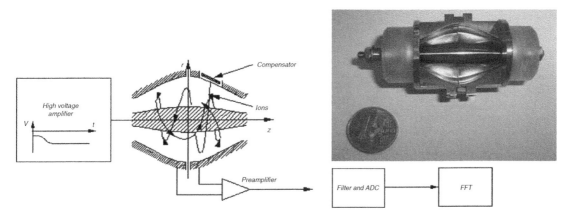

Figure 3.35 Basic principles of orbitrap FT MS. Ions enter the orbitrap at the compensator, lowering the potential slightly, so that the ions can be 'caught' by the potential of the wire like a satellite in orbit round the earth. Without the compensator the ions would either 'overshoot' or fall into the wire. The transit of ions between the left and right electrode in this view is used to measure the induced current. The ADC current is filtered out of the signals and fast Fourier transformation is used to determine the frequencies and thus the m/z values of the ions in the trap. Reprinted with permission from Hardman & Makarov (2003). © 2003 American Chemical Society. Photo: ipg.obs.ujf-grenoble.fr/IMG/jpg/Orbitrap.jpg.

measurement of the current induced by passing ions, followed by FT to measure all ions in a mixture. However, instead of using ICR and strong magnetic fields, the ions are trapped in motions with circular elements by means of an electrostatic field along a central electrode (see Figure 3.35). The basic principle is relatively simple; ions are shot at high speed into an orbit around a charged wire, rather like putting a spacecraft into orbit around a celestial body. Once in orbit and without perturbation (vacuum at a nearly ICR-like 10^{-8} mbar), the ions will happily fly around the wire.

Similar to FTICR MS, groups of ions travelling as a 'packet' will induce a current when moving past electrodes in the outer walls of the trap. These currents are measured and amplified. If a mixture of ions with different m/z is in the trap, the generated signal needs to be Fourier transformed to yield a sine wave for ions of every m/z present in the trap. The frequency of the sine wave is related to the m/z, its intensity to the amount of ions at this m/z value present in the trap. However, the orbitrap is anything but simple; the ions need to be injected in a high vacuum at a precisely arranged speed depending on the m/z value or, better expressed, the ions need to have an energy that is precisely matched to the attraction of the central 'wire' electrode. All ions of a certain m/z need to be delivered at precisely the same time, in a very small physical space ($1\,mm^3$), and if the

ion density is too big space charge effects will lead to deviations of the optimal circular trajectory and ions will be lost to the measurement or be measured at reduced resolution. To increase the space and thus the capacity of the orbitrap the central electrode is not a wire but a 3D spindle shape (Figure 3.35) and ions are 'pushed' from the entry point towards the other end of the barrel shaped orbitrap by RF voltages between the two lateral outer electrodes of the trap. Having this 3D trap allows many more ions to be trapped than the 2D 'ions around a wire' principle would allow; the ions now fly in corkscrew-like movements inside the trap without blocking each other's charges. The spread of ions is achieved as the complex shape of the electrodes allows ions of different m/z to have different flight paths, barely overlapping, filling the trap but still flying with an orbital frequency that is strictly dependent on their m/z. The injection of ions in the trap is achieved by collecting the ions in a specially formed 'C' trap, from which they are ejected and accelerated much like ions in a ToF mass spectrometer; this enables good focusing of the ions but also ensures that all ions have the same kinetic energy, as they are accelerated in the same field (Figure 3.36). Thus small ions have a higher speed than larger ions, perfectly matched to be trapped in the axial field of the orbitrap, where small ions travel at a higher speed, the same axial diameter and thus higher frequency around the centre electrode. However, trapping

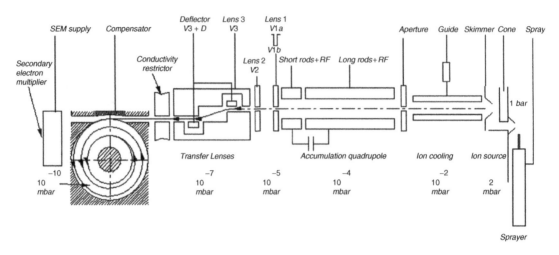

Figure 3.36   The orbitrap as a practical mass analyzer. Overview of the first orbitrap. Ions enter the mass spectrometer on the right and are moved from right to left, towards the orbitrap, which is depicted in an optical section in the plane of the ion entry on the left. The accumulating quadrupole (middle) is an LIT to enrich/focus the ions before injection. An optical section of the orbitrap in a perpendicular plane as well as a picture of an actual orbitrap analyzer can be seen in Figure 3.35. The circular motion in this plane is only used to trap the ions, not for measurements. Reprinted with permission from Hardman & Makarov (2003). © 2003 American Chemical Society.

is a tricky business – too fast and the ions fly out of the trap, too slow and they are attracted too strongly to the centre electrode and hit it. For efficient trapping, ions are pushed towards the electrode as they enter, while at the same time the field is changed (lowered) to allow catching of the ions. After injection the potential on the centre electrode is raised again to about $-3\,200\,V$ ($+3\,200\,V$ for positive ions) – and here is one limitation of the orbitrap, it can only trap either positive or negative ions, not both at the same time. Ions move along the axis of the trap due to RF applied between the lateral outer electrodes; this, together with the shape of all three electrodes, allows some control of the ion trajectories. Using this RF it would also be possible to fragment ions, but in commercial instruments this is not done; similar to FTICR MS, the orbitrap has more potential if it is used 'just' as the final analyzer in combination with other mass analyzers, which take on the role of fragmenting devices. In combination with several other traps, orbitrap instruments can also use ETD, as the negative ions can be stored in a dedicated trap (see below). The use of the orbitrap in a fully fledged mass spectrometer for proteomics will be discussed in Section 'Other hybrid mass spectrometers'.

In terms of performance the orbitrap is very close to FTICR instruments. Resolution does diminish less rapidly with increased mass than in the FTICR, so the very good resolution of 60 000–100 000 is available for a relative wide m/z range, limited to about 4 000 Th by other hybrid instrument components. Accuracy is also very close to FTICR values, around 2–5 ppm when externally calibrated and below about 2 ppm when internally calibrated. The dynamic range is larger than 4 orders of magnitude, comparable to other traps, but smaller than in quadrupoles. The sensitivity is also just shy of what FTICRs can achieve, in the sub-femtomolar range. As in FTICR analyzers, the performance values for sensitivity, resolution and accuracy depend on cycling times, or more precisely on the duration of the transient measurement. However, even at short cycle times of about 0.2 s, the resolution and accuracy are on a par with any other ion trap. At about 0.5 s measurement time the resolution of orthogonal ToFs is reached. At 0.9 s cycle time the resolution is clearly better than that of any ToF analyzer. This is usually the time used for analysis of peptides with high resolution (40 000–60 000) and accuracy. The full performance, rivalling FTICR MS at larger m/z, is achieved at 1.6 s analysis time and is mainly used for proteins. Compared to FTICRs the orbitrap is a robust and practically maintenance-free analyzer, more stable than ToFs without internal calibration, and has no external magnet and a small footprint. In other words, the orbitrap is an almost ideal instrument, the only disadvantage being the capital cost; this is higher than most traps and is in the region of high end ToFs but below FTICRs with strong magnets.

It is fair to say that since their commercial introduction in 2005 orbitraps have become very popular and versatile instruments in proteomics. They have easily overtaken other FT MS analyzers in terms of sheer market volume and practical applications, and one can safely predict that they will become a mainstay of proteomic applications with their mix of versatility, accuracy, robustness and excellent performance at acceptable to high speed measurements, particularly in the form of hybrid instruments with other mass analyzers (Section 'Other hybrid mass spectrometers').

3.3.6 Hybrid instruments used in proteomics

In previous sections the main mass analyzer types used in proteomics have been introduced. In the last decade an ever confusing array of new commercial mass spectrometers combining different mass analyzers has appeared. Furthermore, high end instruments often allow the optional addition of MALDI to many primarily ESI based instruments as well as the optional addition of new features such as ETD or ion mobility MS to high end (and high price) instruments. There are about half a dozen global manufacturers that offer a variety of instruments, as well as many smaller manufacturers that specialize in certain mass analyzers. It is very easy to get confused by this variety, and of course, by the marketing claims of the manufacturers. The obvious lesson from this variety is that there is no such thing as the optimal mass spectrometer for proteomics – a given instrument will have advantages over others in some fields and disadvantages in other fields. Often very different mass spectrometers will perform in everyday use to a similar level, provided that the strength of each analyzer is incorporated into the experimental layout and the data analyses, while strategies need to be designed in order to avoid hitting a 'brick wall' due to the limitations of the mass spectrometer used. Often it comes down to personal preferences and experience; this might be a good enough reason to prefer a certain instrument.

It is impossible to introduce every single available instrument combining different mass analyzers in a textbook such as this, but we will explain the most important analyzers used for proteomics with their main characteristics.

Quadrupole-ToF hybrid instruments

We begin our exploration of hybrid instruments in this section with quadrupole-ToF mass analyzers. The first commercial q-ToFs were introduced in 1996, and at the time of writing they are in widespread use for proteomic applications, due to their very attractive mix of features and performance. Q-ToF mass spectrometers (or, more precisely, tandem quadrupole ToF instruments) combine the versatility and controlled CID capabilities of the triple quadrupole with the high resolution, accuracy, sensitivity and speed of ToF mass analyzers. This holds true for MS measurements and some of the MS/MS analysis modes that are typical for triple quads. Other instruments, such as ToF/ToFs and hybrids of quadrupole and ion traps, are now close rivals, but for half a decade the combination of speed, resolution, accuracy, sensitivity and versatility that q-ToFs offer was unrivalled. Replacing the third (scanning, 'one ion at a time') quadrupole with a 'measure all ions simultaneously' ToF analyzer makes the instrument a hundred times more sensitive than a quadrupole analyzer for MS and product ion MS/MS measurements (Figure 3.37). This increase in sensitivity already takes into account all possible problems with orthogonal ToF principles (see below). Modern q-ToF instruments deliver a resolution in the region of 20 000, some instruments achieving up to a staggering 40 000. Accuracy is routinely better than 10 ppm under optimal conditions, and internally calibrated down to about 2 ppm. This means that the third digit behind the point of a peptide mass is correct within one to several units! Moreover, the high accuracy is maintained for the MS/MS fragments; for reasons we will discuss further below, it is even easier to generate more accurate MS/MS spectra, somewhat limited by increasing signal to noise ratios. However, the MS/MS data is not as good for database searches as the data from ion traps; although the resolution and accuracy of a q-ToF will be much better than an ion trap, ion traps tend to produce fragments that are easier to interpret and lead to a higher score in searches, as there are also fewer fragments. The reason for this is that, just like in quadrupoles (identical fragmentation principle), the initial fragments in the MS/MS mode undergo further collisions/fragmentation and produce a variety of secondary fragments. These can, on the other hand, be very useful for the characterization of PTMs.

Next to all the great advantages, there are three disadvantages to combining a quadrupole and a ToF analyzer.

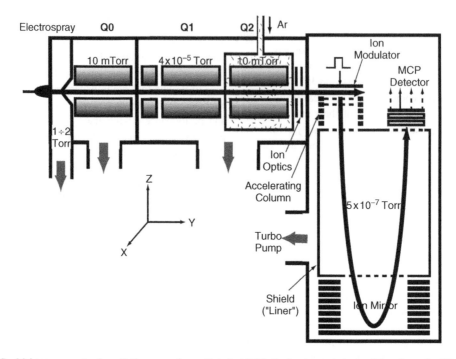

Figure 3.37 Main components of a q-ToF mass analyzer. Note that Q0 is for ion transmission and focusing only, Q1 can be used for scanning or transmission of one ion only for MS/MS, and Q2 is only used for transmission (MS) or for collision (MS/MS). For CID, gas is introduced into Q2 and ions are accelerated between Q1 and Q2 to give the collisions enough energy for fragmentation. Changes in acceleration voltage are usually used to switch between MS and MS/MS mode. A 'W'-shaped double reflector can be used to enhance resolution. For more details see text and Figure 3.38. Reproduced with permission from Chernushevich *et al.* (2001).

First, ToF analyzers are not mass limited, that is, there is no in-built upper mass range and ions up to a million Th can be measured in a MALDI ToF. Since the ions in a q-ToF hybrid have to pass through the quadrupole first, the limit for MS operations is about 20 000 Th and for MS/MS about 4 000 Th, which are the mass limits for transmission on the quadrupole in transmission mode and effective scan/ion selection in the scanning mode. However, using ESI, which easily produces charge states higher than 15 on larger proteins, this is not really as much of a limitation as it might be when using MALDI, with its poor ionization of mainly +1 and up to +3 or +4 in large proteins. Given the very good resolution of q-ToFs, medium sized proteins can be analyzed in top-down approaches, reaching isotopic resolution of fragment ions for proteins up to about 20 kDa in size. Also very useful is the ability to analyze protein complexes and multimers under 'native' conditions in ESI up about 60 kDa with accuracies of several dalton or around 20–30 ppm by analyzing multiple charge states (higher than +10).

When analyzing a protein by ESI MS, all the signals from the ions in the various charge states are integrated and used to calculate the true molecular weight of the protein by deconvolution (see Figure 3.16 and Section 4.2.2). A spectrum of a single protein in high resolution ESI is very complex, as different charge states exist in parallel.

The second area of possible limitation of the triple quad hybrid concerns different MS/MS modes in which the triple quadrupole can operate. While the hybrid offers superb sensitivity, resolution and accuracy when able to take full advantage of the ToF rather than a scanning analyzer, in MS/MS this is only possible for the product ion scan (compare Figure 3.27). The precursor ion scan offers reduced sensitivity compared to a quadrupole (but still better resolution and accuracy), and the neutral loss scan is not possible on the hybrid instrument as such, but can be simulated at reduced sensitivity by data analysis of the complete ToF data set.

Finally, the third potential limitation of q-ToFs is that the amazing dynamic range of quadrupoles is curtailed to the still respectable dynamic range of ToF instruments, starting at much lower signal intensities than, for example, MALDI ToF instruments.

In order not to give the wrong impression, it should be said that using a triple quadrupole to the best of its abilities on a complex sample eluting from an LC would result in far fewer proteins being detected from their peptides by MS and MS/MS than can be expected using a q-ToF; enhanced speed, sensitivity resolution and accuracy would allow identification of several times more peptides compared to a triple quad. This advantage of using the q-ToF would, however, pale into insignificance for high accuracy quantifications or if a precursor ion scan were to be performed to identify phosphorylated peptides.

Most q-ToFs for proteomic applications routinely use ESI and have a front end that is in principle identical to triple quadrupoles. The similarities extend to the second quadrupole (see Figure 3.37). This is where the hybridization starts; after the second quadrupole the ions enter into a region where they are focused and where the vacuum changes dramatically; quadrupoles have relatively slow moving ions in a medium vacuum, which is slightly lower in the second quadrupole as this quadrupole may contain gas (nitrogen or argon) needed for CID (see Section 3.1.2). ToF analyzers, on the other hand produce, very fast moving ions and only work efficiently in a high vacuum, as the ions fly long distances and cannot be allowed to have any collisions at the high speeds at which they travel. For maximal sensitivity and resolution, pressures are around 10^{-6} mbar. At this point the ions also change from an environment of constant ion flow to pulsed ion flow; ToFs only work when they extract ions in very small time frames (i.e. nearly instantaneously) and need the ions highly defined in space and speed for an accurate measurement. The ions are extracted from a constant stream with a maximal pulse rate in the range of 30 kHz (Figure 3.38). This rate is effectively the acquisition rate of the ToF. It has to be changed depending on the size of the largest ion that is to be measured (up to around 20 000 Th in MS mode, 4 000 Th – and in some instruments 8 000 Th – in MS/MS mode). The larger the ions, the more the frequency has to be reduced to allow a cycle of ToF measurements to be completed before the next one starts, as the larger ions move most slowly in the ToF flight tube. A slower pulse rate or ToF measurement rate results in more 'dead' time, and thus a lower sensitivity. The result of this alone is that larger ions can only be measured with reduced sensitivity and lower

sampling rates, unless the pusher cell involves some sort of 'ion trap' feature (see below). Thus, although it is a ToF instrument, with potentially unlimited m/z range, it only measures the range of m/z that is really necessary (for other reasons too – see below), a situation not encountered in offline MALDI ToF instruments. The ions are accelerated orthogonally to their movement away from the second quadrupole, which is very slow in terms of ToF ion movements. Orthogonal ToFs can achieve amazing performance and the resolution is on a par with – if not better than – that of any MALDI ToF. A so-called pulser/pusher injects the ions into the ToF stage. For the sake of performance the ToF stage is usually a reflectron stage, sometimes even a double reflectron. This results in a 'W'-shaped flight path (see Figure 3.38) and allows high resolution with a modestly sized, relatively compact instrument with high (thermal) stability. Detectors can be similar to the analogue transient recorders discussed in Section 3.2.1. However, far fewer ions per ToF analysis reach the detector (and the acquisition rate is 10–1 000 times higher!), so time to digital converters can be used. These still use MCPs, but transform the arrival of a signal in a digital fashion. They produce less noise but are also saturated more easily. With a time to digital converter even signals of four to seven detected ions in an MS/MS spectrum can measure a peak with sufficient accuracy. The reflectron geometry takes into account the lateral motion of the ions and, in addition to the considerations in our discussion of the ToF analyzer, has to account for this movement and a better focusing of ions with different speeds in the lateral direction.

Ideally all the ions coming from the second quadrupole should be measured by the ToF, but the pulse mode necessary for ToF analyzers does not permit this. Allowing a high pulse rate of orthogonal extraction can help to get a high level of ion transmission and thus sensitivity, but the ions arriving at the ToF extractor in between pulses cannot be measured. The ions coming from the quadrupole region travel relatively slowly from the quadrupoles; if they can be delayed in between pulses, while the ions of the previous pulse are being accelerated and fly in the ToF tube, they can accumulate in the pusher region, 'waiting' for the next pulse. With a little care nearly all arriving ions can thus be measured. However, as manipulation of ion movement is mass dependent, this 100% duty rate can only be achieved for a small range of m/z values (typically 100 Th) and not for the whole mass range normally

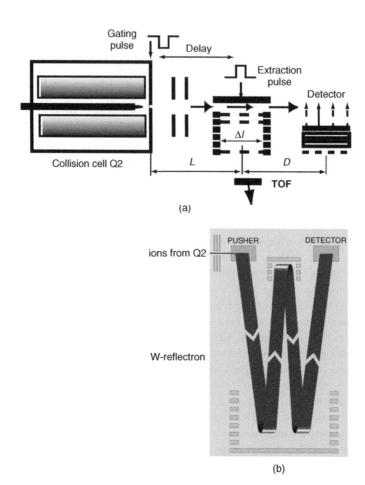

Figure 3.38 Ion transmission in the ToF stage of a q-ToF. (a) Careful design of the components in the ion modulator (pusher) and between Q2 and the ToF stage can enhance the duty cycle, and changes in potential can be used to 'trap' ions for short times before they reach the pusher, so that a certain mass range (e.g. 100 Th wide) can be transmitted with 100% efficiency at this stage. (b) A 'W'-shaped double reflectron allows resolutions in excess of 20000 in compact instruments with minimal thermal instability. (a) Reproduced with permission from Chernushevich *et al.*, J Mass Spectrom, 2001; 36: 849–65. (b) Micromass, http://www.protein.iastate.edu/seminars/micromass.ppt, slide 16.

measured by the ToF. The duty cycle of the ToF analyzer discriminates against small ions (Figure 3.38); if the ions are pushed towards the ToF extractor, small ions are faster than large ones. As soon as the largest ion of a packet arrives at the far end of the extractor, the ions are extracted. At this time part of the smaller ions are already outside the extractor, so they are lost to the analysis. Now the extractor starts to fill again and the cycle starts all over again. It would also be possible to discriminate against the larger ions, but usually their signals are weaker anyway (for several reasons, see also the discussion on ToF) so it is more practical to discriminate against smaller ions. Due

to the need of all ions to pass through the quadrupoles, there is a loss in sensitivity if the m/z range defined in the set-up of a particular measurement exceeds the range that the quadrupoles can let pass at a given RF setting. This loss in sensitivity is around 50% as two different RF settings need to be used on the quadrupoles alternatively. Such an loss in sensitivity becomes inevitable when the ratio of the m/z of the smallest to largest measured ion exceeds a factor of about 5.

This alignment of ion arrival and pusher frequency can be used to perform 'precursor' or 'neutral loss' ion scanning on a q-ToF with sensitivities closer to those on a triple quadrupole than under normal q-ToF operations. The overall ion transmission (i.e. from entering the instrument to the recorded spectrum) in a q-ToF is around 2%, compared to a triple quadrupole's 5–20% (both instruments in transmission mode). Losses due to low duty cycle and pusher/orthogonal acceleration efficiency (efficiency around 20%), loss to acceleration/reflector grids of the ToF (efficiency around 10%) entrance and detector and efficiency of the MCP/detector (around 65%), all add up to a total transmission of 2%. This 2% overall ion transmission is still 100 times better than that of a quadrupole, scanning at unit resolution over a range of 1 000 Th in about 5 000 steps, and an overall efficiency/transmission of 10% (when not scanning), as the scanning quadrupole only spends one-thousandth of its good transmission time on any one ion with the width of 1 Th. For q-ToF, transmission depends on the mass of the largest ion that is to be measured. The highest sensitivity is achieved at the highest pusher frequency, typically 30 kHz. As the ions get larger the time they need to drift in the pusher aperture increases, as well as the time they spend in the ToF stage, as they fly more slowly than small ions. For example, if the largest ion is 1 000 Th the periodic time for the pusher is about 40 μs and for 16 000 Th around 200 μs. This means the instrument can perform 30 000 measurements per second on ions up to 1 000 Th, but 'only' 5 000 for ions up to 16 000 Th, and will thus be less sensitive. Remember that the q-ToF in MS or MS/MS mode does not need to operate a quadrupole in scanning mode, unlike the triple quad. If the quadrupole of the q-ToF is in scanning mode (e.g. classical precursor ion mode) than it gets less sensitive by a factor of 1 000, so the triple quad is 10 times more sensitive now! Employ the trapping before the pusher, the q-ToF might still only be half as sensitive as the triple quad in precursor ion mode, albeit with better resolution, measuring all fragments with better accuracy and thus with results inspiring more confidence.

We observed earlier in this chapter that the ToF acquisition rate is several kilohertz; nevertheless the q-ToF has a cycle rate of several hertz only. The cycle rate is defined as the number of different operational modes the instrument can switch per second. If the instrument is to take a simple MS measurement, it needs to stay in a given mode a certain amount of time to acquire enough data for a sensitive, highly resolved and accurate measurement. During this time all the signals from

the ToF analyzer are added up, so a 0.05 second cycle of MS could be made up of the summation of several hundred to more than a thousand individual ToF measurements. If then a certain ion is chosen for fragmentation the instrument again needs to spend a certain time acquiring data for this, (say) 0.2 s, or several thousand ToF measurements. So for weak ions only a handful of these thousands of ToF measurements will actually contain a signal. The huge number of ToF analyses summed up for the read-out (together with low-noise digital detectors) allows a higher dynamic range than is typically achieved in MALDI ToF analyzers. In most proteomic applications cycle times of 0.1–0.4 s will be realistic and will deliver good data, while most q-ToFs will be capable of even shorter cycle times. With its high sensitivity and short cycle times, q-ToFs are a good match to the short peak elution times in uHPLC applications. Cycle times can be seriously affected if the collision gas in the second quadrupole is to be removed or introduced. However, the presence of gas in this quadrupole does not adversely affect MS measurements in the ToF analyzer, on the contrary – gentle collisions of the ions in this quadrupole can 'cool' the ions down, result in a more uniform distribution of energy (and thus speed) and minimize the spatial spread of the ion beam, resulting in better focused ions, higher resolution, higher sensitivity (as more ions contribute to the peak) and accuracy. All this is true at low acceleration voltage between Q1 and Q2 (typically below 10 V), while the switch to MS/MS measurements and CID just requires an increase in the potential between Q1 and Q2 (to perhaps 30–50 V), allowing fast switching between MS and MS/MS mode without the need to evacuate the collision gas out of Q2.

The efficiency of the transfer between the stream of ions and the orthogonal ToF can in theory be close to 100% if the first mass analyzer is an ion trap instead of a quadrupole; all ejected ions could then be measured in the ToF analyzer (see the paragraph on MALDI quadrupole ion trap instruments in Section 'Other hybrid mass spectrometers'). The q-ToF was also one of the first commercial non-ToF instruments to be equipped in large numbers with a MALDI source; the timed, discontinuous generation of ions is a perfect match for a ToF (see Section 3.2.1). However, development of orthogonal MALDI source technology allowed ion traps, FTICR and quadrupole instruments to be equipped with very efficient MALDI sources (Figure 3.39).

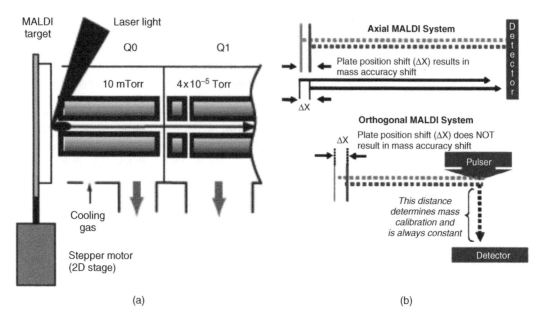

(a)   (b)

Figure 3.39 Orthogonal MALDI ToF. (a) Orthogonal MALDI ToF is very different from high vacuum MALDI, as the source contains gas and the ions are focused in time and in their energy spread by quadrupoles. This can result in a near constant ion stream. (b) Orthogonal MALDI is not prone to inaccuracy or loss of resolution due to differences in the sample's x-axis, be it due do differences in the target plate at various positions or the 3D structure of the sample preparation (crystals). Speed (energy) distributions along the x-axis as a result of the violent MALDI explosion are also less problematic.

Orthogonal ToFs as used in q-ToFs have advantages over typical MALDI ToF instruments with their axial geometry. These advantages come from uncoupling ionization and mass analysis. First of all, it is easier to achieve high mass accuracy with real samples as the changes in LASER intensity do not cause problems with resolution, detector saturation and calibration. There are no raised crystals that would influence the spatial distribution of ions prior to the ToF stage. There is no negative influence of the 3D topology of the sample or the sample plate to take into account. In axial MALDIs some form of DE has to be used to achieve high mass accuracy, as the MALDI process leaves the ions with a wide energy, speed and spatial spread. No DE is needed for high mass accuracy and to counteract the effects of the MALDI process in orthogonal ToFs, as the beam of ions is well defined. All these advantages come from the fact that the ionization is uncoupled from the ToF stage. The ions undergo multiple collisions in an intermediate pressure environment (usually N_2), which cools the ions down. At the same time they are accelerated gently away from the MALDI source (e.g. with a field of 400 V) into

a quadrupole, which is used as ion guide. Inside this quadrupole the ions undergo further gentle collisions, cooling them down further. All these processes together spread the 'packet' of ions resulting from the nanosecond LASER pulse into a stream of several milliseconds; at high LASER repetition rates a nearly constant stream of ions is thus generated. The DE mechanisms in axial MALDIs always have an optimal mass region; ions outside this region are detected less sensitively, and DE makes a multipoint calibration curve necessary. In a q-ToF the quadrupoles need calibrating once a year (they are very small and stable, with comparatively low accuracy of about 500 ppm) and the ToF stage can be easily calibrated with only one or two masses. Just as in a MALDI ToF, the calibration of the ToF stage of the q-ToF is affected by temperature and drifts considerably from day to day (typically 150 ppm) as well during the usage of the instrument (50 ppm over several hours) as the components warm up. Typically the drift in mass due to temperature changes is around 100 ppm per degree Celsius and can be automatically compensated to about 10 ppm/°C. However, recalibrating a measurement with known ions during or after measurements is easy enough, as there is no DE, and 5–10 ppm over the entire mass range in MS and MS/MS mode can routinely be achieved, whereas high end axial MALDIs struggle with 10 ppm for real-world samples.

Other hybrid mass spectrometers

It is impossible in a book like this to discuss all the hybrid mass spectrometers available and in use today for proteomic analyses. Here we will do no more than glance at the most popular instruments. This does not reflect the usefulness of the instruments for proteomics, but a combination of complexity, space considerations, the novelty of the instruments and sometimes the complexity of operation or the sheer cost of purchase that makes it unlikely that beginners in proteomics will get access to these instruments.

MALDI quadrupole ion trap instruments

Historically this set of commercial hybrid instruments began with an instrument that is a combination of MALDI source with a quadrupole IT and that uses a reflectron ToF as final mass analyzer. However, additional commercial instruments are now available that combine MALDI with a linear quadrupole IT or with a linear quadrupole IT and an orbitrap as final analyzer. These quite unique combinations allow sensitive offline measurement with MS^n capabilities. The strength of this approach for proteomics lies in the ability to analyze complex modifications such as glycosylations in the sensitive offline mode. Obviously it is challenging to 'catch' all the ions from a MALDI source at high efficiency and to eject the ions in an efficient way with high temporal, spatial and energetic homogeneity into the acceleration stage of the ToF analyzer. The offline mode of MALDI allows the experimenter to perform some data analysis during acquisition; an informed choice can then be made for the next step, be it the MS^2 of a certain molecular ion or the MS^3 (or higher) of ions that contain complex modification such as branched chains of glycosylations. The sample consumption is minimal, as the MALDI is only operated when sample is needed, while the ToF analyzer allows for relatively high resolution and accuracy of the results. Using the orbitrap as final analyzer for MS and tandem MS allows sensitive measurements with resolutions in excess

of 50 000 and accuracies of below 1 ppm – levels difficult to achieve with axial MALDI ToF instruments. These instruments are rather new at the time of writing, but it is clear that this kind of performance will be beneficial for many proteomic applications, including identification of proteins from gels and imaging mass spectrometry (Rietschel *et al.*, 2009).

Q-Trap

This instrument is the combination of ESI quadrupole-quadrupole followed by an LIT (Figure 3.32, although the Q-Trap has axial ejection). It allows the use of the unique scanning modes of the triple quadrupole with the higher sensitivity and resolution achieved by an ion trap. The instruments can run either as triple quadrupoles or with the third quadrupole used as an LIT with axial ejection. These configurations can be changed during an ongoing measurement in a data dependent fashion, so that complex measurements can be made, for example precursor ion scan with the emphasis on identifying PTMs (using the third quadrupole as just that) alternating with product ion scans to identify the PTM-carrying peptide (using the third quadrupole as LIT). For best results the ion trap is not used for MS/MS fragmentations; these are induced by CID in the second quadrupole. This allows the trapping of fragment ions within a wide mass range. In typical (2D or 3D) quadrupole ion traps the smallest fragment that can be trapped must not be smaller than about three to four times the size of the largest fragment. This limitation arises from the need to excite the parent ion while trapping the fragments at the same time. For measurement of iTRAQ quantifications, which produce ions around 100 Da (always with a single charge) for quantification purposes, the largest peptides that can be analyzed are around 400 Th in size in quadrupole ion traps, and using different excitation modes even the latest ion traps cannot go much beyond 600 Da. The Q-Trap does not have this limitation. It also produces ions by CID, so the fragments are generated in a 'harder' way than in ion traps, which can be a blessing (for small fragments) or a curse for peptide identifications. The trapping efficiency of the LIT is enhanced due to collision and thus diminishing energy of the ion in the entry axis of the LIT. The Q Trap is also capable of MS^3, the second MS/MS event taking place in the axial ejection LIT. All classical triple quadrupole MS/MS modes are possible, and the sensitivity of the product ion scan is enhanced significantly as the trap has less dead time than a scanning quadrupole. The dynamic range of the ion trap is enhanced as the first two quadrupoles make sure that only ions of interest fill the trap (down to only one specific peptide ion); the limited number of ions that can 'fit' in a 'stand-alone' quadrupole trap is always taken by chemical noise. In most proteomic applications there will be more contaminants than ions of interest in the sample at any given time.

Interestingly, ions can be trapped before the first quadrupole in a guidance quadrupole called Q0. This trapping is performed while the third stage (used as an LIT) is expelling/measuring ions. The measurement step is the longest step, particularly in an axial ejection LIT, and particularly during high resolution measurements, because this means prolonged dead times. Trapping before the first quadrupole is possible in all measurement modes where the first quadrupole is not scanning, that is, it is either transmitting all ions or just a certain one. Applications using these unique modes are well suited for the analysis of multiple PTMs, as we will see in Chapter 5, allowing the sensitive detection/quantification of hundred of phosphorylation sites using MRM (Section 5.4).

Hybrid FTICR instruments

FTICR analyzers offer unique potential for high resolution, accuracy and sensitivity. They can be used to perform MS/MS experiments, but this might interfere with consequent measurements as, for example, CID needs gas and high performance ICR measurements need ultra-high vacuum. The longer the ions are measured in the ICR cell, the better the sensitivity, resolution and accuracy. So to get the best out of this analyzer it is a great advantage to uncouple fragmentation and measurement, using quadrupoles or ion traps before the final analysis is performed in the ICR cell. Such combined instruments also offer the advantage that more than one measurement can be carried out simultaneously; another ion trap can perform one or several scans while the ICR cell is measuring parts of the sample at high resolution (which may take easily more than a second). Mass analyzers in front of the ICR cell also allow a more efficient filling of the cell (which only has a limited ion capacity); only ions of interest are let into the cells and chemical noise or unnecessary ions are kept out, thus allowing a higher sensitivity, higher dynamic range and shorter measurement times. The biggest advantage must be that ion fragmentation is moved outside the ICR cell. The most successful fragmentation method for proteomic applications is CID. However, CID depends on the presence of gas molecules and ICR measurements need ultra-high vacuum. If CID is performed before the fragments are transferred into the ICR cell a relatively short cycle time of 1 second can be achieved. FTICR instruments are very large, rather complicated and depend on very strong magnets. The addition of other mass analyzers does not add significantly to the bulk of these instruments and adds only a small fraction to the overall cost. This addition actually makes them simpler to use, as part of the measurements and fragmentations can be performed independently of the FTICR operation. Commercially successful hybrid instruments with FTICR analyzers are combinations of LITs or tandem quadrupoles (Figure 3.40). The ultra-high resolution and sensitivity of FTICR MS are somewhat counteracted by the lack of speed for proteomic analyses, as a high quality measurement in the ICR cell takes more than a second. However, new measurement strategies are possible because of the performance available; at such high accuracy (around 1 ppm in practical measurements) MS/MS is not always necessary to identify a single peptide. The accurate mass tag approach is a strategy in which the identity of a peptide is derived from its very accurate mass measurement and its retention time in LC. With this information alone proteomics analyses can be performed with a limited amount of sample (see Chapter 5). A high accuracy FTICR MS scan lasting 2 seconds can thus identify more peptides from a limited amount of sample during an LC run than the typical combination of survey scan followed by three to eight MS/MS scans, as at low sample concentration these steps take longer than 2 seconds with standard mass analyzers. Also, by avoiding MS/MS at all, the loss of signal intensity is prevented that comes from dividing the intensity of one parent ions onto many fragment ions. Some hybrid FTICR MS instruments can generate data from both mass analyzers in parallel, for example ion trap MS/MS data and a high resolution/accuracy MS of the ions used for MS/MS. This approach leads to higher confidence data in measurement times that are compatible with LC MS separations and tries to combine the best of both worlds; sensitive but not very accurate MS/MS data combined with ultra-precise measurements of the masses of the molecular ions.

The high resolution and accuracy as well as sensitivity offered by FT MS are perfect for the requirements of

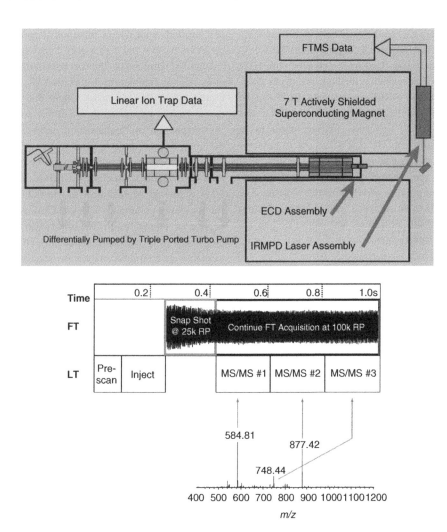

Figure 3.40 Hybrid FTICR mass spectrometers. The QIT FTICR is in fact built out of two mass spectrometers, which can deliver data in parallel measurements. The linear ion trap (QIT) in the middle of the sketch performs a quick pre-scan to make sure an optimal amount of ions is used for the next stages (see lower panel). The non-destructive nature of the FTICR data acquisition allows an overview to be produced at lower resolution, which is than used to direct the MS/MS capability of the QIT. The QIT can thus perform three independent MS/MS experiments on different peptide ions eluting from an LC run, while the FT is still acquiring data for high resolution and high accuracy MS of the eluted peptides. Thus after about 1s of measuring time, three peptide data sets with very high MS accuracy (around 1–2 ppm) and very good MS/MS patterns from the QIT are produced, allowing high significance identifications (see also Figure 5.3). The QIT FTICR can be used in a variety of different schedules, involving ECD or infrared multiphoton dissociation fragmentations or extended FT measurements with ultra-high resolutions (greater than 300 000). Reproduced with permission from articlesFile_754 at http://www.thermo.com/com/cda/product/detail/1,22172,00.html. © Thermo Fisher Scientific Inc. All rights reserved.

top-down proteomic approaches; here complete proteins are analyzed, usually from more or less highly enriched samples, and the lack of sub-second analysis time is not a limitation. It is more of a limitation that the resolution goes down with increasing m/z values. As with

other hybrids, the combination with quadrupole analyzers imposes limits on the m/z range that can be injected into to ICR cell.

FT MS (i.e. FTICR and orbitrap) is the instrumentation of choice for top-down proteomics of proteins larger than about 30 kDa, due to the extremely high resolution. As the best results for protein fragmentation (best coverage of all amino acid bonds) are achieved with ETD and ECD, these fragmentation methods are made available in many hybrid FTICR instruments. ECD can be performed directly in the ICR cell, while for ETD a special source before the first mass analyzer needs to be built into the instrument. The ICR cell is one of the view mass analyzers that can be used for ECD directly, as it allows the low energy electrons to be trapped for sufficient times along the positively charged proteins. The instrument needs to be equipped with a special electron emitter (Figure 3.40). Fragments and molecular ions are stored together with the electrons in the ICR cell during fragmentation, which takes around 5–20 ms to allow sufficient numbers of events to occur. ETD needs a special nCl source right after the ions enter the instruments. Here reactive nCl is combined with the analyte ions and transfers its single electrons. ETD reactions take 20–200 ms to create sufficient fragmentation events. The MS measurements have to be performed discontinuously, as the sample and the ETD reagents need to be brought together from different flight paths in the instruments. ETD and ECD can deliver backbone fragmentations in proteins that can cover between 60% and over 80% of all peptide bonds. Together with ultra-high FT MS resolution, this is often enough to assign particular modifications to the purified proteins, as well as identify the exact amino acid at which PTMs are present. FT MS is also ideally suited to analyze protein/protein interactions by ESI MS. The advantage of FTICR MS over the orbitrap is that ions can be stored for prolonged time without loss. So diverse gas phase chemistry experiments can be performed to study 3D interaction either on the same protein or between different proteins by, for example, deuterium exchange reactions. In these experiments it is observed how deuterium from one position is exchanged intra- or intermolecularly over time to different positions, while the ions are in the ICR cell, taking full advantage of the non-destruction mass analysis in ICR cells.

Orbitrap hybrid instruments

When orbitraps became commercially available they were actually only available as hybrid instruments. A commercial instrument with a 'pure' orbitrap analyzer became available only recently for ESI MS with ultra-high performance. Orbitraps are available in a variety of combinations with one or more ion traps and additional octapoles for low and high energy collisional dissociation reactions (Figure 3.41). Orbitrap hybrid instruments thus offer several fragmentation mechanisms and avoid the limitations in the low mass range, due to limitations of the LIT to trap small ions together with large ions. This expansion of fragmentation capabilities allows at the same time a sensitive identification of peptides (via CID) and a reliable quantification using small fragment ions with iTRAQ. Similarly to hybrid ICR analyzers, ETD is offered as additional fragmentation mechanism for top-down proteomics and the PTM analyses at the peptide or protein level. The somewhat more efficient and faster ECD mechanism used in FTICR MS is not available, as orbitraps cannot store negative and positive ions at the same time.

With their ease of use, analysis-speed compatible with LC MS and high accuracy, IT orbitrap hybrids offer superior performance to ion traps in terms of reproducibility, if the instruments are kept tuned and used in the best possible way. Increasing the accuracy from 50 ppm for the precursor ion mass to around 1–2 ppm usually leads to more peptides in complex LC MS run being identified correctly. Very significantly, the reliability of the identifications increases and the grey zone between 'shaky' or uncertain identifications and highly reliable database hits becomes considerably narrower (see also Chapters 4 and 5). To keep the duty times high during LC MS separations, the orbitrap can perform high resolution MS scans while the hybrid ion trap used to transmit ions into the orbitrap can perform 5–20 MS/MS scans (Figure 3.41).

This mode allows high resolution and high accuracy MS scans combined with MS/MS scans of the three to eight most abundant ions in cycle times below 2 seconds. The enhanced accuracy offered by the FT MS scan is especially helpful in reliably detecting PTMs like phosphorylations in automated surveys, where CID can be used as first MS/MS and higher energy collision dissociation as 'quasi' MS[3], for example if the MS/MS results show fragments indicative of PTMs.

While FTICR instruments can outperform orbitraps in terms of resolution for ions below 1 000 Th, the situation changes with larger m/z, and above about 2 500 m/z the orbitrap performs as well as if not even better than ICR analyzers in terms of resolution and sensitivity.

Although hybrid orbitraps have only been commercially available for a couple of years they have already

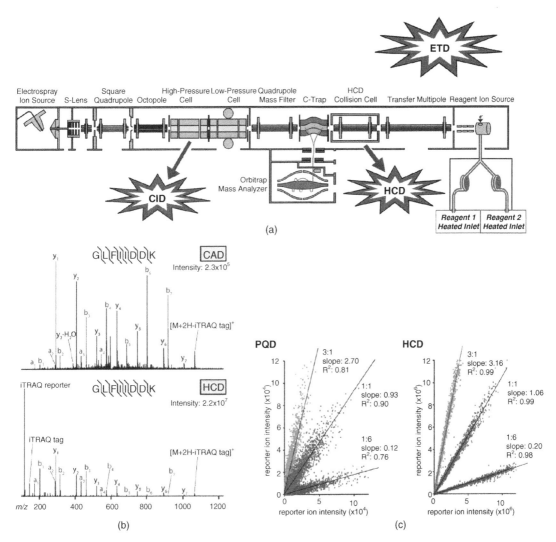

(a)

(b)

(c)

Figure 3.41 Orbitrap hybrid mass spectrometers. The principal components of a hybrid orbitrap in combination with a dual pressure LIT (see Section 'Linear ion traps') are shown in (a). The higher pressure LIT is used for CID. The instrument has an additional C-Trap, used for controlled injection of ions into the orbitrap analyzer, and a quadrupole collision cell, used for CID under typical quadrupole conditions, which transfer more energy than ion trap CID, hence it is named higher energy collision dissociation (HCD) here. In addition ETD can be performed, using ion generators (towards the right-hand end of the sketch). The instrument has various modes of operation. As an example, ions are passed from the left-hand ESI source to the LIT. Here a short pre-scan measures the ion content for optimal filling of the orbitrap. Now the linear trap is used to feed into the C-Trap. This trap injects the ions into the orbitrap for an initial low resolution scan. The results are used to direct several consecutive CID MS/MS measurements in the LIT, independent of the ongoing high resolution scan in the orbitrap. Once the orbitrap is finished with the high resolution scan, ions from the LIT can be directed into the C-Trap, and from here into the HCD collision cell. The fragments of this collision (without the loss of small ions typical of ion trap CID) can then again be transferred to the C-Trap and into the orbitrap for analysis. This ion ping-pong makes the instrument very versatile and allows high throughput and accuracy as well as the measurement of iTRAQ reporter ions (b). LITs allow a special excitation protocol that circumvents the 1/3 rule for low mass ion loss (see Section 3.3.3). This PQD process cannot compete with HCD on the hybrid orbitrap, which delivers a 100 times higher intensity for the iTRAQ reporter ion (c). The quantifications of iTRAQ test-samples derived from HCD data (using the orbitrap as mass analyzer) are much closer to the expected signal intensities than the data from PQD measurements on the same instruments, allowing more reliable quantifications. (a) Olsen *et al.* (2009), Figure 1. (b), (c) Reprinted with permission from McAlister *et al.* (2010). © 2010 American Chemical Society.

Table 3.1 Overview of some performance parameters of mass spectrometers used in proteomics. Specifications of instruments are often higher than values in this table, which reflects performance under optimal conditions with real proteomic samples. All values are just for comparison and might be different under certain measurement conditions (scan speeds, sample preparation methods, etc.).

Mass analyzer	Resolution	Accuracy (ppm)	Speed cycle/s	Dynamic range	m/z range (Th)	Comments
ToF lin	5 000	circa 200	10–1 000	$<10^3$	700–10^6	No MS/MS! good for proteins, large molecules
ToF ref & ToF/ToF	15–40 000	5–50	10–1 000	10^3	700–20 000	Precursor selection resolution from 100 (ToF ref) to 1 000 (ToF/ToF)
Quad	2 000	500	0.7–4	$>10^5$	100–4 000	Precursor selection resolution 1 000, several unique MS/MS modes, e.g. for PTM analysis
Quad IT	4 000–25 000	20–200	0.7–10	$<10^5$	50–4 000	MS^n, very informative MS/MS data for proteomics
q-ToF	10–40 000	2–10	2–20	10^4	50–12 000	MS/MS only below 4 000 Th
Orbitrap (IT hybrid)	15–100 000	1–5	0.5–4	$<10^5$	50–4 000	Performance dependent on analysis time
FTICR MS	20 000–1 000 000	<1–3	0.1–4	$<10^5$	50–200 000	Performance dependent on analysis time, very strong magnet required, resolution strongly m/z dependent

proven their value for proteomic applications and new developments will enhance performance and usability even further in the future, allowing high performance FT MS to be used for more proteomic applications. They have in recent years set a new standard for mass spectrometry performance in proteomic applications (see the examples in Chapter 5).

3.4 CONCLUDING REMARKS ON MASS ANALYZERS FOR PROTEOMICS

From what we have seen in this chapter, mass spectrometry instrumentation and data acquisition strategies are rapidly developing fields. Mass spectrometry is a prime example of how proteomics is driven and enabled by technology. At the same time the field is diversifying more and more, as different approaches have different needs for data generation in terms of speed, resolution, accuracy, dynamic range and reproducibility. This is matched by different data analysis strategies (see Chapter 4), and it is very difficult to directly compare the capabilities of different approaches to answer biological questions (see Chapter 5). It is thus impossible to summarize these performances into simple numbers. Table 3.1 is meant simply as a reminder of some of the performance indicators of the previously discussed instruments.

Inevitably one ends up comparing apples and pears, as the different analyzers have different operational

requirements. For the performance real samples are considered, and instrument specifications with standards are usually higher. Sensitivity is a major issue and is addressed in the description of the different analyzers. As sensitivity id dependent on operational parameters such as ionization efficiency or resolution and cycle time settings or sample quality and operational mode (e.g. triple quadrupoles are hundreds of times more sensitive for MRM than for scanning applications), it is not considered here. All mass analyzers are capable of working with samples in low femtomole amounts, most even below this.

From the sheer length of this chapter it is clear that there is no optimal mass spectrometer, ionization method or even fragmentation principle; it is strictly horses for courses, and it is important to identify a good strategy starting with a biological question, a good biological system to work with and then a good MS data acquisition and data analysis strategy.

REFERENCES

Barker, J. (1999) Mass spectrometry, Wiley.

Baumgaertel, A., Becer, C.R., Gottschaldt, M. and Schubert, U.S. (2008) MALDI-TOF MS coupled with collision-induced dissociation (CID) measurements of poly(methyl methacrylate). *Macromol Rapid Commun*, **29**, 1309–1315.

Bogdanov, B. and Smith, R.D. (2005) Proteomics by FTICR mass spectrometry; top down and bottom up. *Mass Spectrum Rev*, **24**, 168–200.

Bondarenko, P.V., Second, T.P., Zabrouskov, V., Makarov, A.A. and Zhang, Z. (2009) Mass measurement and top-down HPLC/MS analysis of intact monoclonal antibodies on a hybrid linear quadrupole ion trap-orbitrap mass spectrometer. *J Am Soc Mass Spectrom*, **20**, 1415–1424.

Chernushevich, I.V., Loboda, A.V. and Thomson, B.A. (2001) An introduction to quadrupole-time-of-flight mass spectrometry. *J Mass Spectrom*, **36**, 849–865.

Chhabil, D. (2001) *Principles and Practice of Biological Mass Spectrometry*, Willey.

Cotter, R.J., Grifith, W. and Jelinek, C. (2007) Tandem time-of -flight (ToF/ToF) mass spectrometry and the curved-field reflectron. *J Chromatogr B*, **855**, 2–13.

Desiderio, D.M., (Ed) (1991) *Mass Spectrometry of Peptides*, CRC Press.

Elias, J.E., Haas, W., Faherty, B.K. and Gygi, S.P. (2005) Comparative evaluation of mass spectrometry platforms used in large-scale proteomics. *Nature Meth*, **2** (9), 667–675.

Fan, X. and Murray, K.K. (2010) Wavelength and time-resolved imaging of material ejection in infrared matrix-assisted laser desorption. *J Phys Chem A*, **114** (3), 1492–1497.

Flensburg, J., Haid, D., Blomberg, J., Bielawski, J. and Ivansson, D. (2004) Applications and performance of a MADI-TOF mass spectrometer with quadratic field reflection technology. *J Biochem Biophys Methods*, **60**, 319–334.

Hager, J.W. (2004) QTRAP™ mass spectrometer technology for proteomics applications. *Targets*, **3** (2), S31–S36.

Gross, J.H. (2004) Mass Spectrometry: A Textbook, Springer.

Hardman, M. and Makarov, A.A. (2003) Interfacing the orbitrap mass analyzer to an electrospray ion source. *Anal Chem*, **75**, 1699–1705.

Hillenkamp, F. and Karas, M. (2007) The MALDI process and method. In MALDI MS: A Guide to Instrumentation, Methods and Applications, (eds. F. Hillenkamp and J. Peter-Katalinić), Wiley-VCH Verlag GmbH & Co. KGaA.

de Hoffman, E. and Stroobant, V. (2007) Mass Spectrometry: Principles and Applications, 3rd edn, John Wiley & Sons, Inc.

James, P. (Ed.) (2001) Proteome Research: Mass Spectrometry-Principles and Practice, Springer.

Jonscher, K.R. and Yates, J.R. III (1997) The quadrupole ion trap mass spectrometer- A small solution for a big challenge. *Anal Biochem*, **244**, 1–15.

Krutchinsky, A.N., Loboda, A.V., Bromirski, M., Standing, K.G. and Ens, W. (1998) A MALDI ion source with collisional cooling for orthogonal-injection TOF. Department of Physics, University of Manitoba, Winnipeg, Canada R3T 2N2. http://www.physics.umanitoba.ca/~ens/Krutchinsky.pdf.

Kussmann, M., Nordhoff, E., Rahbek-Nielsen, H., Haebel, S., Rossel-Larsen, M., *et al.* (1997) Matrix-assisted laser desorption/ionization mass spectrometry sample preparation

techniques designed for various peptide and protein analytes. *J. Mass Spectrom*, **32**, 593–601.

March, R.E. (1997) An introduction to quadrupole ion trap mass spectrometry. *J Mass Spectrometry*, **32**, 351–369.

McAlister, G.C., Phanstiel, D., Wenger, C.D., Lee, M.V. and Coon, J.J. (2010) Analysis of tandem mass spectra by FTMS for improved large-scale proteomics with superior protein quantification. *Anal Chem*, **82** (1), 316–322.

Moskovets, E., Preisler, J., Chen, H.S., Rejtar, T., Andreev, V. and Karger, B.L. (2006) High-throughput axial MALDI-TOF MS using a 2-kHz repetition rate laser. *Anal Chem*, **78**, 912–919.

Olsen, J.V., Schwartz J.C., Griep-Raming, J. *et al.* (2009) A dual pressure linear ion trap Orbitrap instrument with very high sequencing speed. *Mol Cell Proteomics*, **8** (12), 2759–2769.

Perry, R.H., Cooks, R.H. and Noll, R.J., (2008) Orbitrap mass spectrometry: instrumentation, ion motion and applications. *Mass Spectrom Rev*, **27**, 661–699.

Rietschel, B., Baeumlisberger, D., Arrey, T.N. *et al.* (2009) The benefit of combining nLC-MALDI-ToF/Orbitrap MS data with nLC-MALDI-ToF/ToF data for proteomic analyses employing elastase, *J Proteome Res*, **8**, 5317–5334.

Schwartz, J.C. and Senko, M.W. (2002) A two-dimensional quadrupole ion trap mass spectrometer. *J Am Soc Mass Spectrom*, **13**, 659–669.

Simpson, R.J. (2003) Proteins and Proteomic, a user manual, CSH Laboratory Press.

Suckau, D., Resemann, A., Schuerenberg, M., Hufnagel, P., Franzen, J. and Holle, A. (2003) A novel MALDI LIFT-TOF/TOF mass spectrometer for proteomics. *Anal Bioanal Chem*, **376**, 952–965.

Takats, Z., Wiseman, J.M., Gologan, B. and Cooks, R.G. (2004) Mass spectrometry sampling under ambient conditions with desorption electrospray ionization. *Science*, **306**, 471–473.

Tang, N., Perkins, P., Miller, C. and van de Goor, T. (2007) Protein phosphorylation sites determination using a microfluidic chip interfaced with ETD ion trap and Q-TOF mass spectrometry. Poster P03-59-M at MSB-2007. Agilent Technologies, 5302 Stevens Creek, Santa Clara, CA 95052, USA. http://bit.ly/dDn4nb.

Wilm, M., Neubauer, G. and Mann, M. (1996) Parent ion scans of unseparated peptide mixtures. *Anal Chem*, **68**, 527–533.

Wuerker, R.F., Shelton, H. and Langmuir, R.V. (1959) Electrodynamic containment of charged particles. *J Appl Phys*, **30** (3), 342–349.

Yates, J.R., Cociorva, D., Liao, L. and Zabrouskov, V. (2006) Performance of a linear ion trap-hybrid for peptide analysis. *Anal Chem*, **78**, 493–500.

Zubarev, R.A., (2004) Electron-capture dissociation tandem mass spectrometry. *Curr Opin Biotech*, **15**, 12–16.

4

Analysis and Interpretation of Mass Spectrometric and Proteomic Data

4.1 INTRODUCTION

Data analysis is one of the most important aspects of any proteomic study and can be expected to take much longer than the actual experiments or the data acquisition. In fact, the capacity to study proteomics at the experimental level is much more advanced than the ability to analyze the data completely in a proteomic sense. This is true for gel-based approaches at the level of image analysis (Section 2.3.2), applies even more so for gel-free approaches at the level of mass spectrometric data analysis, and is certainly most obvious in making biological sense of proteomic results. One can distinguish between different levels of data analysis, for example a basic level and an advanced level, with a considerable overlap between them. The basic level of data analysis includes the interpretation and analysis of mass spectrometric data and the identification and analysis of peptides and proteins with the help of such data (or peptide and protein informatics as it applies for proteomics). At the advanced level the already interpreted data are stored in dedicated databases where they can be analyzed and related to other data (e.g. in the same or other public databases). We will refer to the latter as bioinformatics as it applies to proteomics, for example the analysis of protein function based on similarities to known proteins, or automated deciphering of which signalling pathways are affected by differentially expressed/modified proteins using cluster analyses or automated text mining for annotations.

There is a distinction to be made in the way proteomic methods are used; most research is undertaken using proteomics-related methods, but is not performed on a proteomic scale. For example, one can analyze the proteins of cerebrospinal fluid in a case study, by using 2D high pressure/performance liquid chromatography (HPLC) coupled with online electrospray ionization (ESI) MS/MS and identify several thousands of proteins in the process. On the same level of a case study, one could equally well use the same cerebrospinal fluid from healthy animals with those that had some form of trauma, run the samples on 2D gels and compare them. By comparison with some other data both these approaches can deliver important and valuable clues about certain proteins and their relation to the well-being of our most complex organ, the brain, even without the use of extensive bioinformatic analyses. To transform this research into proper proteomic studies, they would have to be performed much more rigorously, optimizing the potential to deal with most proteins in the sample. This would involve statistically relevant sample sizes and, for example, up to 10 independent 2D LC MS runs from every sample. Evaluations would have to be made of how much chance there was of catching small changes, or changes of less abundant proteins (especially for LC MS/MS based quantifications), and it would have to be ensured that the analysis was performed under circumstances that allow direct comparisons with data from other labs (best in international, multi-site studies). The difference between using proteomics tools and working on a proteomic scale is not only the amount of work put into such studies; it will mainly be the amount of data analysis performed at the bioinformatic level and the documentation necessary for this. Every aspect of a proteomic study needs very detailed advance planning with data analysis in mind. If the main data

Introducing Proteomics: from Concepts to Sample Separation, Mass Spectrometry and Data Analysis, Josip Lovrić
© 2011 John Wiley & Sons Ltd

from proteomic studies are to be made publicly available in standard formats agreed by international bodies, publication in a journal alone is not sufficient (Taylor *et al.*, 2007; Gloriam *et al.*, 2010). Dedicated databases with easily accessible user interfaces are needed to make the data available; a publication in a printed research paper can only describe the 'highlights' of the study. Only with this effort can proteomic information be made available to other researchers in the field and provide data, for example, for clinical researchers trying to develop new diagnostic tools.

A basic data analysis is needed for a case study as well as for a proteomic study, but the proteomic study will involve a lot more bioinformatics. When it comes to returns, many questions can be answered with the help of a case study; often a full-blown proteomic study is not even justified before we know much more about a particular subject through one or more case studies using proteomic approaches. We will in this book mainly discuss basic data interpretation and only very briefly outline data analysis requirements for studies on a truly proteomic scale.

There are different objectives for mass spectrometry data analysis in proteomics. Not all analyses have the same targets; some studies are carried out to get a quantification of as many proteins in the proteome as possible, others get qualitative data for comparisons. Some studies might want to quantify a subset of well known proteins, others focus entirely on detection or even quantification of posttranslational modifications (PTMs). All these questions can be addressed at the peptide level, or at the level or proteins. They can be part of strategies in which the single protein is (more or less) pure before the MS analysis or based on complex mixtures being presented to the mass spectrometer. Listed roughly in the order in which the data analysis proceeds (and in order of complexity), we may want to:

1. create a mass list of peptides or proteins present in the sample;
2. correlate the elution times from LC with the mass spectra;
3. create a list of fragments from the peptides or proteins identified under 1;
4. analyze which database entry corresponds (matches best) to an MS data set (peptide mass fingerprint, PMF);
5. identify possible modifications of peptides and database conflicts from MS data;
6. quantify single proteins from purified samples;

7. analyze which peptides in database entries correspond best to a set of MS and MS/MS data from a peptide;
8. identify possible modifications and database conflicts from our MS and MS/MS data;
9. identify which proteins are present in a complex mixture based on MS and MS/MS data;
10. quantify the proteins present in our sample from a complex mixture (perhaps to establish candidates for biomarkers);
11. establish the modifications present on the protein level.

The next steps in the analysis are based on the results of previous steps, but only deal with heavily interpreted data and are clearly in the realms of bioinformatics.

12. compile a complete list of all components in a complex system (e.g. serum or single- cell organism cell line or tissue or even whole multicellular organism under different conditions), incorporating information from different sources and technologies;
13. make qualitative/quantitative comparisons of the composition of complex systems (see 12);
14. list all possible PTMs of interest compared to which ones are actually observed under certain conditions (e.g. phosphorylation during mitosis, or inflammation).
15. examine the interaction of single proteins or all proteins of a cell/species (interactom) in different conditions;
16. identify candidates for biomarkers rationalized from data derived under 12–15.

The analysis goals 1–11 are based on single measurements and their interpretations, while 12–16 are examples of advanced analyses on truly proteomic scales. We will show examples of the latter in Chapter 5 but will not discuss the analysis requirements in detail in this book.

4.2 ANALYSIS OF MS DATA

The analysis of MS data in proteomics is often only the first step towards more specific MS/MS analyses. However, it is a very important first step that will introduce us to the more complex analyses. Although the higher specificity provided by MS/MS technologies is often desirable, it is not always the best way to analyze a sample; if time and sample are precious, often the simpler, more sensitive and considerably cheaper MS analysis, with its often higher resolution, can do the job on its own if the experimental strategy takes all appropriate factors into consideration.

4.2.1 Analysis of MS data for PMF

In high throughput studies most of the issues in this section dealing with MS data analysis will be dealt with automatically by a piece of silicone. However, to understand the potential problems with any of these settings and procedures it is a good idea to go over them manually with a sample of a known protein and optimize them for the specific laboratory and the practices/instruments used. We have seen some typical mass spectra in Chapter 3. The easiest mass spectra to interpret are those of matrix assisted LASER desorption/ionization (MALDI) based MS of peptides, as they are typically singly-charged. A typical context for taking such a spectrum is the identification of proteins from gels after tryptic digestion, using PMF, as introduced in Sections 2.1.1 and 3.1. A typical spectrum of a tryptic digest from a spot of a 2D SDS PAGE gel is seen in Figure 4.1.

The overview shows a dozen strong peaks, indicative of a good overall result. All instruments will offer automatic processing of spectra and peak recognition software that will determine the monoisotopic mass of each peptide and even discard peaks that are contaminants, such as trypsin (normally present as 50–150 ng per sample), or any mixture of keratins (which are strongly dependent on local factors such as the experimenter). The spectrum is usually processed, often so that the experimenter does not even notice; the spectrum is smoothed and the background is subtracted. This improves the accuracy of the measurement, if it is done correctly. With very weak spectra the settings for normal spectra can 'overdo' it, so it

is good to be aware of this, as weak peaks or shoulders in peaks might be erased in the processed spectrum but visible in the original, and it is advisable to adjust the peak processing parameters according to the sample quality. The next step is the peak picking, that is, determining the monoisotopic masses of all significant peptides in the sample. It takes about 2–10 minutes per sample to perform a manual peak picking, and depending on the software used the manual peak picking can easily be better than an automated one. Manual peak peaking may have the advantages of resolving peaks that are close in mass to each other (below 1 m/z) or in discarding peaks that are obviously not peptides but, for example, polymers (recognized as peak groups with characteristic/constant mass differences). Manual peak picking can also deal with detecting peaks that are 'riding' on one another (mass difference 1, 2 or 3 m/z) by choosing the correct monoisotopic masses as well as delivering masses for peptide signals with poor signal to noise ratio that are more accurate than those derived from automatic peak picking. In medium to high throughput analysis this step has to be automated. It is a good idea to compare the results of automatic peak picking with a manual approach, and adjust the parameters for the automatic picking accordingly. Typically the sample has been calibrated externally (see also Section 3.1.1). In most cases the software also allows the user to calibrate the sample internally based on know contaminants, which will increase the accuracy from about 150 ppm to about 50–10 ppm, depending on sample preparation method and quality of preparation/mass spectrometer. Calibrations can be performed at any stage after the measurement, should the need arise (e.g. calibration is obviously out, as seen on known contaminants with deviating masses). For every spectrum the question arises which of the peaks should

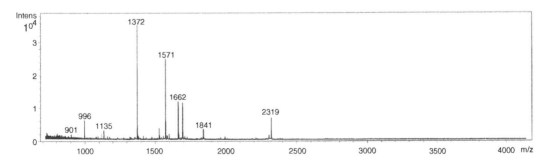

Figure 4.1 Peptide mass fingerprint with MS data of tryptic digest. See text for details. The generation of PMFs is outlined in Figure 2.15.

be picked for later PMF searches. The quality of peaks depends partially on their intensity or, better, their signal to noise ratio, thus peaks with low signal to noise ratio will be measured less accurately. Furthermore, every sample contains many unwanted contaminants (e.g. background proteins from the gels, smeared proteins, keratins, chemical noise), and the lower the signal intensity the greater the chance of picking up more of these signals. Therefore, it is a good idea to use a cut-off for either absolute intensity or the signal to noise ratio, or even just the number of peaks (e.g. choose the 20 most intense peaks) used for the database search. A search can easily be repeated with more peaks, but this first filtering allows the clear identifications to be separated from the non-significant ones.

After the best possible calibration and choosing the peptides, it is time for a database search. Some prior knowledge is important to make the search successful:

1. Is it known from which species the protein is derived?
2. Is it likely that (or known whether) there are any chemical modifications on the peptides?
3. What size was the original protein? What is its pI?
4. How accurate is my measurement?
5. Am I likely to have one or more proteins in my sample?

A number of observations are in order. Concerning point 2, PMF is (like many other MS based identification methods) error tolerant; if the sequence of the protein of interest is not known in the species that is analyzed, the homologue from a related species might be, so one would not identify the exact protein but a close relative (see later).

Again concerning 2, it is advisable to prevent oxidation by treating the sample with reducing agents followed by irreversible modifications of the reduced residues. This allows peptides containing Cys and Met to be detected in one or two peaks, rather than in a confusing and data-diluting multiplicity of unknown mass (e.g. multiple oxidations or adduct formation). Are acrylamide adducts observed regularly? Urea adducts?

Turning to 3, the protein size is just an estimate; the database entry might have a longer coding sequence, as the protein in question might be a fragment, or shorter, as the databases might not contain this protein's splice variant or the database entry might contain a truncated sequence. It is also important to remember that SDS

PAGE only delivers an estimate of the real molecular weight, variations of 10–20% are to be expected and PTMs can change this estimate considerably. Some proteins also just do not follow the marker protein separation at all and are consistently detected at masses as much as several hundred per cent out of the calculated values. The isoelectric point is more reliable, but it is changed dramatically by unpredictable modifications (e.g. at the N-terminus, PTMs).

With regard to 4, this is a parameter based on experience in this type of experiment in the particular laboratory: typical assumptions range from 50 to 150 ppm for MALDI ToF generated data. ESI quadrupole MS or ESI IT data might be out by several hundred parts per million and are usually not usable for PMF analyses.

Finally, regarding 5, some software is particularly written to take this problem into account (e.g. PepMAPPER; see Figure 4.2) and it might be a good idea to use several database search engines in case of ambiguous results.

Other search parameters usually asked for in the graphic user interface (GUI) of the search engine include the protease used, the numbers of missed cuts per peptide allowed, and additional information such as the species from which the protein is derived. The species can be often be chosen by limiting the database entries against which the search will be performed to a specific species or a higher evolutionary unit (e.g. mammals or primates). The search engines also allow the definition of a set of variable or fixed known chemical modifications of certain amino acids and a choice of mass value input formats (e.g. singly-/doubly-charged, theoretical masses M or measured masses $M + 1$, monoisotopic or average masses). These questions are likely to be asked in the process of data entry for the database search. Figure 4.2 shows some of the database search software (or search engines) publicly or commercially available.

Figure 4.3 shows the result of a typical reliable PMF identification. As PMF relies on statistical results, it is very important to be able to evaluate the search results. All search engines allow the experimenter such an evaluation in different ways. Typical parameters are the difference between masses of peptides expected and observed for this particular 'hit', experimental masses, the amount of trypsin cleavage sites present in the peptide sequence, indications of any PTM or artefactual modification, the 'sequence coverage' of the matching peptides and, most important of all, a score for the hit.

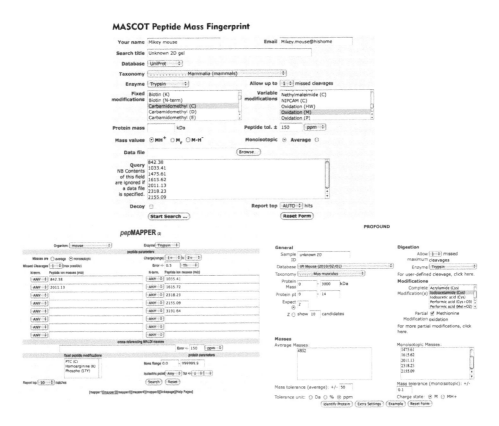

Figure 4.2   Using search engines for PMF. The screenshots of some popular search engines for PMF data – MASCOT, PepMAPPER and ProFound – show that similar parameters are needed to fill in the search forms. PepMAPPER has different versions, depending on how many proteins might be in the digest. ProFound has more additional settings and also asks how many proteins are expected. See details in text.

Ideally the score gives a very good indication of the reliability of the match between the experimental data and the database entry. This score is in turn related to the probability of such a hit occurring at random. This probability depends on various factors: accuracy of the measurement (or assumed accuracy), the number of peptide masses used for the search, the number of masses that actually match to the database entry and the number of amino acids in the matching database entry. The more peptide masses are used for a search, the more likely is a false positive hit with some of the many database entries, in particular if the accuracy is not high. The larger the database, the better the match must be to be a rare event and to allow a significant sore. The shorter a sequence in the database, the less likely is a false positive hit, as a random match with a long amino acid sequence which contains many tryptic peptides is more likely. The higher

the mass of a peptide matching a database entry, the more this particular peptide adds to the overall significance of the hit; at larger masses there are more (theoretical) sequences that could result in such a mass. However, not all theoretically possible sequences are present in the databases; the discrepancy grows the longer the peptide, so hits with larger peptides are more specific, as they occur rarer 'by chance'. Based on all these factors, a certain expectancy can be calculated for a match to occur; this is in turn related to the significance of a match. The combination of database size and expectancy for a hit to occur result in a score for a certain match. The score is the most important indicator for the quality of the result. Figure 4.4 shows some of the relationships between search parameters and scores. Some search engines deliver different scores, with different meanings, and it is important to familiarize yourself with the

{MATRIX} Mascot Search Results
{SCIENCE}

User : Mikey mouse
Email : Mikey.mouse@hishome
Search title : PMF
Database : UniProt 0 (6462751 sequences; 2097290313 residues)
Timestamp : 1 Apr 2010 at 15:30:23 GMT
Top Score : 103 for LAT_HUMAN, Linker for activation of T-cells family member 1 OS=Homo sapiens GN=LAT PE=1 SV=1

Probability Based Mowse Score

Protein score is -10*Log(P), where P is the probability that the observed match is a random event.
Protein scores greater than 81 are significant (p<0.05).

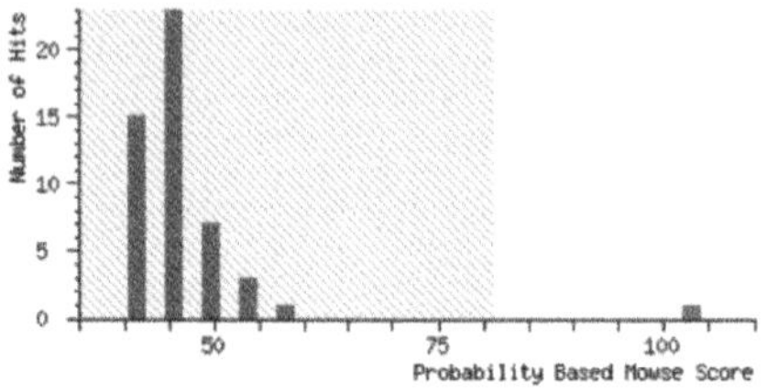

Protein Summary Report

[Format As] [Protein Summary ⬍] Help

  Significance threshold p< [0.05] Max. number of hits [AUTO]

(Re-Search All) (Search Unmatched)

Index

Accession	Mass	Score	Description
1. LAT_HUMAN	28140	103	Linker for activation of T-cells family member 1 OS=Homo sapiens GN=LAT PE=1 SV=1
2. B1VHG7_CORU7	25347	58	Putative uncharacterized protein OS=Corynebacterium urealyticum (strain ATCC 43042 / DSM 7109) GN=cu1248 PE=4 SV=1

Results List

1. LAT_HUMAN **Mass:** 28140 **Score:** 103 **Expect:** 0.00032 **Queries matched:** 6
  Linker for activation of T-cells family member 1 OS=Homo sapiens GN=LAT PE=1 SV=1

Observed	Mr(expt)	Mr(calc)	ppm	Start	End	Miss	Peptide
1615.7200	1614.7127	1614.7692	-34.96	120 - 134		0	R.GAQAGWGVWGPSWTR.L
1998.9000	1997.8927	1997.8562	18.3	100 - 119		0	R.DSDGANSVASYENEGASGIR.G
2155.0900	2154.0827	2153.9573	58.2	99 - 119		1	R.RDSDGANSVASYENEGASGIR.G
3191.6400	3190.6327	3190.3949	74.5	234 - 262		0	K.TEPAALSSQEAEEVEEEGAPDYENLQELN.-
3460.1800	3459.1727	3458.8503	93.2	53 - 83		0	K.RPHTVAPWPPAYPPVTSYPPLSQPDLLPIPR.S
3586.2100	3585.2027	3584.8963	85.5	1 - 31		0	-.MEEAILVPCVLGLLLLPILAMLMALCVHCHR.L

  No match to: 1035.4100, 1475.6100, 2398.2400, 3602.2700

Figure 4.3 Search results from PMF. Example of PMF search result with peptides from Figure 4.6. Search parameters were set as shown in Figure 4.2, except that the peptide tolerance (accuracy) was set to 100 ppm. Note that most of the identified peptides have a carboxy-terminal R (Arg), which is preferably detected by MALDI based MS over peptides ending with a Lys (K). One peptide is carboxy-terminal in the protein, and thus has no tryptic end. One peptide has an expected missed cleavage (amino terminal R—D bond was not cleaved, see Figure 4.5), as can be seen by the amino acid preceding the tryptic peptide, separated by a dot. Mr(expt) is a mass assigned to the experimental mass by MASCOT, and is used to make sure the mass falls into a certain mass 'bin'. Mr(calc) is the expected real mass of the peptide shown on the right, and thus 1 Da lower than the observed (experimental) mass, which carries an additional proton.

definitions of the scores and how they are controlled (e.g. they might search your masses against the same database you used, but with all sequences read backwards). In some search engines the score is defined as an 'absolute' score, so it defines the significance of the match. With such a score it is possible to make an informed decision whether to trust a certain match (i.e. the identification

of a specific database entry from PMF data) or not. The cut-off value for the significance is usually set at 5%; this means that the chance of any given identification being wrong at this significance level is 5%. This in turn means that if there are 200 measurements with identification at this significance level, then 10 of them are likely to be false positives. Depending on the proteomic application,

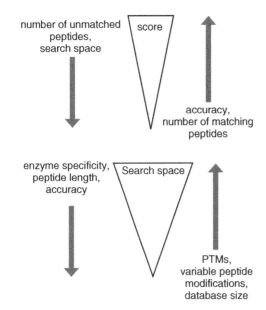

Figure 4.4 Parameters influencing the outcome of a PMF search.

this might be acceptable, but usually a good, reliable match is much better than this, more in the region of 0.01% or one false positive in 10 000 experiments. It is very important to be familiar with the scoring system of the software used to understand what the scoring means.

It is also important to remember that the absolute score or the significance of a hit is only reliable if the mathematical models used to calculate this score are correct and take all experimental factors into consideration. If, for example, you do not know the oxidation state of your peptides or whether a significant portion carries PTMs, the models applied are usually prepared to take this into account. Another often encountered inaccuracy is the prediction of the protease specificity; trypsin, for example, is not an exopeptidase (see Section 2.1.3), and it does not like to cut if the bulky, charged amino acids (Glu or Asp) are right in front of Arg/Lys or if phosphorylated Ser or Thr are one amino after the bond to be cut (Figure 4.5). Also, in MALDI based mass spectrometers, Arg is preferably detected. The only trypsin limitation fully accounted for in most search engines is the fact that it does not cut the sequences Arg-Pro and Lys-Pro. So if we still find a match, and all these parameters are not accounted for in the search engine in the first place, the significance of a match is usually underestimated, leading

to false negative results (Siepen *et al.*, 2007). If all the aforementioned experimental conditions were allowed for in the search, the search space (i.e. the size of the database against which the search is performed) would be so large that no significant search results could be returned. With all statistical methods there can be no absolute confidence in the results, but only a cut-off at which the experimenter has to decide if a true match is achieved or not. It must be the aim of all strategies to make sure to avoid many matches in the grey area of the scoring algorithms.

To stay with the example given in Figure 4.4, let us see how the search results change when we change different parameters (Figure 4.6). At 150 ppm search tolerance, without any further information the 11 strongest peptides picked deliver a score of 101 for human LAT. This score would still be significant at the 0.07% level (p value = 0.0007, i.e. there is a 1 in 1 400 chance that this hit can be achieved by chance alone). If we had not picked the smallest peptide, the score would drop to 84, significant at the 1% level, and if we had not picked the largest peptide, the score would drop to 80, which would be regarded as non-significant ($p = 0.7$, i.e. 1 in 14 experiments would produce a false positive). One peptide more or less can make this difference! If we had not picked one or two of the peptides that did not match, the score would have improved to 104 and 108, respectively. This would have brought the significance to the 1 in 10 000 level. However, if in doubt, it is advisable to have more rather than fewer peptides in the search, as the loss of a single matching peptide can make all the difference. Removing two additional peptides resulting from trypsin auto-digestion brings the score up to 119, a really reliable score. The best practically achievable score at this accuracy would have been 108, as we 'knew' the two peptides were from trypsin, and they are observed in nearly every sample (for further peak-list editing, see Figure 4.8). Searching in a smaller database does not change the score, but the score of the next best hit is further away from the score of the real hit. This shows how it is beneficial to start searching in reasonably large databases; it does not ruin the score and enhances the chance of a match, if the protein is not present in the database of the 'right' species, but only in a close or distant evolutionary homologue form. Including many peptides in the search is a double-edged sword; including another nine (not matching) peptides brought the score from 104 down to the edge of significance at a score of 81. Increasing the perceived/expected accuracy

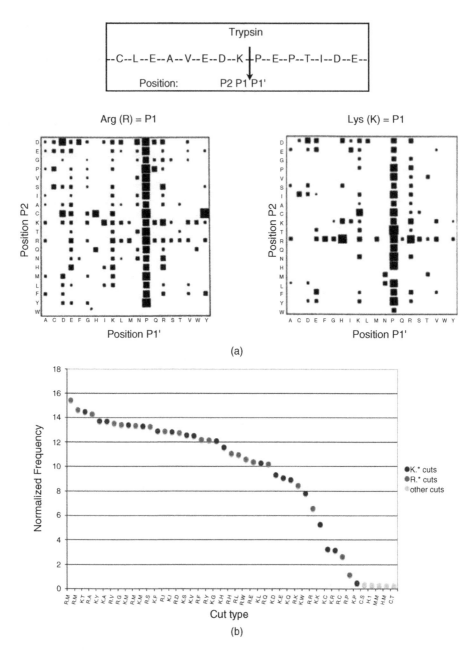

Figure 4.5 Sequence specificity of trypsin. (a) Amino acids that have a negative influence at a certain position are shown as black squares. The bigger the square, the more negative influence the amino acids have. For example, a Pro at the P1′ position inhibits a cut completely. The same is true for a Tyr (Y) at the P1′ position, but only if the cut is initiated by Arg (R in P1) and if Cys precedes the Arg in the sequence (C in P2). Smaller squares lead to less efficient cutting by trypsin. Despite cuts after Arg being inhibited by many amino acid combinations, Arg is preferably observed in MALDI MS based PMF (see Figure 4.3). Cutting efficiency is also influenced by residues further away from P1, but the influences are smaller. (b) Based on nearly 30 000 peptide identifications by ESI MS/MS, the relative frequencies of observed cuts are shown, in pairs of P1 = P1′, supporting that K/R-P cuts are very rare, as are K/R-C cuts, which is not obvious from the data in (a). (a) Reprinted from Keil, B., 1992. Specificity of proteolysis. With kind permission of Springer Science+Business Media. (b) Reprinted with permission from Rodriguez *et al.*, *J Prot Res*, 2008; 7: 300–5. © 2008 American Chemical Society.

1) 3602.27
2) 3586.21
3) 3460.18 LAT_Human
4) 3191.64 LAT_Human
5) 2155.09 LAT_Human
6) 2318.23 LAT_Human
6a) 2390.24
7) 2011.13 Trypsin
8) 1615.72 LAT_Human
9) 1475.61
10) 1035.41 LAT_Human
11) 842.38 Trypsin

HAc measurement

1) 3601.896
2) 3585.996
3) 3459.869 LAT_Human
4) 3191.425 LAT_Human
6) 2318.143 LAT_Human
8) 1615.782 LAT_Human

Search	Score	sig. level	expect
A	101	0.07%	1: 1 400
B	84	3%	1: 33
C	80	7%	1: 14
D	104	0.03%	1: 3 300
E	108	0.01%	1: 10 000
F	119	0.001%	1: 100 000
G	108	0.01%	1: 10 000
H	95	0.20%	1: 500
I	81	5%	1: 20
A*	87	2%	1: 50
A**	87	2%	1: 50
A***	— –-	— –-	— –-
A+	62	— –-	— –-
A++	66	— –-	— –-
HAc	86	2%	1: 50

Figure 4.6 Variations in PMF results with data quality and search strategy. A peptide list generated from a gel separated spot was used to search a database with settings as in Figure 4.3 and slight variations of some of the settings. The protein was identified as protein UniProtKB/Swiss-Prot O43561 (LAT_HUMAN). Search A was performed with all peptides from the peaklist in the upper box; B without no. 10; C without no. 3; D without no. 1; E without nos 1 and 2; F without nos 1, 2, 7 and 11; G without nos 7 and 11; H with two additional non-matching peptides; I with nine additional non-matching peptides; A* with peptide tolerance set to 250 ppm; A** with peptide tolerance set to 100 ppm; A*** with peptide tolerance set to 70 ppm; A+ with peptide 6 in phospho form (+80 Da) and phosphorylation as variable modification in search parameters; A++ as A+ but measured at 100 ppm; and HAc with peptides 1, 2, 3, 4, 6 and 8 measured at 10 ppm accuracy, search set to 10 ppm. Note that A and A* have six peptide matches, A** five and A*** only three peptide matches, as peptides were measured with lower accuracy than the search tolerance. Scores shown in bold are matches that would be regarded as significant. G is shaded, as this is the best possible match without the high accuracy data and specific prior knowledge of the sample. For further improvements of matching with peptide 6a, see Figure 4.8.

of the measurement to 100 or even 70 ppm brings the score down or loses the hit altogether, as fewer peptides match (i.e. some of the 'real' matching peptides are measured at a lower accuracy than assumed) and the significance is lost. Actually performing the experiment at a higher accuracy drastically improves the score. Measured at 10 ppm only four matching peptides are needed for a reasonably reliable score of 86, which is achieved even when two non-matching peptides are included in the search (Figure 4.6, HAc measurement).

In other words, measure as accurately as possible, but do not overestimate your accuracy; if in doubt, it is better to underestimate it for best search results. If we allow as variable modification the phosphorylation of Ser/Thr and Tyr (a reasonable assumption for a signalling molecule) the score goes down below the significance level at the 5% threshold. This holds true even if one phosphopeptide is included in the search and now seven instead of six peptides match. The search space is too big and the accuracy would have to be much better than 100 ppm until the search space is small enough to achieve a hit at the modest 5% confidence level.

Adding the mass of the protein (e.g. because you found it at 27 kDa on a gel) does not change the scores dramatically, as MASCOT uses a 'sliding window' of the indicated mass and not enough entries get excluded from the search for a significant improvement of the score.

If you try to repeat the above procedure, it will usually not be possible to generate exactly the same search results; using free search engines on the internet has (among other disadvantages) the problem that the underpinning databases are changing, usually from month to month. They are updated and corrected. That also offers the opportunity to get results with old data sets that did not deliver a good hit; the newer version of the database might contain the right sequence and deliver a hit. So it is worth trying again later or taking advantage of automated searches that inform you via e-mail of a successful search.

The search space is defined as the number of database entries that are actually used for a PMF. For instance, if the accuracy is given as 100 ppm and the mass as 2 318.14, then every sequence that generates a peptide in the mass window 2 318.14 Da $\pm$ 100 ppm or 2 317.91–2 318.37 Da is part of the search space created by this peptide. This will constitute several thousands of potential peptide sequences, and this is true for each of the 10 peptides searched in the above example. Increasing the accuracy to 10 ppm makes the search space smaller, 2 318.12–2 318.16 Da, but this will not result in reducing the matching potential sequences to a tenth of the sequences, as the sequence distribution follows the so-called peptide mass rule. This rule results from the addition of the sub-decimal masses of bio-relevant atoms and states that the masses of peptides are clustered; for example, thousands of peptides will be in the range 2 318.06–2 318.28 Da, but only a handful in the range 2 317.91–2 318.06 and 2 318.28–2 318.37, although the combined mass range (0.27 Da) is larger

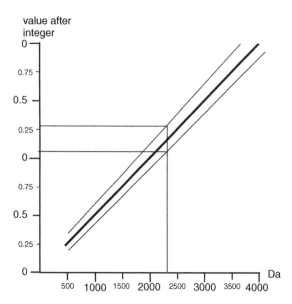

value after
integer

Figure 4.7 The peptide mass rule. All atoms have a mass different from their nominal mass (except carbon 12, by definition); for example, H weights not 1 but 1.007825 Da. Since peptides have similar formulae with similar ratios of all atoms (C/H/O/N/S roughly 3 : 4 : 1 : 1 : 0.3), certain values for the mass after the integer are rare or 'forbidden'. Thus most peptide masses cluster around the regions indicated in the figure. The bold line represents the mean of the clusters, the thin lines represent the range into which most of all peptides fall.

than the one adjacent to the statistical average (0.22 Da); see Figure 4.7.

Still higher accuracy makes the search space smaller, and at less than 1 ppm only a single or a few peptides will match each measurement. This can be used to develop entirely different search strategies with MS data (see later this chapter). On the other hand, using variable modifications (i.e. a modification might be there, but is not necessarily there) makes the search space very large. In the above example the sequence LPGSYDSTSSDSLYPRGIQFK at 2 318.14 Da was found. Allowing for variable phosphorylations, all combinations of possible phosphorylations of this peptide (and any other peptide in the database) have to be implemented in the database; at seven possible phosphorylation sites this equates to 49 possible phosphorylation patterns with eight different potential masses (i.e. from unphosphorylated, through 1×, up to 7× phosphorylated). Looking at it the other way round, allowing phosphorylations, there will be about an order of magnitude more

peptides that match the mass bracket, around 2 318.14 m/z, at any given accuracy. As phosphorylations occur in three amino acids and we have (currently) no way to predict reliably which might be phosphorylated, it is easy to see how allowing variable modifications can make the search space too big to be searched with data that would have delivered good results if we had omitted the potential modification.

To achieve the fewest possible matches in the grey zone of the scores it is important to define the parameters of the search engine in the best way and use an optimal number of peptides for the search (Berndt *et al.*, 1999); however, as we are searching for the unknown, this is not always possible. If in doubt one should always start with settings which will deliver more false positives, but where the best hit is still hidden under perhaps irrelevant false positive hits. Why not start with the most specific settings and the largest number of peptides possible in the first place? Because it might be far too time-consuming! It is usually possible to go back to the original data and to improve it – for example, by removing peaks that are obviously not from peptides (because they do not follow the peptide mass rule, or their isotopic pattern is wrong for a peptide), removing peaks from contaminants, manual recalibration, or trying to find masses to corroborate the best hits. This can be done by using the sequence from the hit and performing a theoretical tryptic digest using tools such as 'FindPept' from ExPASy (Figure 4.8). Under these circumstances, one can allow a multitude of PTMs and see if there is any indication to support such peptides in the measured data. The signals for these masses might be very weak, but if they fit in with the hit, one can assume that the protein (or a related one) is present in the digest. Also one can initiate a search for more than one protein in the sample; remove all peaks for the hit with the highest score and see if the rest of the peaks can deliver a hit with a reasonable score again. All these manipulations have in common that they are very time-consuming when it comes to medium or high throughput measurements. At a more systematic level one can initiate another search strategy to identify the proteins; either use MS/MS on a post source decay equipped MALDI ToF instrument (often limited by sample amount/peak intensity and quality of data from post source decay) or use full MS/MS on ToF/ToF instruments. Another way forward is to use the remainder of the sample for LC MS/MS analyses; the nature of the MALDI procedure does allow one to get very good results with a small portion of the entire

Post-Translational/Artefactual Modifications for the protein Linker for activation of T-cells family member 1:

Count	Nature	Position	Source	Occurrence	Mass	Description
1	PHOS	35	Feature Table	Optional	79.966	Phosphoserine.
1	PHOS	39	Feature Table	Optional	79.966	Phosphothreonine.
1	PHOS	40	Feature Table	Optional	79.966	Phosphoserine.
1	PHOS	41	Feature Table	Optional	79.966	Phosphoserine.
1	PHOS	43	Feature Table	Optional	79.966	Phosphoserine.
1	PHOS	84	Feature Table	Optional	79.966	Phosphoserine.
1	PHOS	101	Feature Table	Optional	79.966	Phosphoserine.
1	PHOS	106	Feature Table	Optional	79.966	Phosphoserine.
1	PHOS	109	Feature Table	Optional	79.966	Phosphoserine.
1	PHOS	110	Feature Table	Optional	79.966	Phosphotyrosine (Probable).
1	PHOS	156	Feature Table	Optional	79.966	Phosphotyrosine (Probable).
1	PHOS	161	Feature Table	Optional	79.966	Phosphotyrosine (Probable).
1	PHOS	220	Feature Table	Optional	79.966	Phosphotyrosine (Probable).
1	PHOS	224	Feature Table	Optional	79.966	Phosphoserine.
1	PHOS	240	Feature Table	Optional	79.966	Phosphoserine.
1	PHOS	241	Feature Table	Optional	79.966	Phosphoserine.
1	PALM	26	Feature Table	Optional	238.230	S-palmitoyl cysteine.
1	PALM	29	Feature Table	Optional	238.230	S-palmitoyl cysteine.
4	CYS_CAM	C	Artefactual	Mandatory	57.021	Iodoacetamide

http://www.expasy.ch/ tools/findpept.html

Peptides resulting from the cleavage of contaminants:

User mass	DB mass	Δmass (ppm)	type	contaminant	peptide	position
1475.610	1475.749	94.000	Keratin	P04264 Keratin, human (KRT1)	(K)/WELLQQVDTSTR/(T)	212–223
1475.610	1475.785	118.700	Keratin	P04264 Keratin, human (KRT1)	(R)/FLEQQNQVLQTK/(W)	200–211
3602.270	3602.744	131.600	Keratin	P04264 Keratin, human (KRT1)	(R)/QFSSRSGYRSGGGFSSGSAG IINYQRRTTSSSTR/(R)	4–37
1475.610	1475.785	118.700	Keratin	P35908 Keratin, human (KRT2A)	(R)/FLEQQNQVLQTK/(W)	198–209
3586.210	3585.960	-69.700	Keratin	P13645 Keratin, human (KRT10)	(R)/LKYENEVALRQSVEADINGL RRVLDELTLTK/(A)	236–266

Matching peptides for specific cleavage:

User mass	DB mass	Δmass (ppm)	peptide	position	modifications	missed cleavages
1035.410	1035.533	118.700	(R)/SPQPLGGSHR/(T)	84–93		0
1615.720	1615.776	34.800	(R)/GAQAGWGVWGPSWTR/(L)	120–134		0
1998.900	1998.863	-18.200	(R)/DSDGANSVASYENEGASGIR/(G)	100–119		0
2155.090	2154.965	-58.100	(R)/RDSDGANSVASYENEGASGI R/(G)	99–119		1
2318.230	2318.130	-43.100	(R)/LPGSYDSTSSDSLYPRGIQF K/(R)	32–52		1
2398.240	2398.096	-59.900	(R)/LPGSYDSTSSDSLYPRGIQF K/(R)	32–52	PHOS	1
2398.240	2398.096	-59.900	(R)/LPGSYDSTSSDSLYPRGIQF K/(R)	32–52	PHOS	1
2398.240	2398.096	-59.900	(R)/LPGSYDSTSSDSLYPRGIQF K/(R)	32–52	PHOS	1
2398.240	2398.096	-59.900	(R)/LPGSYDSTSSDSLYPRGIQF K/(R)	32–52	PHOS	1
2398.240	2398.096	-59.900	(R)/LPGSYDSTSSDSLYPRGIQF K/(R)	32–52	PHOS	1
3191.640	3191.402	-74.500	(K)/TEPAALSSQEAEEVEEEGAP DYENLQELN	234–262		0
3460.180	3459.858	-93.200	(K)/RPHTVAPWPPAYPPVTSYPP LSQPDLLPIPR/(S)	53–83		0

Figure 4.8 PTM search and PMF refinement. FindPept can be used for refining a search result and matching more of the measured peaks against the sequence of the identified protein. Using all the masses from the peptide list in Figure 4.6 and the database entry for protein UniProtKB/Swiss-Prot O43561 (LAT_HUMAN), FindPept shows all PTMs known to occur in the sequence (upper panel), automatically removes known typical contaminants in proteomics samples (middle panel) and shows how peptide masses found in the experiment could match the database entry (lower panel). Note that there can be no scores, as this is not a statistical analysis. The phosphopeptide (6a in Figure 4.6) matches several possible phosphorylation sites on the same tryptic peptide. Only MS/MS data can identify which of the amino acids actually carries the PTM.

digest, usually below 10%. It is often not productive to analyze the whole sample in the same way; because of the small volumes and sometimes ultra-low content in the protein of interest, any attempt to clean the sample up further and load a substantial larger proportion ends with either sample loss or deterioration of signals due to increased loading of contaminants. However, loading the sample onto a nano-RPLC online coupled to ESI MS/MS is highly complementary, as often other peptides are detected, and it allows MS/MS analysis of some peptides or the removal of other contaminating proteins from the search. This approach is likely to consume the entire remainder of the sample in one experiment and the RP acts as both sample clean-up and sample concentration device. If the sample of the tryptic digest was 10 μl and the peptides of this sample elute from the HPLC in peaks of 30 s length at a flow rate of 300 nl/min, they will elute in 150 nl, which is a concentration factor of 67. However, the intensity increase is greater than that, as chemicals interfering with the ionization process (e.g. impurities from gels/stains) are separated from the peptides. We will go into the MS/MS analyses in Section 4.3.

PMF analyses are extremely sensitive; the sample manipulation can be kept to a minimum, and since no MS/MS is performed the time spent per sample is very short. Also, avoiding MS/MS keeps the signal of every peptide in a single peak; with MS/MS this is split into dozens of peaks with concomitant decrease in sensitivity, which needs to be dealt with. PMF is in several ways a limited approach. It cannot be used for quantification; this is usually achieved at the gel-image level (Section 2.3.2). With PMF of protein samples from gels it is often difficult to obtain enough high quality peptide masses or clear spectra (without too much mixture) for a clear identification. This is often seen in proteomic studies based on 2D gels; thousands of spots are seen, often more than a thousand are analyzed as images, but only a fraction of these are actually identified by MS, due to great losses on the way from gel to mass spectrometer. This fraction of identified spots can be enlarged by several approaches; often just using only half of the gel for a digest with trypsin and the other half for another protease can improve results significantly using PMF and MS data alone. Using two alternative matrices for MALDI also increases the peptide coverage and the chance of identifying proteins without the need for splitting the sample.

Other approaches aim to get higher amounts of the sample by using larger, preparative gels, or combining spots/bands from several individual gels (up to about three to five gels before chemical/protein noise gets too big) and performing LC ESI MS/MS or ToF/ToF analyses with the digested samples (see Section 4.3).

It is usually not possible to use data from ion traps or triple quadrupoles for PMF, due to the lack of accuracy. These instruments perform better when used in MS/MS mode. Offline PMF from higher resolution mass analyzers such as ESI ToFs or hybrid instruments is possible in principle, but not often used as the MS/MS approaches of these instruments are more powerful. Ultra-high precision MS measurements on some orthogonal ToFs, orbitraps or FTICR instruments are an exception. Here proteins can be identified based on very precise measurement of the masses of their tryptic peptides and their elution times from LC columns, called the accurate mass and time (AMT) approach. The approach needs about two peptides to identify a protein from a local database, combining 1–3 ppm accurate mass measurements with the elution time in RPLC. Together this information is about as specific as an MS/MS measurement of these peptides (Figure 4.9). The disadvantage is clearly the need for very expensive mass spectrometers, to allow measurements at ultra-high resolution and thus precision of several parts per million or lower. The advantage is that this approach works with as few as 5 000 cells per measurement, which is several orders of magnitude more sensitive than MS/MS approaches, since in MS the signal of one peptide is kept together and not split into many fragments as in MS/MS and measurements are also faster. Quantification in AMT can be achieved by several labelling approaches, using ^{18}O water during the tryptic digest or afterwards, by using stable isotope labelling with amino acids in cell culture (SILAC), isotope coded protein labelling (see Chapter 5) or run to run comparisons without labels.

In a typical proteomic analysis with PMF, only a fraction of the peaks experimentally detected can be assigned accurately to either the protein of interest or a known contamination. This fraction can be as low as 20% (even for a significant identification) and rarely gets any better than 60%. Looking at it from a different angle, the sequence coverage, that is, the fraction of the sequence of the identified protein covered by the matching peptides, for significant matches starts at around 18% and rarely exceeds 60%. It is important to bear these figures in mind when analyzing PMF data; the data cannot prove that the protein is pure, as not all peaks can be explained, and we do not know with certainty if the protein is identical to the sequence identified in the database, or if it is somehow modified or just a related protein. In fact these statements are also true, for example, for data derived from a Western blot using a monoclonal antibody. The difference is the setting against which Western blots and PMF are performed; Western blots analyze single well characterized proteins and are often used to confirm or test a certain hypothesis. PMF can be performed on thousands of proteins with often very limited knowledge about most of them. Sometimes the interpretation only delivers

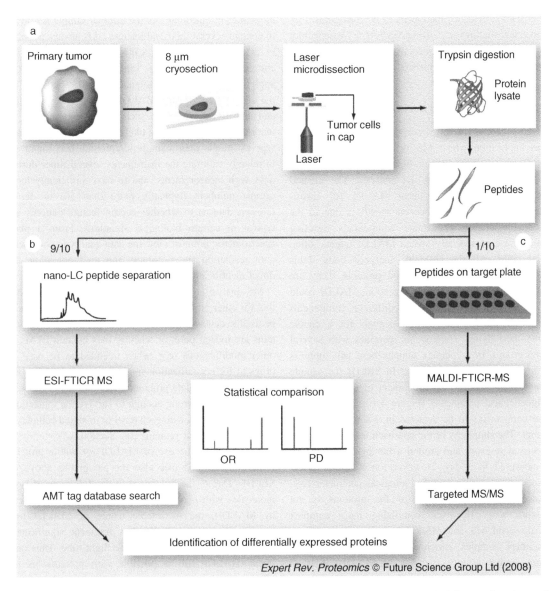

Figure 4.9 AMT strategy or sensitive and fast protein identification. The AMT strategy is best suited for samples where only minute amounts can be generated, such as clinical material. To increase the specificity LASER microdissection is used, as about 80% of the sample does not represent tumour cells but, for example, normal tissue, blood vessels and immune cells (a). In LASER microdissection the tumour cells are selectively cut out and collected using a LASER, so the sample amount is even lower than that from a complete cryosection. 90% of this highly enriched material is used in nano-LC separation of tryptic peptides, where the long LC gradients and ultra-high resolution MS allow the reproducible characterization of thousands of peptides by their mass and elution time, which are entered into a local database (b). In order to identify the differential peptides and their parent proteins between cancers responding to treatment (OR) and the progressive diseases (PD), MS/MS analyses have to be performed. This can be done offline using MALDI FTICR MS from LC separated peptides (c) or separate nano-LC ESI runs from related material (e.g. cultured tumour cells) which are likely to contain the same proteins, albeit in different quantities. Umar *et al.* (2008). Reproduced from *Expert Rev. Proteomics* 5(3), 445–455 (2008) with permission of Expert Reviews Ltd.

circumstantial evidence that needs to be corroborated by independent findings (e.g. using MS/MS or Western blot data). This applies, for instance, when highly homologous proteins need to be distinguished.

4.2.2 Analysis of MS data from complete proteins

Surface enhanced LASER desorption/ionization (SELDI) is an application where MS of complete proteins or peptides is used to detect biomarkers. Body fluids are a natural target for SELDI analysis. The method depends on rather simple linear MALDI ToF instruments and the name 'surface enhanced' is due to the enrichment/isolation of sample components resulting from the use of specially coated MALDI targets, often called chips. All the normally analyzed body fluids contain thousands of peptides and proteins. They are analyzed without prior sample digestion. MALDI could not cope with this variety of different components and would usually only deliver signals for a couple of the 'best flying' components: peptides with several amino groups. In a complex sample these will suppress the ionization of other analytes. In SELDI the sample complexity is reduced by enrichment of proteins based on biochemical parameters (Figure 4.10), thus allowing more components to be detected in a semi-quantitative manner. The simplicity of the approach makes it popular in clinical research and around a hundred samples can be measured in a day. However, once certain peaks are correlated with a certain biological parameter (e.g. a cancer), the identification of the proteins is not straightforward. However, combining more complex and powerful MS and MS/MS approaches as well as sometimes complex enrichment and separation steps (using similar affinity steps as in SELDI), it is in principle possible (if not straightforward) to identify the original component(s) that casued the differential SELDI signals. The data generated by SELDI are typically separated into two size ranges and are measured from the same target/spot. The mass ranges are typically from 1 500 Th to 10 000 Th for components of low molecular weight and 10 000–40 000 Th for larger components. One can expect to detect reliably some 20–200 features per sample in each mass range. Using up to 10 different chips/binding conditions and allowing some overlap between the surfaces on the chips, one can expect several hundred features as candidates for the

biomarker in questions. As the original sample is known to contain several thousand features, it is always possible to extend the number of candidates by new chips or enrichment steps in the sample preparation before applying the sample (colloquially called 'spotting') on the differentially treated MALDI targets. The specificity and sensitivity are often such that a combination of markers (a pattern) is more likely to be linked to a certain type of sample than a single ion/protein. It is important to reliably calibrate the instruments several times during days with measurements, and to have sufficiently large sample numbers (typically more than 50) to derive relevant data as to whether certain feature patterns are typical of certain biological situations. From a mass spectrometric point of view the data quality is poor, with poor resolution and accuracy, and does not allow the direct identification of the components (see above).

MS analysis of complete proteins is also a powerful tool for other proteomic applications. For example, it can be used to determine whether single isolated/purified proteins are indeed pure, or whether they contain PTMs or other modifications (e.g. when intended to be used as drugs or for crystallization studies) . Another application of MS from complete proteins is the analysis of associations (which are strong enough to survive the ionization) of purified proteins with each other or in mixed complexes containing different proteins (see Section 5.5).

Unfortunately the use of MALDI for whole proteins is severely limited (see also the paragraphs above on SELDI). Due to the low speed in the ToF stage of large molecules with one to three charges only (as produced by MALDI), the metastable nature of MALDI ions becomes limiting and the ions fragment significantly during the time they spend in the flight tube. Thus only linear MALDI ToF is possible with proteins, as the effect of metastable decays is not detrimental in linear MALDI ToF (see Section 3.3.1). For reasons discussed in Sections 3.2 and 3.3, the resolution for large proteins is well below 1 000, typically in the region of 80–300 (see Figure 4.10), so the accuracy of measurement can only be expected to be low. Furthermore, different proteins need very different LASER energies for ionization, so it is also difficult to calibrate the ToF analyzer better than 0.1% or 1 000 ppm, and the situation gets even worse for larger proteins (not to mention the problems of getting clean and well defined standard proteins). At these low resolutions, resulting in substantial peak width,

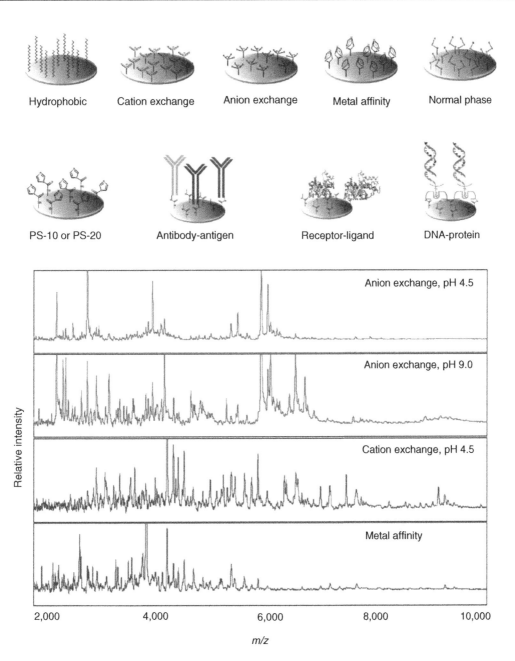

Figure 4.10 SELDI for the analysis of full-length proteins. Different surfaces on MALDI targets can be used to enrich for various proteins in SELDI. Next to surfaces which bind proteins differentially due to their physicochemical parameters (first row) there are several surfaces that can be loaded with specific factors (second row; PS-10 and PS-20 are materials with a reactive epoxy group, so the user can immobilize reagents on them). The lower panel shows some typical results for the low molecular weight range. The resolution of the peaks at 6 000 Th in the anion exchange chips is about 160. Reprinted with permission from Issaq *et al.* (2003). © 2003 American Chemical Society.

single modifications can only be detected if they result in substantial mass shifts, for example deletion of amino acids or phosphorylations (loss of 57–186 Da and gain of 80 Da, respectively) glycosylations (about 300 Da each) or ubiquitinations (8.5 kDa each). Modifications that cannot be detected and simply add to the peak width are deamidations (−1 Da), reduction of cysteine/disulfide bridges (−2 each) or single oxidations (+16 Da). Whole proteins are not homogeneous; they contain a very wide distribution of isotopic peaks as well as a mixture of modifications. At the low resolution offered by MALDI ToF for proteins these go over into a continuum, and it is impossible to determine with certainty if, for example, a trailing peak is due to modifications or an artefact of the measurement (LASER power too high, bad sample preparation). The difficulties are also dependent on protein size, so while in small proteins (20–30 kDa) point mutations might be easily detected, in larger proteins this becomes increasingly difficult.

Using ESI it is possible to analyze even large proteins with very high precision. This can be done in offline or online mode (see Sections 3.3.4–3.3.6). Depending on the mass analyzer, there are relative low upper values for the largest m/z that can be analyzed (e.g. if quadrupoles are involved, 8 000–12 000; see Section 3.3.6), but proteins are multiply-charged by ESI (on average one charge per 2–3 kDa), so even a complete IgG antibody can be measured at 3 000 Th with a charge state of +50. Ultra-high resolution analyzers such as orthogonal reflectron ToFs (up to resolution 40 000), orbitraps (up to resolution 100 000) or FTICR (up to resolution 1 000 000 for commercial instruments) can distinguish single isotopic peaks as well as modifications, if the sample is not too complex a mixture of different proteins or modifications. The multiple charges work in favour of accurate measurements also in another way; orbitraps and especially FTICR analyzers have lower resolutions with increasing m/z values and multiple charges keep the m/z values of large proteins relatively small. However, it is impossible to interpret ESI spectra of complete proteins directly. While ESI derived spectra of peptides may contain two to three different charge states of each component, the spectra of complete proteins contain dozens of different multiply-charged species of all proteins (ions with different m/z but the same mass); they need to be deconvoluted first. Deconvolution is the mathematical transformation of all (or most) of the features of a spectrum to the original

components, so that it looks comparable to a MALDI spectrum with only one charge state (+1). In this (usually software based) interpretation of the raw data, every peak of the original spectrum is assigned a charge state, based on interpretation of all its neighbouring peaks in the raw data (Figure 4.11). If the sample is too complex for deconvolution, or the original data do not have sufficient resolution, the assignments could be incorrect. If the resolution is high enough to distinguish isotopic peaks of the ions in the different charge states, the deconvolution can be very reliable (Figure 4.11), as the charge state can simply be deduced from the distance between monoisotopic peaks, similar to the situation with ESI spectra of peptides. The highest resolution FT analyzers can achieve isotopic resolution for proteins up to about 100 kDa. With the availability of the charge states it is still impossible to determine the monoisotopic mass, as only a very low fraction of the total population is monoisotopic (this fraction is very close to 0%), and the signal is correspondingly weak. However, in typical top-down approaches, the amino acid sequence, and thus the chemical formula of the protein in question, is known; the aim of the study is to detect/quantify modifications and verify the sequence of an isolated protein. Under these circumstances the theoretical isotopic distribution can be compared with the observed pattern, and any discrepancies would suggest a wrong interpretation of data and hint at chemical or post-translational modifications, or adducts or fragmentation of the major ion on the skimmer, all problems encountered with pure protein samples and even more confusing in mixtures. Of course this also means that an overlap of theoretical and experimental data supports the interpretation. This is particularly strong if the isotopic pattern can be observed and compared to calculated values based on the formula. However, to achieve isotopic resolution of complete proteins, very high resolutions are needed; the resolution needs to be equal to or, even better, several times higher than the size of the protein in question in dalton. Thus a resolution of 20 000 will deliver isotopic resolution for proteins up to about 18 000, and for larger proteins larger resolutions are needed (see Figure 4.12).

4.2.3 Analysis of MS data from LC applications

MS analyses from LC separations are rarely useful for proteomics applications on their own (with some exemptions such as AMT; see Section 4.2.1), but of course they are the basis for MS/MS measurements and will thus be

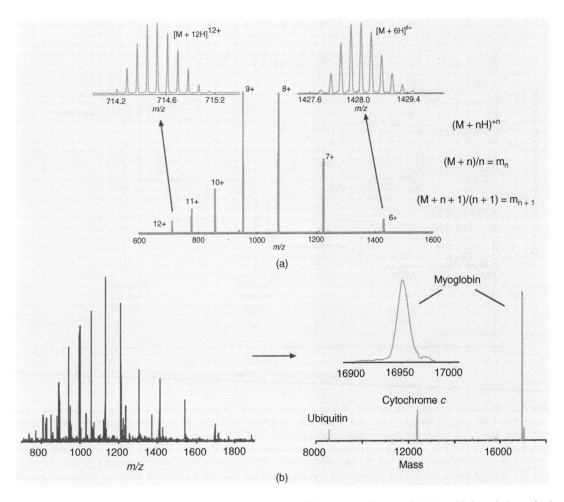

Figure 4.11   Deconvolution of ESI MS from complete proteins. (a) ESI spectrum of bovine ubiquitin, with isotopic ion series in several charge states. The upper term on the right shows the formula of each peak; M is the mass of the analyte, n is the charge state and H is the mass of a proton. The middle term shows which equation has to be solved for any given peak to determine its mass. The last equation is for the next lower charge state. Assigning the masses to all different charge states so that there is a solution for all peaks in the spectrum according to this formula is called deconvolution and is usually software based. With isotopic resolution and single proteins it is relatively easy to solve the equations as the masses can be determined by the charge state alone. (b) High resolution ESI spectrum of a mixture of three proteins, and the resulting deconvoluted spectrum. Deconvolution may be impossible if there are too many components and the isotopes are not resolved. Reprinted with permission from Marshall *et al.* (2002). © 2002 American Chemical Society.

discussed first. LC MS data provide a wealth of information about a sample and its separation over time. There are several ways to access this information, and we have already seen examples of single MS spectra from LC runs. However, the only way to access this information completely is the so-called contour plot (Figure 4.13). This is very important to remember when we come to derive quantitative information from LC MS runs as we

will see shortly. In a contour plot three pieces of information are combined: the intensity (colour) of every ion at a particular m/z (vertical axis) over the time of the LC run (horizontal axis). For purposes of interpretation different information formats are often chosen, but it is essential for a meaningful data interpretation to see this reduced information in the context of the contour plot. A 2D interpretation of the contour plot is given by the total ion

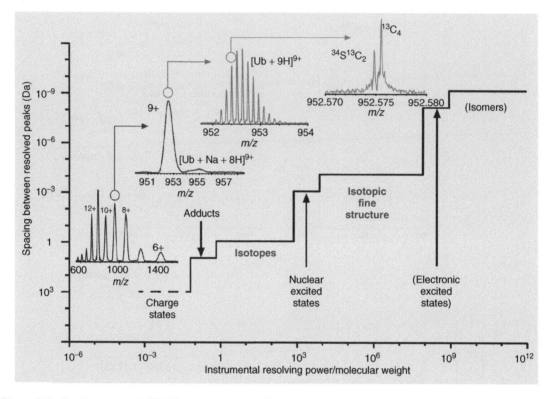

Figure 4.12 Resolution at work. This diagram gives an excellent overview of the power of resolution in MS. The Y axis gives an indication of the absolute distance between resolved peaks in dalton, the X axis the resolution in relation to the molecular weight of the analyte. The graph then shows which feature can be resolved. The resolution of adducts includes differences between proteins/peptides due to PTMs. At isotopic resolution the mass accuracy takes a great leap forward. The isotopic fine structure allows us to distinguish between atoms with the same nominal mass, and to determine with certainty whether a peptide contains no, one or two atoms of sulfur, as it not only changes the isotopic pattern but also shows a minimal difference in the mass after the integer. Resolution up to 1 000 000 for proteins are possible with commercial FTICR MS. Reprinted with permission from Marshall *et al.* (2002). © 2002 American Chemical Society.

current (TIC) over time (Figure 4.13), sometimes together with the information from UV detectors (see Figure 2.22). However, in the TIC lane we only see the addition of all MS signals together; we have no way of knowing whether a strong signal at any given time is made up of one ion, a series of ions or a burst of chemical noise, or even changes in the background chemical noise due to changes in the ESI source. The TIC lane presentation compresses the data in the dimension of m/z. A single spectrum is like a vertical slice through the contour plot; the contour plot is composed of thousands of individual scans, each lasting about a second, and one slice at a certain time is shown in each single mass spectrum from a certain time (see Figure 4.13). Of course, several single spectra can be added up. If the summed up spectra are measured over the time that covers the elution of a peptide, they contain all the MS information regarding this peptide. Whether the summed spectra really cover all the data for a certain peptide can be assessed easily by looking at the contour plot. Alternatively, if looking at the data in spectrum mode, single spectra right before and after the summation border can be checked for the presence of the signal for the peptide in question. If the summation time is extended, several peptides that eluted earlier or later start to appear in the spectrum; their intensity in relation to the main peak can only be judged in the context of the contour plot, not by analyzing single spectra or summation of spectra on their own (see also Figure 5.15(d)).

The intensity for a peptide is best reflected by summing the signal at this mass over the time frame the peptide

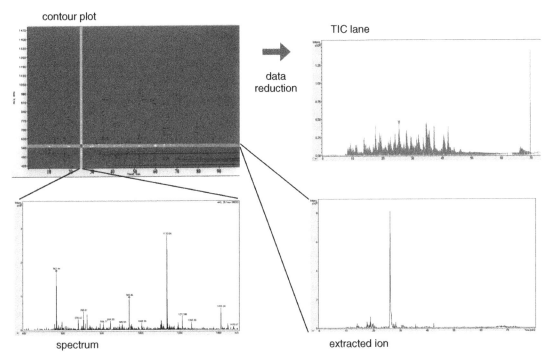

Figure 4.13   Data presentation from LC MS runs. The contour plot is the most comprehensive way to present LC MS data (not MS/MS data). The Y axis is mass, the X axis is time, and the colour represents the intensity of the signal at this point in time and mass. This representation requires intensive computation, but is the best source of data to compare different LC runs. Data reduction, that is, loss of the mass information, leads to the TIC lane presentation. One 'slice in time' leads to a single spectrum, one 'slice in mass' to the extracted ion presentation, which is also very useful for peptide quantification (see also Figures 4.24 and 4.25). The extracted ion lane is also called an extracted ion current (XIC) chromatogram. See also Figure 4.17 for a more detailed view of a contour plot.

eluted; this is the most precise information to be used for the quantification of a certain peak. The peak can also be extracted from the entire chromatogram; the software is set to display the intensity of the mass right at the centre of the peak plus or minus the peak width (e.g. 586 m/z $\pm$ 0.2). The resulting chromatogram is called an extracted ion current (XIC) chromatogram (Figure 4.13). It is the area under such an extracted peak in the XIC that is the best representation of the ion's the intensity, and if other possible charge states and isotopic peaks of the same peptide are also considered, the combined areas under the XIC peaks represent the total amount of the peptide (see also Section 4.4). Looking at all the spectra that belong to a certain peptide, one can see that the intensity of the ions for this peptide is low at the beginning of the elution of the peak. The intensity of the ions gets stronger up to the point in time where the bulk of the peak elutes and

from there on the ion intensity gets weaker before the signal is finally lost in the noise (Figure 4.13).

Apart from XIC we can also use spectral counting for quantification; instead of extracting every peak and making all the complicated calculations to determine the area below a peak or the volume of a peak over time, the number of spectra in which the peak is used for MS/MS analyses is counted. The more intense the peak, the more spectra will show its presence, as weak peaks will not be visible in their initial slopes due to low signal to noise ratio and high chemical background. However, these are not the only relevant considerations for peptide quantification as opposed to peak quantification; a peptide might be detected in several charge states, and these would need to be considered as well. There are variations in the measured data over time; for example, in one run most of the peptide might be detected as doubly-charged, while in the next run it might be detected as triply-charged. Depending on the resolution of the mass spectrometer, a substantial fraction of a peptide might be detected in a charge state that cannot be precisely determined, and this peak might be lost from the quantification. Protein quantifications are different again; for this the quantification of every peptide is taken in to consideration. Based on possible errors (some of which we mentioned above) not all peptides of a protein will show the same quantitative properties; some may be weak due to poor recovery, poor tryptic digest, heterogeneous modifications (including PTMs) or simply their co-eluting peaks, which might 'overshadow' them. In semi-quantitative or relative quantitation approaches each peptide might thus react in a different way. For more reproducible quantification (e.g. comparing one biological situation with another) the average of all the changes is taken, usually omitting peptides whose changes deviate too much. This is a good workaround in a situation where thousands of peptides are quantified; some will always show technical problems. However, the deviation of a peptide in quantifications compared to other peptides derived from the same protein might also be the result of a wrong annotation to a certain protein of any of the peptides involved; the peptide might in fact belong to a different (but homologous) protein all together. This will be discussed in more detail in Sections 4.3 and 4.4.

Next to relative quantifications, absolute quantifications can also be performed with LC MS signals. Standards of known concentrations are separated together with the sample. The standards have to have identical physicochemical parameters to the sample to guarantee the same level of detection (mainly ionization, but also stability and separation behaviour and efficiency of digest by trypsin). To achieve this similarity and still be able to separate sample from standard, isotopically labelled peptides are used, with amino acids or chemical compounds labelled by heavy hydrogen, carbon or nitrogen atoms (Section 4.4). ESI is ideal for these measurements, as it is robust and ionization is not suppressed by standards. From the LC MS point of view it is most important to measure the whole intensity of the peak of a peptide, rather than a single spectrum, and the same applies to the standard peptides. In this regard it can be slightly counterproductive that some heavy compounds elute a very short time after their unlabelled counterparts in RPLC. The most critical step in isotopic labelling approaches is the quantitative incorporation of labelled

amino acids or high stoichiometry and reproducibility of the labelling of different samples, with different compositions and protein concentrations. We will look at these parameters further in Sections 4.4 and 5.3.

4.3 ANALYSIS OF MS/MS DATA

Pure MS is limited to measuring the mass of the entire protein (top-down proteomics) or the mass of peptides derived by digestion (bottom-up proteomics). In MS/MS (also called tandem MS) or MS^n measurements, the masses of proteins or peptides are measured, just as before, but in addition the ions are further fragmented in the mass analyzer and the masses of these fragments are also determined. Often the term MS is used imprecisely to describe all kinds of MS measurements.

MS/MS is performed for top-down and bottom-up approaches, using a wide variety of fragmentation principles and mass analyzers. One of the most important features of MS/MS is that the fragment masses have to be connected to the parent masses. Thus additional 'structural' information can be derived from the measurements. The ideal outcome for any fragmentation study is to find fragments that support and cover the complete amino acid sequence (primary structure) of a peptide or protein. In reality most of the time all the fragment ions measured can only support a part of the amino acid sequence. This is also usually only possible after a great deal of interpretation of the data. For most experiments in proteomic settings this interpretation is not possible and often not even necessary. It is enough to get a high statistical probability that a certain fragment pattern is derived from a certain sequence in a database. In principle, MS/MS studies can be used to determine elements of secondary, tertiary and quaternary protein structure, for example by measuring intra- or inter-molecular contact points using deuterium exchange. However, in practice these experiments are currently too complex to be performed at the proteomics level.

4.3.1 MS/MS analysis of peptides

By far the most common application of MS/MS in proteomics is the fragmentation of tryptic peptides in bottom-up approaches. This strategy is very successful on purified proteins as well as on complex protein mixtures. In most cases a single peptide can be identified in the largest known databases without interpretation of MS/MS spectra, provided the quality of the data is good enough. The data acquired from peptides must almost always be transferred to the level of proteins. Depending on the analytical target, a couple of peptides may be enough to determine the identity of a protein, its presence in a certain organelle or its changes in expression level during an experiment. Although peptide based proteomics using MS/MS is the most powerful proteomic tool in terms of throughput, sequence coverage, sensitivity and versatility, it is important to realize the degree of information that can be derived from peptide based approaches. While MS/MS data on single peptides can show, for example, the presence of a member of a protein family, it can be very difficult to determine exactly which member of the family is present (compare Figure 5.8). Equally a single peptide can identify the presence of a PTM on a protein; but it might be impossible in peptide based proteomics to establish the phosphorylation isoforms of the protein that are actually present, as they arise from combinations of phosphorylations placed across different peptides. Finally, peptide based protein quantification has to be carefully examined. Often the limited information can be combined with other information to allow extended interpretation (e.g. identifying PTM sites from peptides of proteins separated on 2D gels).

MS/MS analysis of peptides from purified proteins

MS/MS data from purified proteins are the easiest to interpret; most, if not all, peptides in such a sample will derive from one protein or a simple mixture of a couple of proteins and some contaminations. If the 'purified' protein came from a band of an SDS PAGE gel, the sample can soon become 'complex' (see Section 'MS/MS analysis of peptides from complex mixtures'). The easiest spectra to interpret (from a beginner's point of view) are ToF/ToF spectra; all ions are singly-charged, there can be no confusion between Th and dalton units, and all fragments appear actually smaller than the parent ion (Figure 4.14). Such spectra are typical where gel (1D or 2D) separated proteins are digested in the gel and then 5% of the digest is loaded onto a MALDI target.

The chosen example shows how important a high resolution for the precursor ion selection is; a double oxidized form was chosen (as it turns out after the database search) for MS/MS analyses, surrounded by other peptide forms differing in mass by ± 16 Da (oxygen). A resolution of about 110 would have been enough to separate the

precursor cleanly. The instrument was actually capable of a resolution of about 2 000 for the precursor selection. The data quality in this example is good enough to identify the protein without further interpretation and two such MS/MS sets form 2 different peptides would be enough to identify a protein with high statistic significance. Lets have a closer look at the data and perform a limited interpretation of the MS/MS spectrum. Starting from the strongest fragment ion (986.34 Da) the next reasonably strong smaller ion is 873.03 Da, $\delta = 113.31$, reasonably close to the mass of Leu/Ile, which cannot be distinguished based on their mass (thus 'N' is used to denote either a Leu or an Ile in the interpretation of the spectrum). There is also a signal at 86 Da, corresponding to the immonium ion of Leu/Ile. The next mass on the 'ladder' is 743.03, $\delta = 130$, within 0.8 Da of what would be expected for Glu (or 'E'). When this iterative process is continued, the sequence of 10 amino acids (this unfortunately leads to the second N in the sequence) can be derived. Together with the accurate mass of the parent (ToF typical at 41 ppm accuracy as it later turns out) this would match with several sequences from the database. Together with the 'knowledge' (based on the distance of 32 Da from a much stronger ion in the MS spectrum) of the double oxidation, the correct peptide could have been manually assigned. As it was, there was a 'hunch' from PMF data as to which protein it might be, so the automatic assignment was good enough.

The MS/MS data in Figure 4.14 are not quite sufficient for a complete de novo sequencing of the peptide; some parts of the sequence are not supported by fragments (missing). But from this example you can see that de novo sequencing is a bit like a crossword puzzle; practice makes better but it still takes some time, and you never know at the beginning if you can finish it, even if it looks good to start with. The fragmentation pattern has nothing to do with chance; it is strictly determined by the amino acid sequence, the instrument settings and the type of fragmentation. The amino acid sequence determines of course which fragments are possible at all; however, it also determines which fragments are preferred. Certain fragments will always be strong, others weak or absent. This is ruled by mechanistic aspects of the fragmentation process. For example, rigid Pro (the backbone bond can not rotate) fragments more often, so much so that it terminates all the nice series of fragments we like to see. It is akin to putting a stiff link into a flexible chain; if you

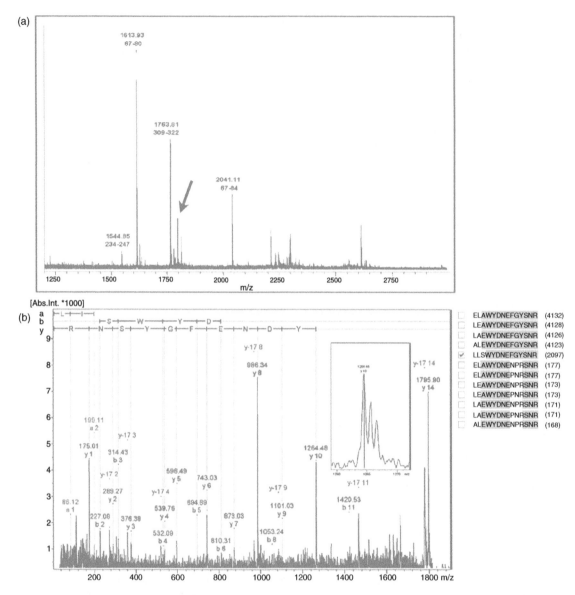

Figure 4.14 Peptide analysis by ToF/ToF MS/MS. A protein from a 2D gel was digested with trypsin and used for PMF (see detail of MS in (a)) followed by ToF/ToF for identification. The PMF resulted in 3-phosphate dehydrogenase as a candidate. To confirm this, ToF/ToF of the ion at 1 795.8 Th was performed (b). Automated sequence determination resulted in the putative peptide sequences indicated in the box on the right (scores in parentheses). Note the sequence matching to the 3-phosphate dehydrogenase peptide 309–322 (LISWYDNEFGYSNR) was only found among the putative sequences after oxidation of Try was specified as modification. From the pattern in MS it was obvious that the peptide at 1 795.8 Th was the twice oxidized version of LISWYDNEFGYSNR, which was found at 1 763.81 Th. This is a good example of how several lines of analysis come together to identify a protein. Note that the a1 fragment is the immonium ion for Leu/Ile. Reprinted from Suckau *et al.* (2003). With kind permission of Springer Science+Business Media.

twist the chain slowly to breaking point (as in CID) it will always break near the stiff link (Pro). If you apply a bolt cutter at a random position of the chain (ECD or ETD fragmentation) it does not care for the stiff link and all fragments will appear. Also phosphorylations on Ser/Thr will suppress fragments two to three amino acids away in CID, but not ETD/ECD, while phosphorylation of Tyr does not show this effect at all. Of course if a lot of material is available and the sample preparation is good, the accuracy and intensity of fragments will improve, and assigning a Glu (E) at a delta mass of 130 Da is not as confidence inspiring as assigning it at 129.2 Da, when its real mass (the theoretically expected mass difference) is 129.1155. Although the ToF/ToF in question showed a typical accuracy for the measurement of fragment ions of around 0.2–0.4 Da (according to the manufacturer) this call at 0.8 Da accuracy is correct, but just outside the assumed accuracy range. Modern ToF/ToF instruments can show far better accuracies – the example is from 2003. However, because the assignment even at low accuracy makes the fragment part of a long series, it is very trustworthy. Other instruments show higher accuracies for fragments, and if the fragment is 'out' for a particular assignment by 0.09 Da, then this assignment is not correct (see next example).

As we will see in Chapter 5, such a high accuracy does not make an instrument automatically better for proteomics as an instrument with lower accuracy, if this one tends to produce better ion series. The big enemy of MS/MS spectra interpretation is a multitude of fragments; combined with a certain inaccuracy it is very daunting to not mis-assign several fragments, and end up with a similar but wrong sequence in the end. Often sequences stay ambiguous, because of isobaric amino acids (Ile/Leu, Gln/Lys) or the fact that the masses of two consecutive amino acids are isobaric with a single one (e.g. Ala + Gly = Lys), or because of amino acid modifications (e.g. homoserine); see also Figure 3.8. Note that from the assignment of 10 amino acids to a series, we still would not know which series it is, and thus the order of the sequence (where is the n-term?). We do not know what the absolute modification (state of N- and C-terminus) of the fragments is, as long as they are all the same they form a series with the delta masses corresponding to the exact mass of the amino acid. This example shows how tricky it can be to interpret MS/MS spectra. If you were to generate several hundred of them per day,

uninterpreted search against a database that leaves you with a score at the end is a very welcoming option. It is also noticeable that not all ions in the MS/MS spectrum were assigned; a typical situation, even for confident matches like this one. The interpretations/assignments in this case are also consistent with the combination of other data (PMF, ToF/ToF data of different peptides), which adds more confidence to the assignments made.

The second example of an MS/MS spectrum (Figure 4.15) shows the MS/MS spectrum of a peptide found by a q-ToF in a 2D spot, measured at 1 093.91 Th, with an error of 290 ppm. Notice that this was a doubly-charged ion (perfect for MS/MS) and the error on the real mass (2 185.161 Th from the database) stays the same when expressed in ppm. This is far off from the accuracy (50 ppm) that can be achieved with such an instrument, and shows how calibration in a ToF instrument can drift off during measurements (this one was performed in the offline ESI mode). All the fragments also show an offset of +300 ppm, showing the precision of the instrument. The delta between the peaks is only 0.03–0.06 Da out, making manual annotation quite reassuring and allowing a nearly complete de novo sequencing in this particular case, leaving only two gaps of three amino acids in total. Modern instruments with best accuracies (ESI q-ToFs, hybrid ToFs or FT instruments) can distinguish between Gln/Lys or Phe/Met (delta of 0.036 or 0.033 Da, or below 18 ppm for fragments of around 1 000 Da), making assignments even more reliable.

The b and y ion series are nearly complete, except for gaps at the C-terminus and around two of the three prolines (one of the prolines also produces the strongest fragment at 737.72 Th). Notice how weak the b_{18} ion is, and how it is still included in the 'hit parade', owing to the superb low background noise (intensities of 0–2) of q-ToF instruments; an intensity of 3 can be an accurate peak and b_{19} has an intensity of 6 on the arbitrary instrument intensity units. Also notice the absence of any immonium ions in this milder CID than in the ToF/ToF. Singly-charged peptides are not fragmented very efficiently in such an instrument, as opposed to the ToF/ToF (see also Chapter 3 for the reasons for this). This peptide comes from a phosphorylated protein (shown by metabolic labelling with radioactive ^{32}P), but shows no sign of phosphorylation on its lone Tyr. It belongs to casein, a known heavily phosphorylated protein. It is typical not to be able to see the phosphopeptides, as long as they are not targeted (see

(a)

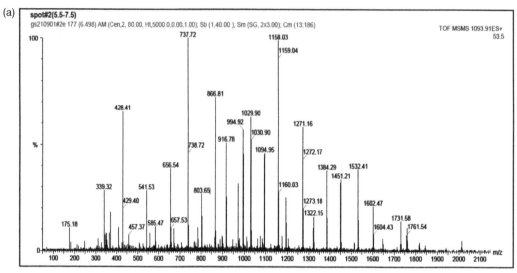

(b)

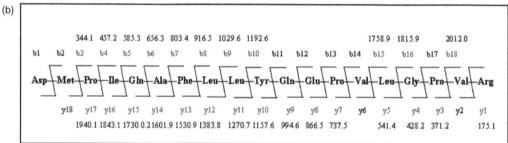

(c)

Tryptic fragment	Parention mass (Da)	Charge state	Sequence (observed fragments in red)	Casein isoform
106–115	634.4	2+	YLGYLEQLLR	alpha
38–49	693.4	2+	FFVAPFPEVFGK	alpha
199–219	1093.9	2+	DMPIQAFLLYQEPVLGPVR	beta

Figure 4.15 Peptide analysis by q-ToF MS/MS. The tryptic digest of a 2D gel separated protein was analyzed by offline ESI on a q-ToF. (a) MS/MS spectrum of an ion at 1 093.9 Th, which was doubly-charged. (b) Interpretation of the spectrum. (c) Summary of all peptides analyzed by MS/MS from this sample. Casein could be identified clearly, but no call can be made on the isoforms, or if both isoforms were present in one spot at pH 5.3 and 25 kDa. The database mass for the above peptides is 2 185.161 Da and 1 093.588 Da, respectively, and the peptides were measured as doubly-charged ions. See text for more details.

Section 5.4). Note also that the parent ion at 1 093.91 m/z (too weak to be seen in this overview) is in the middle of the size range of its fragments, as most of the fragments are singly-charged.

ESI MS/MS can also be performed online, coupled to RP LC or to CE. Offline measurements are very good for detailed analysis of (partially) purified proteins, for example in the search for PTMs, but they allow only a very low throughput. However, a trained operator can

often achieve a lot more offline than online, if the sample is pure enough. Using precursor ion scans targeting fragments specifically for e.g. Ile/Leu immonium ions, a trained operator can detect peptides whose intensity would be too low to be picked up in a standard MS measurement and can even identify fragments of these peptides by follow-up parent ion MS/MS analyses. Precursor ion scans targeting phosphorylation specific fragments of 80 or 98 Da (−79 Da in negative mode) can be used to identify cryptic phosphorylation sites. However, the sample has to be cleaned up before offline analysis. This is best achieved using C18 RP beads or, even more conveniently, mini-columns, prepared in pipette tips or directly in the spraying needles. Then the peptides are eluted directly into the spraying capillary. While this clean-up leads to a loss of much of the sample (as can be evidenced in MALDI applications on clean samples), without it the chemical noise and the ionization suppression of contaminants would not allow a useful measurement at all.

Another special application for offline spray is the use of an ion trap for MS^3 or even Ms^n. In the example in Figure 4.15 of the casein peptide, an MS^3 of the b15 ion (1 758.9 Th) could have been performed to try and get a more complete sequence coverage. While samples in an ESI offline experiment can last 20–50 minutes, a MALDI IT would allow even more time to deduce partial sequences and plan the next measurement. This is important for the complete elucidation of complex PTMs such as multiple glycosylations or complete sequence coverages.

While online ESI MS of single proteins can produce a lot of data, it rarely covers 100% of the protein sequence (neither can offline ESI MS/MS do this all the time; see above). If a specific PTM is known to be on a protein from other lines of analysis and needs to be localized/identified, offline might offer advantages, just as it does when it comes to explaining unaccounted-for peptides due to, for example, modifications during sample preparation or other, unknown reasons which often stay unknown even after extensive analysis. If a general overview or 'just' a protein identification is required, online ESI is faster, and can work automatically without any data interpretation. Online ESI is also very powerful for identifying all the contaminant in a sample, and 'clean' 2D spots (one identified protein according to MALDI ToF PMF) usually yield about two to six proteins, apart from ubiquitous contaminants such as keratins.

Using online ESI MS, all peptides of one or several proteins (e.g. 1D gel or spot from 2D gel) are spread in time through elution from a nano-RP column. This results in high sensitivity without precursor mass scan as most peptides are clearly separated and highly enriched (from, say, 2 µl in an offline experiment to 150 nl elution volume of one peak from an RPLC run). Although every single peptide might not be assigned a high score, the protein identifications scores, combined from several

peptide scores, usually are much higher than for PMF of the same sample. Even when PMF does not deliver a clear identification an online ESI MS/MS analysis can often identify the protein. The drawback is that online ESI analysis will consume all of the sample, while MALDI usually consumes less than 5% of it. Figure 4.16 shows a typical protein identification from an online ESI MS/MS analysis. The TIC lane shows a serrated intensity profile as every MS scan is followed by an MS/MS scan of the two strongest ions (thus not generating any MS data at this point). Using dynamic exclusion of every ion that was fragmented twice for the following 60 seconds guarantees that even weak ions get a chance of being analyzed by MS/MS. The resulting fragment spectrum is not very detailed. The reason for this is that during an LC MS run there is no time to play around with fragmentation conditions to optimize the parameters for every peptide and to wait long enough to see as many fragment ions as possible, because the next peptide is eluting as well and needs to be measured. As the database search results show, eight peptides produced reasonable fragmentation patterns. Most of the peptides were measured 2–3 times, either in different charge states or even in the identical charge state. This shows that dynamic exclusion did not work properly, presumably the software decided that ions derived from different peptides, as they were measured at slightly differing masses. The IT that was used did not have a chance to determine charge states higher than +3 and the software has no chance during acquisition to see that two ions are related; this only becomes obvious after the database search. The total sequence coverage is 50%; this value varies between 30% and 80% in a typical measurement. Such a sequence coverage would have the chance to detect some, but not all, PTMs in a second round search, and targeted searches would have a much better chance of identifying more PTMs (see Section 'Targeted mass spectrometric approaches for the detection and quantification of PTMs'). Another approach for MS/MS analysis is to not even choose the ion for fragmentation but to fragment all ions together (Figure 4.17). The assignment of parent and fragment ions to each other is only possible using the coinciding elution times of the parent and fragment ions. A lot of peptides overlap in their elution, but they can still be separated as they start to elute at different times, reach their maximum intensity at different times and also many peptide differ in their 'peak length'. A recent approach is to use a q-ToF and switch

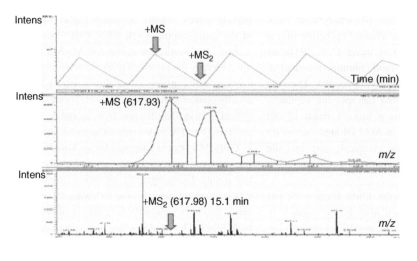

(a)

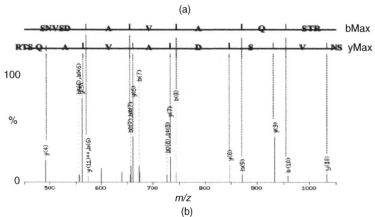

(b)

Start - End	Observed	Mr(expt)	Mr(calc)	Delta	Miss	Sequence
33 - 52	1096.5170	2191.0194	2191.0620	-0.0425	0	K.VPADTEVVCAPPTAYIDFAR.Q Carbamidomethyl (C) (Ions score 66)
33 - 52	1096.5530	2191.0914	2191.0620	0.0295	0	K.VPADTEVVCAPPTAYIDFAR.Q Carbamidomethyl (C) (Ions score 29)
33 - 52	731.4870	2191.4392	2191.0620	0.3772	0	K.VPADTEVVCAPPTAYIDFAR.Q Carbamidomethyl (C) (Ions score 21)
69 - 84	811.4110	1620.8074	1620.8181	-0.0107	0	K.VTNQAPTGEISPGMIK.D (Ions score 25)
69 - 84	811.9620	1621.9094	1620.8181	1.0913	0	K.VTNGAPTGEISPGMIK.D (Ions score 35)
69 - 84	811.9750	1621.9354	1620.8181	1.1173	0	K.VTNGAPTGEISPGMIK.D (Ions score 40)
85 - 98	793.4600	1584.9054	1585.7307	-0.8253	0	K.DCGATWVVLGHSER.R Carbamidomethyl (C) (Ions score 45)
85 - 98	529.5930	1585.7572	1585.7307	0.0264	0	K.DCGATWVVLGHSER.R Carbamidomethyl (C) (Ions score 21)
85 - 98	793.9440	1585.8734	1585.7307	0.1427	0	K.DCGATWVVLGHSER.R Carbamidomethyl (C) (Ions score 25)
100 - 112	729.9160	1457.8174	1457.7151	0.1024	0	R.HVPGESDELIGQK.V (Ions score 26)
113 - 130	904.4640	1806.9134	1806.9662	-0.0527	0	K.VAHALAEGLGVIACIGEK.L Carbamidomethyl (C) (Ions score 51)
113 - 130	904.4800	1806.9454	1806.9662	-0.0207	0	K.VAHALAEGLGVIACIGEK.L Carbamidomethyl (C) (Ions score 76)
113 - 130	904.5010	1806.9874	1806.9662	0.0213	0	K.VAHALAEGLGVIACIGEK.L Carbamidomethyl (C) (Ions score 81)
160 - 174	801.9570	1601.8994	1601.8817	0.0177	0	K.VVLAYEPVWAIGTGK.T (Ions score 37)
194 - 205	617.8100	1233.6054	1233.5949	0.0105	0	K.SNVSDAVAQSTR.I (Ions score 79)
194 - 205	617.9300	1233.8454	1233.5949	0.2505	0	K.SNVSDAVAQSTR.I (Ions score 54)
206 - 218	663.3380	1324.6614	1325.6649	-1.0035	0	R.IIYGGSVTGATCK.E Carbamidomethyl (C) (Ions score 25)
206 - 218	663.9650	1325.9154	1325.6649	0.2505	0	R.IIYGGSVTGATCK.E Carbamidomethyl (C) (Ions score 23)

```
  1 APSRKFFVGG NWKMNGRKQS LGELILILNA AKVPADTEVV CAPPTAYIDF
 51 ARQKLDPKIA VAAQNCYKVT NGAFTGEISP GMIKDCGATW VVLGHSERRH
101 VFGESDELIG QKVAHALAEG LGVIACIGEK LDEREAGITE KVVFEQIKVI
151 ADNVKDWSKV VLAYEPVWAI CTGKTATPQQ AQEVHEKLRG WLKSNVSDAV
201 AQSTRIIYGG SVTGATCKEL ASQPDVDGFL VGGASLKPEF VDIINAKQ
```

(c)

between MS and total fragmentation MS/MS and thus analyze all peptides in a run by MS/MS. The 'poor man's' version of this approach is skimmer fragmentation, which can be enhanced by increasing the skimmer voltage. Skimmer fragmentation is a process that happens in all circumstances to a certain degree and some peptides show very strong skimmer fragmentation in any experimental circumstances. The disadvantage of using systematic skimmer fragmentation is that the fragmentation conditions are not well controlled; on the positive side, it is possible to perform this analysis with any MS only instrument, as the fragmentation occurs (just about) outside the mass spectrometer.

The skimmer fragmentation raises the issue whether your sample does get efficiently into the mass spectrometer in the first place. Many instruments differ in the data they produce and deliver overlapping results. Next to fundamental considerations (e.g. ToF analyzer versus IT), the design of the ESI source is an important factor in the overall performance of a mass spectrometer.

Before MS/MS data are used for database searches the original data are reduced automatically in several ways. Instead of analogue signal intensities, these are interpreted automatically and reduced to a few numbers: the m/z value of the monoisotopic peak, intensity (in arbitrary instrument units) of the ion, and elution time when it was measured (time into experiment). The ion intensity will be calculated in different ways by different manufacturers (either taking the monoisotope alone or the other isotopic peaks as well) after the peaks have been smoothed and the background has been subtracted. Some mass analyzers can identify charge states, others cannot; if they cannot, they simply assume it is any of the ones from +1 to +4. Then a similar data reduction is performed on the

MS/MS ions and this data set is related to the parent ion. Cut-off values for low intensity, which are different for MS and MS/MS data, are defined, so chemical noise does not contribute too much to the search results. This is the minimum automatic data reduction, it can be a lot more complicated. All settings can be adjusted to user values (e.g. noise cut-off values) to adapt them to special applications or the sample (amounts) used in a certain lab. From this data file (an example can be retrieved in form of an 'MGF' file from the website of this book) is created and exported. This file is a lot smaller than the original data file (e.g. down from 40 MB to 300 kB) and is in principle a text file. The file is then uploaded into a database search program, and the right parameters need to be set in this program for an efficient search. These parameters are much like the ones for PMF (see Section 4.2.1) with some all-important additions. These include the type of analyzer and fragmentation energy/method used, as different fragments are considered for the MS/MS search. Most fragmentation types generate b and y ions, and higher energies introduce additional fragment types. The more fragments types are considered in a search, the bigger the search space and thus the bigger the chance of spurious hits and the smaller the significance of any single matching fragment mass. On the other hand, if the observable fragments are not included in the search list, they will be missed, and the chances of a specific identification will be reduced. Another setting is the accuracy for precursor and fragment ions, which they might be different. The setting for how many proteins you expect should make you suspicious; it makes it obvious that there is no easy answer as to which proteins are present or not and the various algorithms and software programs will deliver slightly different results. However, the results should be

Figure 4.16 Nano-HPLC ESI IT MS/MS for protein identification from a protein separated by 2D gel electrophoresis. (a) The MS TIC lane from a short interval of the LC run (upper panel), the MS signal at the base peak (the strongest peak) in the spectrum in positive mode at 15 minutes (middle panel) and the MS/MS spectrum from this ion at 15.1 minutes (lower panel). (b) Automatic database search result. The example is only one of several ions that were found for this protein (triosephosphate isomerase (c)). The example in (b) is one of the only two peptide MS/MS data sets that were significant hits on their own (i.e. ion scores higher than 76) for a peptide of triosephosphate isomerase. However, together with the information from the other peptides (although not significant on their own), the match is very reliable, with a sequence coverage of 50% (bold in sequence in (c)). Three MS/MS spectra per peptide in (c) may indicate that the peak elution time is slightly longer than the dynamic exclusion time of 60 seconds. Note that the peptide with the calculated Mw of 1 585.7307 Da was measured in the +2 and +3 charge state and the relatively great variations in accuracy, up to 1Da. Note also how many peptides show the carbamidomethylation of Cys; without this treatment Cys would have been oxidized in one of several stages, and thus the peptides not detected because of their deviation from the expected mass. For more details see text.

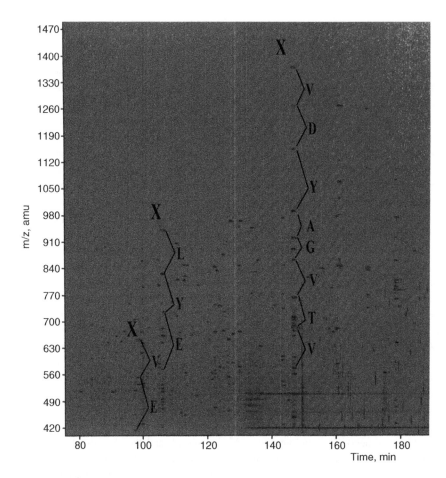

Figure 4.17 Skimmer fragmentation during nano-LC MS for peptide identification. The zoom into a larger contour plot shows three peptides marked by 'X' and fragment ions that perfectly co-eluted with these peptides. The indicated amino acids could be deduced from the b-ion series. Together with data from some other peptides, murine RKIP (Raf inhibitory protein) was identified from a 2D gel spot using a single quadrupole and skimmer fragmentation.

similar for the most part, differing only in the details (which isoforms, which species, which contaminants). If this is not the case, then the data or your search parameter settings need to be significantly improved.

The database search itself for a single experiment can take several minutes (depending on the number of MS/MS spectra and database size), compared to seconds on the same sized server for a PMF. The data output is also increased compared to a PMF. Here we encounter some challenges that will be more important the more complex the samples is (see Section 'MS/MS analysis of peptides from complex mixtures'). In fact, the problems are similar to those encountered in PMF, but this time our tool is sharper, and we are actually able to see the problems.

If we encounter two peptides of a protein in PMF, it will not show up as a hit, as too many other proteins could fit the data. In MS/MS analysis every single peptide can indicate the presence of a specific protein and the presence of two peptides with significant matches is often one critical criterion chosen by most experimenters to define the identification of a protein. Thus two peptides could identify a contaminant missed with PMF. Even if a hit for a single peptide is significant (according to the search software), it only means a hit for a peptide, not a protein. Several proteins share peptides, even if they have different functions. This has a lot to do with the evolution of life; beginning with the first organisms, life has always used modules to create new genes, even new functions. Remains from this evolutionary mechanism are still visible in the conservation of short stretches of amino acids. On the positive side, this allows the analysis of proteomes of species with unknown genomes using MS/MS analyses (see Section 5.3.2 for an example of newts with uncharacterized genomes). For every single peptide, several database entries will show up as the 'hits' in complex mixtures. In a typical spot from a 2D gel, some 100–500 MS/MS spectra are generated. About 30% of these will have been generated from the same peptides (see Figure 4.16), fragmented several times (e.g. in several charge states). Of the 30–100 peptides, 4–8 might be summarized to the most significant protein hit, another 4–12 might belong to another 2–3 proteins co-migrating with the protein on the gel. The rest will belong to a variety of keratins (highly conserved proteins), trypsin and other environmental contaminants. The majority of fragmented peptides will not be assigned to any protein at all; the data are not of a good enough quality. If you go to a single band from an SDS PAGE gel, depending on the sample complexity and gel size, some 10–30 proteins can be reliably identified. Just multiply the number of MS/MS spectra from a 2D gel spot by 3–20 to see the how much information there is in one gel band! In the case of spots from 2D gels the major protein is usually identified by many more peptides than other significant hits, often allowing it to detemine the protein whose changed abundance has caused the spot intensity to increase. On SDS PAGE gels this is often only possible for highly purified proteins, and not for lysates or gels with more than dozens of bands on them. Highly accurate FT or ToF instruments can reduce the amount of unassigned peptides and increase the overall sequence coverage of the identified proteins. But even with less accurate instruments, positive identifications typically have very high scores, and stay significant at a p value of 1×10^{-8} (see Figure 4.6). Although combining scores from peptides towards scores for proteins is not straightforward, it is possible with some approximation (hence, for example, the more conservative 'at least two peptide for one protein' rule). The combined scores for several peptides are astronomical; to achieve this result 'by chance' you would have to repeat the experiment a million times, thus exceeding 'court room' significance, which is several orders of magnitude higher than what PMF can achieve typically (see Section 4.2.1). Despite all this 'significance', if some peptides are missing, we still might not know which isoform we identified (see Section 'MS/MS

analysis of peptides from complex mixtures'). Because there are so many possible caveats (e.g. if three possible internal fragments have the same weight within the specified accuracy, one detected ion may count three times and thus result in an erroneously inflated score), most researchers would consider a protein as identified if at least two peptides with a significant score are identified; they would disregard a 'one hit wonder', a protein identified by only one peptide, no matter how good the score is. The example in Figure 4.18 identified human G3P with great confidence from a 2D spot, and as minor contamination also HNRNPA2B1, from one peptide in position 4 and from an additional three peptides in position 5 of the 'hitparade'. Closer inspection showed that the fourth hit is for an incomplete sequence in the database. Hence the sequence is shorter and the hit was deemed more significant, as it is more likely to get 'by chance hits' the longer the sequence. The peptide for this hit was allocated to both protein hits, a typical solution to the uncertainty problem, although we have no indication that the partial sequence is even expressed as such. HNRNPA2B1 and G3P have the same theoretical molecular weight (within 1 kDa) and pI (within 0.03 pH units), so it is easy to see why they should 'share' one spot on a 2D gel.

MS/MS analysis of peptides from complex mixtures

MS/MS analysis from samples of complex protein mixtures can be performed using MALDI or ESI ionization; MALDI is used for offline analysis, ESI for online analysis. The use of MALDI adds another complication to the proceedings; the sample has to be spotted on a moving target plate together with a matrix solution. MALDI based offline RPLC MS is not performed for samples of simple composition, as the potential disadvantages are far greater than the potential benefits; in simple samples the ion suppression by diverse peptides is not a limitation, while it would be in complex samples. The MS/MS data acquisition time is much longer than the actual LC run (see also Section 5.3.2), so why do it in the first place? One reason might be the availability of mass spectrometers, the other is that MALDI offers complementary coverage of peptides compared to ESI and shows advantages for isotopic labelling approaches involving the quantification of small fragments (isobaric tags for relative and absolute quantification (iTRAQ), see Section 5.3.2). It also offers high resolution and accuracy as well as the chance to

1) <u>G3P_HUMAN</u>   Glyceraldehyde-3-phosphate dehydrogenase OS=Homo sapiens GN=GAPDH PE=1 SV=3
2) <u>B2RA01_HUMAN</u>   cDNA, FLJ94638, highly similar to Homo sapiens keratin 1 (epidermolytic hyperkeratosis)
3) <u>Q0IIN1_HUMAN</u>   Keratin 77 OS=Homo sapiens GN=KRT77 PE=2 SV=1
4) <u>Q9BWA9_HUMAN</u>   HNRPA2B1 protein OS=Homo sapiens PE=2 SV=1
5) <u>ROA2_HUMAN</u>   Heterogeneous nuclear ribonucleoproteins A2/B1 OS=Homo sapiens GN=HNRNPA2B1 PI
6) <u>K1C10_HUMAN</u>   Keratin, type I cytoskeletal 10 OS=Homo sapiens GN=KRT10 PE=1 SV=4
7) <u>K2C71_HUMAN</u>   Keratin, type II cytoskeletal 71 OS=Homo sapiens GN=KRT71 PE=1 SV=2

(a)

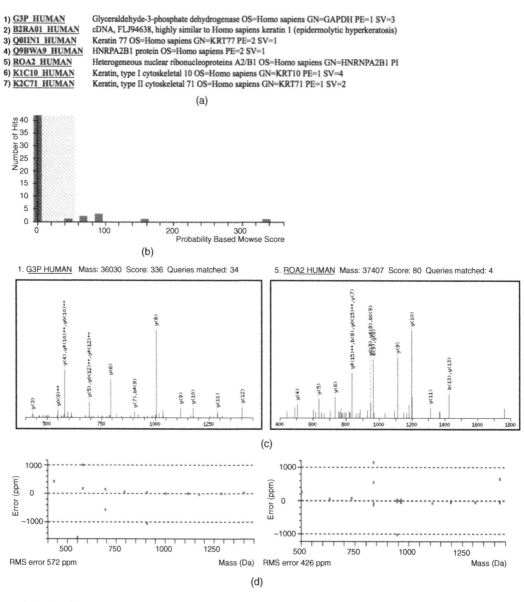

(b)

(c)

(d)

Figure 4.18 Protein identification by database searches with MS/MS data. A spot from a 2D gel was analyzed by nano-HPLC ESI IT MS/MS. (a) Shows the 'hit parade' in descending order of scores from a database search including all MS and MS/MS data of the run. (b) Shows the score distribution, showing the first seven hits (list in a) were regarded as identified with significant scores, which are higher than the scores of insignificant hits, indicated by the shaded area. (c) Shows the detailed results of the matching of fragmentations observed in the experiments with predicted fragmentations 'in silico' from database entries. In (d) the resulting deduced experimental error is shown across the mass range. Note that the score shown in (c) is the total score deduced from all peptide matches, not the individual score for the one peptide only. For the G3P_HUMAN peptide only 16 of the observed 132 fragments matched to any predicted b/y ions. For the ROA2_HUMAN peptide 17 of 134 fragments could be matched. Note that in both cases wrong matches were made, as several times the same experimental fragment was matched to more than one in silico generated fragment. The real average error in this experiment was thus below 200 ppm. More reliable results are achieved if searches are performed assuming and experimental error up to 1 000 ppm as the accuracy can vary (for example depending on the filling state of the ion trap). For more details see text.

reanalyze the data for weeks, if the target plate is stored in a dry and dark place. Ions not chosen for MS/MS can then be re-evaluated, for example.

However, most MS/MS from complex mixtures is performed using online ESI, mainly from nano-RP columns, but recently also using capillary electrophoresis, offering similar flow rates and better separation speed. It is possible in principle to split the samples of one run for MALDI and online ESI, but the complexity of this approach is too great and the same separation conditions are not optimal for both ionization technologies (e.g. MALDI can take advantage of trifluoroacetic acid in LC buffer, for better peptide separation; ESI is not compatible with trifluoroacetic acid and uses formic acid instead).

MS/MS data from complex samples are in principle the same kind of data and use the same analysis of spectra as LC MS/MS from purified proteins. However, instead of several dozen peptides (creating hundreds of MS/MS spectra), several hundred peptides (creating thousands of MS/MS spectra) per LC run have to be measured. The thousands of MS/MS data sets are the tip of the iceberg of even more MS data, which were not strong enough to warrant MS/MS analysis. Time is a critical factor, as new peptides keep relentlessly coming at the mass analyzer. Avoiding redundant measurements (a minor nuisance in simple samples) is absolutely critical for high throughput data generation of complex samples. The goal in high throughput data generation is to find settings that are a compromise between the best quality and the fastest acquisition time; in other words, to obtain the lowest possible data quality one can get away with. In a 100-minute run, with peak length about 30 seconds, 20–100 peptides/ions may be co-eluting at any given time, and many more peptides are hidden in the chemical noise. It is thus extremely important to optimize the data-dependent acquisition (DDA) parameters to get the best out of the combination of LC/mass analyzer for a certain sample and a particular goal. For example, in quantitative analyses reproducibility is less of a factor than in quantitative analyses, where different runs have to be compared and the ion selection procedures for MS/MS have to be more robust, which makes it easier to miss low intensity peptides. Just think how your mass analyzer must feel, measuring all these peptides, when you compare the contour plot of Figures 4.19 and 4.13.

DDA is absolutely essential; constant data acquisition (CDA, for example – one MS scan, one MS/MS scan)

has no chance of obtaining MS/MS data from anything but a handful of the most abundant peptides. High specification mass spectrometers are better suited to the task than instruments with slow acquisition times (at comparable sensitivities/accuracies). In particular, any manual interpretation or analysis of the data would take days for every LC run.

Single LC runs of complex samples are, for example, gel slices from GeLC MS approaches, or fractions from an ion exchange column or a peptide IEF. This means that for one experiment some 10–20 LC runs (like the one shown in Figure 4.19) are combined into a single experiment. For GeLC MS the combination of data can be performed at the peptide level, or (more specific) by analysing the spectra for each gel slice, carrying out the protein identifications and then combining the results of the database searches at the end. Other peptide pre-fractionations have to combine their data at the peptide level. In such a combined set of a complex lysate, several tens of thousands of MS/MS data sets, several thousand identified peptides (often with identical peptides identified in different fractions/RP runs) and more than a thousand identified proteins can be expected. In repeat runs, cell fractionation studies or quantitative approaches, sometimes more than a million MS/MS data sets have to be analyzed in a combined way. It is obvious that even looking at the results is not for the faint hearted, never mind getting to those results in the first place. It is major challenge to deal with the data – it is no longer possible to send some files over the internet to some remote server to get some identifications; dedicated servers on more than one processor are needed and they need hours for the database searches of combined runs. Up to this point the data analysis is straightforward – there is just 'a little' more of it as compared to simple samples. However, the combination of peptide data into protein annotations is a major challenge; the data quality is often lower than on simple samples, a significant proportion of peptides can be shared between several proteins identified in the experiment as a mixture of thousands of proteins was analyzed. Just take vimentin and desmin as an example. They are completely different proteins with very different functions, and yet share 62% protein sequence identity. This results in six identical tryptic peptides (of about 40 tryptical peptides in total). This 15% overlap at the peptide level will influence the MS/MS based identifications or quantifications of the

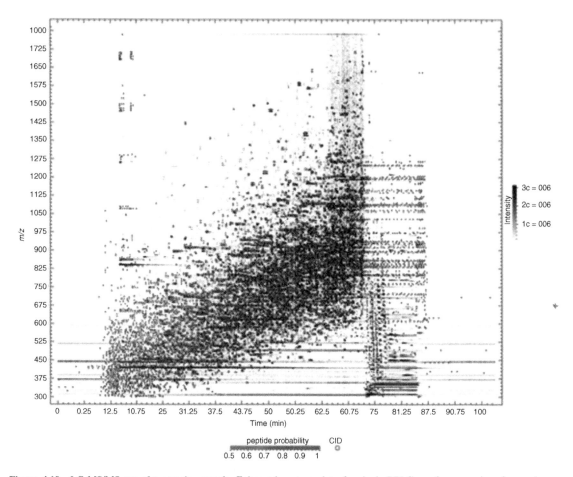

Figure 4.19 LC MS/MS run of a complex sample. Enhanced contour plot of a single RPLC run from a series of several runs comprising a shotgun 2D LC experiment. Such a data set can be used for comparison of different LC runs or for the set-up of an AMT database (see Figure 4.9). This is a high quality data set, as can be seen by the high rate of confident peptide identifications indicated for many peptides. The figure shows the huge amount of data obtained in a single part of a larger shotgun experiment. It is clear that even simple tasks such as peptide comparison (never mind the more complex quantifications) become very complex due to the large number of peptides and the data content. Reprinted from Deutsch *et al.* (2010), courtesy of Elsevier. © 2010 Elsevier.

proteins to some extent, as will the presence of other proteins in the sample with some degree of overlap.

In a typical study results from several runs have to be integrated. For quantitative studies this means a summation of all identifications; for qualitative studies it can mean a much more labour intensive comparison (see Section 4.4).

From all the above it follows that data analysis for complex samples has to be performed automatically; it is simply impossible to achieve this manually. Figure 4.20 shows a typical data handling approach. Note that this example contains manual inspection possibilities at

several levels, which is highly recommended, as it is important to make sure that the data handling produced valid results. Also note that the whole workfloor consists of a variety of software programs that can also work independently of each other. It is important to make sure that import/export of data does not involve losses (e.g. losing the elution time of the peptides), as this also involves losing specificity and manual evaluation potential. Some steps (e.g. converting peptide identifications into protein identifications) can be performed by several algorithms, as the task is not strictly deterministic (data are ambiguous or lost) and there is no 'right or wrong',

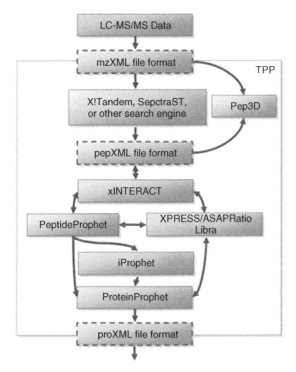

Figure 4.20 Data analysis strategy for complex samples using LC MS/MS. The raw data are simplified by the software of the mass spectrometer, to line spectra, which enter in this example the trans-proteomic pipeline (TPP). The simplified MS and MS/MS data are now generic and can be analyzed by several database search engines. The simplified data as well as the search results can be manually evaluated using Pep3D (see, for example, Figure 4.19). The spectral data are evaluated using PeptideProphet, while peptide identification data are evaluated by iProphet. The final protein level data are evaluated by Protein-Prophet, which is a critical step as there is a loss of information from the peptide to the protein level. Quantifications are then performed with different algorithms, using, for example, XPRESS. The final results are exported in the proXML format, which again can be analyzed by various software. Looking at the complexity and the huge data amount involved in, for example, shotgun quantification, it is easy to see that data analysis takes a lot more time than data generation. Reprinted from Deutsch *et al.* 2010, courtesy of Elsevier. © 2010 Elsevier.

just 'better or worse', and a combination of software for the same task produces overlapping results. Accepting only decisions made by more than one algorithm is conservative but leads to more reliable results. It is very important to control manually all the critical annotations, to make sure all the settings are correct and produce reliable results in an efficient manner. This manual

evaluation is very time-consuming, given the huge data sets. The ambiguity and the different data handling approaches possible between different laboratories are one reason why different studies produce overlapping and complementary data sets (see, for example, Lau *et al.*, 2007, and Section 4.5). Another important reason for ambiguities is that most identifications rely on two peptides only; data for proteins with five or more peptides for identification (more abundant proteins) are more reliable, but represent a small number of the total proteins identified. Even with 2 000 identified proteins in an experiment from mammalian cells, we are usually just scratching the surface of the proteome; repeat experiments will increase this number, but another laboratory, using slightly different equipment, will produce another set of data which will not show a complete overlap. This will be illustrated in several examples in Chapter 5.

Targeted mass spectrometric approaches for the detection and quantification of PTMs

In principle, most PTMs can be picked up by MS/MS approaches with large sequence/peptide coverage of the proteins in question. However, the database search must be error tolerant with regard to the original mass of the peptide. An outright search for PTMs is not a successful strategy under these circumstances. This is mainly because allowing for all possible/presumed PTMs would render the search impossible due to the enlarged search space. Despite having an error tolerant original peptide mass, the MS/MS fragmentation can still identify a peptide, as internal sequence stretches without modification will produce partial fragmentation series. Methylation, acetylation and phosphorylation can be inferred by the observation of common mass shifts (+14, +42 and +80 Da, respectively). The detection of glycosylations by mass shifts is a lot more difficult, as mass shifts involve more than several hundred dalton and a combination of different sugars, the most common ones being GalNAc and GlcNAc, which both produce an increase in mass of 203 Da. In practice a repeated, sequential search pattern is most successful (Figure 4.21).

Generic searches for PTMs are not very likely to find exhaustive amounts of PTMs, as peptides with PTMs are notoriously underrepresented for several reasons. PTMs might interfere with detection (e.g. phosphorylated peptides are often weaker than their

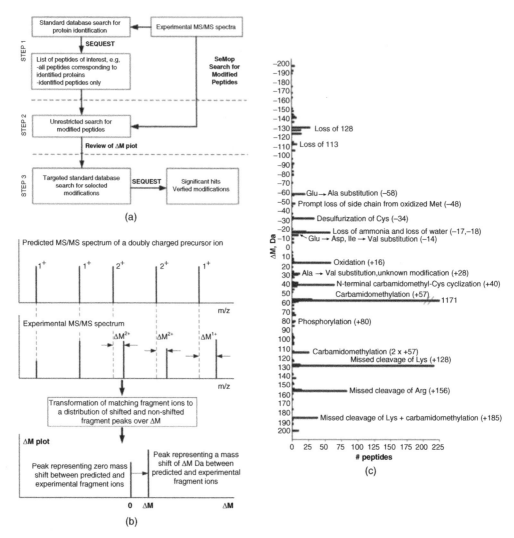

Figure 4.21 Generic strategy for untargeted search of PTMs. In this strategy (a) peptides and proteins are first identified as in any other shotgun approach. Then from the pool of database entries present in the sample an extensive search for all possible modifications is performed (b), taking all fragments into account that do not match the predicted fragment pattern. This is not possible in a first step, as the search space would be too large for any peptide identification. (c) Shows the modifications detected by this approach and the frequency with which they were detected. It is obvious that without enriching, for example, phosphorylations, the vast amount of chemical modifications (artefacts?) is a lot more common than any PTMs, stressing the need for careful sample handling. Baumgartner *et al.* (2008). Reprinted with permission. © 2008 American Chemical society.

non-phosphorylated counterparts in positive mode MS), the stoichiometry of PTM may be low, leading to weak signals, or modifications might be lost during experiments. Targeted searches for PTMs are thus more efficient (see Section 5.4). Targeting PTMs in the MS strategy for data acquisition or data analysis is an important step for the overall targeting strategy.

A typical targeted approach for the detection of phosphorylations is shown in Figure 4.22. A precursor ion scan for the −79 Da ions (phosphate) is performed in the negative mode during the LC separation. If any sufficiently strong signals are detected, a medium resolution scan of the mass range at which the −79 Da fragment originated in the positive mode is started, the peptides are subjected

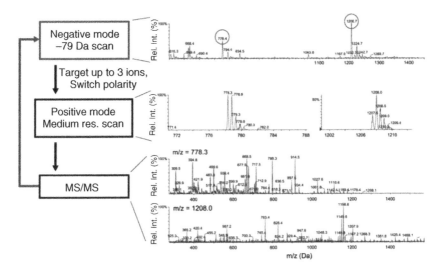

Figure 4.22 Targeted data acquisition strategy for detection and identification of phosphopeptides. Proteins from cellular lysates are separated by anion exchange chromatography and collected in over 60 fractions. Proteins in each fraction are digested with trypsin and the resulting peptides analyzed by two parallel RPLC runs. One is online coupled to a q-trap, for negative mode precursor ion scan searching for a fragment of −79 Da (phosphate), followed by a medium resolution MS scan to determine the exact mass of the peptide in the positive mode. The other RPLC run is online coupled to a hybrid LIT-orbitrap analyzer, to perform MS/MS of the peptides known to carry phosphorylations from the q-trap analysis. This approach takes full advantage of the superb sensitivity of the precursor ion scan on triple quad like instruments, the high specificity of the −79 Da fragment for phosphorylation and the high performance of orbitrap analyzers. For quantification between different signalling situations, contour plots of −79 Da precursor ions are compared, as they are less 'crowded' than positive mode MS plots. Very sensitive analyses reaching down to low expressed phosphoproteins/kinases are possible with this approach. Reprinted from Old *et al.* (2009), courtesy of Elsevier. © 2009 Elsevier.

to MS/MS analysis and the negative ion mode precursor scan for −79 is resumed. This method can allow the identification of hundreds of phosphorylation sites in a complex lysate, together with the identification/sequence analysis of the peptides.

Other targeting strategies include MS³ approaches with ion traps (MS/MS for neutral loss of PTM specific fragments followed by MS³ for peptide identification) and neutral loss scans for PTM specific ions in triple quadrupoles or hybrid trap instruments. In such approaches q-traps are able to scan for multiple neutral losses in a single experiment and thus search for different PTMs at the same time. Targeted approaches can find thousands of modified proteins/peptides in one experiment and can thus involve highly intensive data analysis.

4.3.2 MS/MS analysis of complete proteins

From a data analysis point of view, analysis of MS/MS data of complete proteins is straightforward, as it is only possible for highly enriched/purified proteins in the

first place. In top-down MS/MS, different fragmentation methods than for the MS/MS analysis of peptides are needed. The ubiquitous CID does not transfer enough energy or cannot induce the fragmentations necessary to cover all or even most of the protein sequence. Efficient fragmentation methods for MS/MS analyses of complete proteins are ETD/ECD or infrared multiphoton dissociation (see Sections 3.1.2 and 3.3.6). In Section 4.2.2 we saw that proteins are always multiply-charged in ESI MS and that deconvolution can take all the data from all the charge states to convert it into a de-convoluted spectrum, showing a single charge state. For MS/MS analysis only one charge state (the most intense one) can be used. Usually the sequence of the protein in question is known and the different masses in all the charge states can be assigned to specific modifications (see Figure 4.23). A specific protein isoform from the charge state with the strongest ion intensity is isolated and subjected to ETD/ECD or infrared multiphoton dissociation. Note that the resolution for precursor ion selection is at least in the same range as for high resolution peptide MS/MS

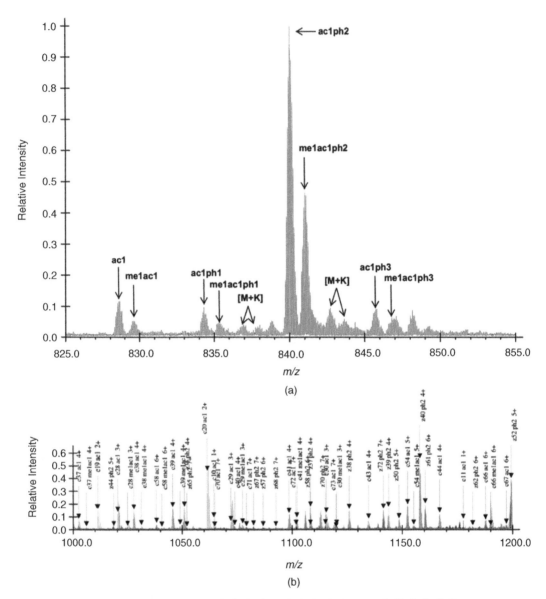

(a)

(b)

ac-S E S S S S K S S Q P L A S K Q E K D G T E K R
     me
  G R G R P R K Q P P V S P G T A L V G S Q K E P
S E V P T P K R P R G R P K G S K N K G A A K T
  R K T T T T T P G R K P R G R P K K L E K E E E E
        2ph
  G I S Q E S S E E E Q

(c)

precursor selection. In a successful fragmentation experiment hundreds of fragments, often in a variety of charge states, are generated. These need to be transformed into a single charge state by deconvolution (automatic with manual evaluation). This is the first hurdle where ultra high resolution (typically 60 000–100 000) is essential for correct interpretation. If the isotopic peaks can be observed, than the assignment to a charge state is very reliable. Now the resulting fragments in the single charge state have to be assigned to certain fragments of the protein. Here again high resolution and the resulting high accuracy are absolutely necessary; only at an accuracy of several parts per million or better is it possible to get correct assignments from the multitude of possibilities. In a sequence with hundreds of amino acids, the most common ion series (c and z) create hundreds of possible fragments; add to this the presumed modifications and internal fragments and you end up with a formidable task. Usually a mixed approach of in silico generated candidates and manual evaluation is followed. Note that in the example in Figure 4.23 with 'only' four PTMs top-down could not localize all of them without ambiguities; here a combination with bottom-up proteomics of peptides can be very useful to support and extend the top-down interpretations. Also note that pure bottom-up proteomics cannot distinguish between protein isoforms with different combinations of PTMs; only top-down can achieve this, based on the accurate masses of the intact proteins. Rare combinatorial isoforms (intensity below about 5% of the total) are difficult to analyze by top-down. Using more material, these rare isoforms can be enriched, for example by free flow IEF, which then makes them intense enough in the enriched fractions for thorough analysis by top-down MS/MS.

4.4 QUANTIFICATION OF LC MS AND MS/MS DATA FROM COMPLEX SAMPLES

MS based quantifications of gel separated single proteins are generally possible but not performed often, as non-MS based methods (i.e. image analysis of stained gels) are well up to the task, usually in steps before the MS analysis. However, for shotgun proteomics, there is no alternative and MS or MS/MS based quantifications of peptides have to be combined to result in quantifications of the underlying proteins. The lack of reliable quantification of shotgun proteomics 'for the masses' has been the single biggest argument in favour of gel-based approaches. The situation is now changing, as different strategies and, most importantly, more and more software tools that are up to this complex task are becoming freely or commercially available. Still, hitherto, most inspiring quantitative shotgun proteomic projects have included some elements of in-house developed software. Differential labelling of proteins or peptides resulting in 'light' and 'heavy' versions of peptides or MS/MS tags have given all shotgun based proteomic approaches the capability of comparative quantification of proteomes. The main advantage of labelling approaches is the ease of data analysis necessary for quantification. However, as software gets better at coping with the enormous tasks of data analysis, labelling (with its inherent disadvantages such as higher cost, experimental set-up complications, higher incidence of variations due to additional experimental steps) is losing ground to the simpler and more robust label-free relative quantification approaches. Absolute quantifications have many advantages for proteomics, as they allow data integration across laboratories (see Section 5.3). They are possible with labelled and unlabelled approaches. All shotgun quantification methods

Figure 4.23 MS/MS on complete proteins for the identification of combinatorial PTMs. HMGA1 is an 11 kDa heterochromatin associated nuclear protein with differential PTMs in malignant cells. ESI MS analysis of the purified whole protein in a hybrid LIT-orbitrap analyzer yields the various modifications carried by the protein, for example in one of its 14-fold charged ions ((a), ac = acetylation, me = methylation, ph = phosphorylation, M + K = Potassium adducts). ETD fragmentation of the whole protein (isolated peak in spectrum (a) from 840 to 843 m/z) was performed yielding hundreds of fragments. In (b) just a small part of the spectrum is shown, indicating the large amount of c and z-type ions in various charge states (+1 to +7). In (c) the coverage of the whole protein sequence by these fragments is shown, indicating for which amino acids within the protein sequence evidence for modifications could be found in the form of c and z fragments derived from ETD. Many more PTMs were found by middle-up and bottom-up approaches on HMGA1 in this study. However, the exact amino acid positions at which the protein was modified could not be established. Note that in (a) the ratio of intensity of the ac + me peaks to the ac peaks stays the same for all of the differential phosphorylation states of the protein: unphosphorylated, once, twice and three times phosphorylated. Reprinted from Young *et al.*(2010), courtesy of Elsevier. © 2010 Elsevier.

work better with lower sample complexity per LC run; less peptides overlap leads to higher quality data, labelling induced 'crowding' of spectra is less of a problem, and DDA picks the same peptides more reliably for MS/MS. For GeLC MS quantifications lower sample complexity can be achieved with more slices from larger gels. For 2D chromatography, using more elution steps in the gradient of the first dimension LC or the addition of a third dimension can reduce the sample complexity and enhance the quantification capabilities. Isoelectric focussing of peptides/proteins such as in OFFGEL is ideal as an additional dimension, since it allows more independent information to be gathered, in form of the pI of the peptide or protein, which might come in handy when PTMs are analyzed. Pre-fractionation of complex lysates (e.g. into nucleus/cytoplasm/membranes) and very shallow elution gradients (90 minutes and longer) for the ultimate RP run are helpful for all approaches.

Successful and popular labelling methods include SILAC, isotope-coded protein labelling (ICPL) as well as isotope-coded affinity tagging for MS quantifications and iTRAQ and tandem mass tag (TMT) for MS/MS quantifications. The different quantification approaches will be introduced in Section 5.3 in more detail.

From a data analysis point of view, iTRAQ and TMT are the easiest to handle in principle. Samples are labelled with different tags, resulting in labelled peptides indistinguishable (isobaric) in the MS mode (Figure 5.10). Only in the MS/MS mode do the labelled peptides produce fragments of different masses. The ratios of the reporter fragments are taken as ratios of the peptides in the different samples. The data are easy to process, as defined peaks in the low mass area of the MS/MS spectrum are compared, and the data can easily be attached to the peptide properties. After assignment of all peptides to a protein (with all the problems mentioned in Section 'MS/MS analysis of peptides from complex mixtures') a mean value for the different peptide ratios is determined to deduce the protein ratios between the samples (see Figure 3.41 c). Up to eight different tags are available, allowing the comparison of just as many samples in one experiment. Peptides with significantly different ratios from other peptides assigned to a given protein are removed or given a smaller weighting factor. This improves the quality of the results in terms of matching control sets of proteins with known ratios. There may be many reasons for these outliers: a peptide

might be wrongly assigned, shared between different proteins, carry a PTM, or (the commonest and most prosaic reason) might only show a low intensity.

The methods are very accurate for peptide ratios close to one, allowing sensitive detection of changes, and become less accurate for larger ratios, often regarded as less important; see the discussion of multiple reaction monitoring (MRM) below. Absolute quantifications are possible, using internal standards, for some but not all proteins, as that would require too many standards (see Section 5.3).

From a data analysis point of view, much of the relative ease of data analysis from labelling approaches comes from the fact that the commercial providers of the compounds used for labelling also deliver software and mass spectrometers for the subsequent analysis and provide a 'one-stop shop' solution. Note that manual evaluation is still needed to get the full benefit of the analysis.

Quantification via 'heavy' labelling produces pairs of peptide peaks, separated in mass by an amount depending on the labelling compounds, usually 4–8 Da (Figure 4.24). Smaller mass differences would cause interference between the different isotopic peaks of heavy and light peptides and are thus avoided; larger differences in the mass of the labelled peptides are more difficult to achieve. From a data handling point of view, co-eluting ions with a fixed distance have to be detected in a shotgun data set. This is usually done after the run; DDA can thus interfere with quantification efficiency. In principle, it would be enough to perform MS/MS analysis only on peptides which (i) are clearly represented by a heavy/light pair of ions and (ii) show different intensities between the heavy and light forms. In reality, the mass analyzer will spend a lot of time fragmenting either peptides that do not fulfil criterion (i) or (ii) (a similar problem occurs with iTRAQ/TMT). This is one of the reasons why only a fraction of identified peptides can also be quantified. In addition, any time spent on MS/MS (essential for identification of peptides) is lost for data needed for quantifications, and (again) a compromise in the data acquisition must be struck. While detection of ions with fixed distances is not in itself challenging, many factors conspire to cause problems. Co-eluting ions can interfere with signal intensities of light and/or heavy peptides; in particular, if the signals are much stronger than those of the original peptide couple, these interferences can spread

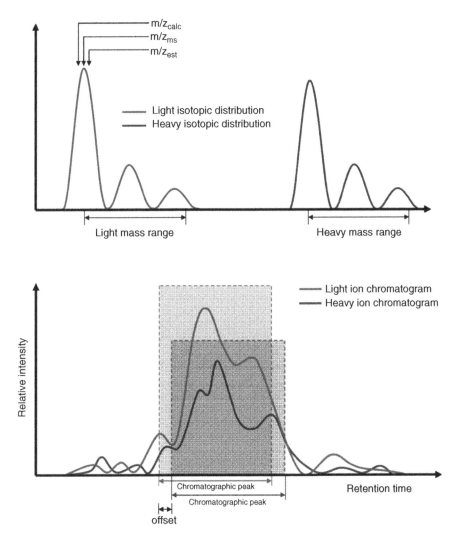

Figure 4.24   Quantification of LC MS sigal intensities from isotopic labelling approaches. (a) Shows the ions of a light/heavy labelled peptide. For XIC based quantification several choices have to be made. Which exact mass is used to mark a peak? Which intensity is used for comparison, the intensity of only the monoisotopic peak or of three or even more isotopic peaks? As each peak has its own deviation from theoretical values each, choice will create its own problems. (b) Shows how an averaged XIC lane for the light and heavy peptides might look; slight variations in elution time have to be considered. It is also not straightforward to determine where the peaks start and the 'background' signal ends, especially for weak peptides. Reprinted from Lau *et al.* (2007), courtesy of Elsevier. © 2007 Elsevier.

over −5 Da from the light monoisotope to +2 Da of the third isotope of the heavy peptide ion. In busy spectra it is quite common to have peaks within the resulting 14–18 Da 'danger zone'. The interfering peaks also do not have to co-elute exactly; overlapping elution is enough, again particularly for strong interfering peaks and weak light/heavy ion pairs (compare Figure 5.15).

That such interference is commonplace can be seen on the elution profile (contour plot) in Figure 4.19. For a meaningful quantifications the elution times and intensities of light/heavy peptides have to be determined, compared and chosen for quantification (Figure 4.24).

With all these possibilities for fooling an algorithm, it is perhaps not so cynical to think it a blessing that usually

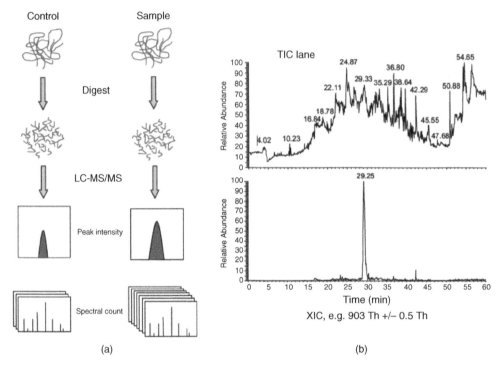

Figure 4.25 Spectral counting and ion extraction for quantification of shotgun data. (a) To compare the expression of peptides (and thus proteins) in shotgun proteomics two label-free methods can be used: either the comparison of peak intensities, or spectral counting. For peak intensity comparisons it is not enough to look at one spectrum and compare intensities of peaks, rather the total amount of MS signals for a peptide as it elutes from LC columns has to be compared between two samples (volume of XIC lane in the lower panel in b). For this purpose the extracted ion current has to be measured (see also Figures 4.13 and 4.24). This can be quite complex for thousands of peaks per LC MS run. If a peak is more intense, it will elute for a longer time and thus also be chosen more often for MS/MS analysis. Thus counting the number of MS/MS spectra for a certain peptide can be used as a semi-quantitative approach. For both methods data for several peptides per protein are averaged. (a) Zhu (2010), (b) Reproduced with permission from Wong *et al.* (2007). © 2007 Oxford University Press.

only hundreds of the thousands of detected peptides are quantifiable and thus need manual evaluation.

Label-free quantification methods are a different kettle of fish. Among their advantages are that no sample manipulations (which might introduce faults) are needed, there is no complex chemistry with the need for quantitative reactions of hundreds of thousands of different components, and these methods are available for every sample type and can easily be expanded to absolute quantifications. The disadvantage is the increased intensity of data handling that is necessary. Two features of a peptide can be compared; its intensity in MS or simply the number of MS/MS spectra that the DDA induced for its analysis (Figure 4.25). While the use of the former appears logical, the latter is 'semi'-quantitative, but works surprisingly well.

For ion intensity comparison the ions need to be extracted from the complete LC MS run; that is, the intensity for the mass of the ion (e.g. 532.5 ± 0.5 Th, Figure 4.25) is determined over the entire time of the run. A peak in elution time indicates the elution time of the ion. The total intensity (in arbitrary instrument units) is determined as the area under the peak, and compared between different runs with mass and time window restrictions, depending on the accuracy of the mass analyzer and the reproducibilities of the LC run, as some data are not easy to define clearly (Figure 4.24).

Spectral counting is a lot easier to carry out, and for low accuracy/resolution mass analyzers like ion traps, is often the better choice. Here MS/MS data exist for peaks in different runs; the ions can thus be relatively easily assigned, for example by their preferred 'hit' in a database search. Despite DDA, stronger ions have the tendency to be chosen more often for MS/MS than weaker ones. So after simply counting how often ions attributed to a particular protein were chosen for MS/MS analyses (spectral count), the abundance of the protein can be estimated. Correction factors for protein length are taken into account, as longer proteins tend to produce more different peptides and thus trigger more MS/MS event. Assuming that on average all peptides are equally well produced, separated and detected, absolute quantities can be estimated using a few internal standards. This 'bootstrap' method has proved accurate over 4 orders of magnitude and works well for proteins that can be detected with more than three to four peptides (which is often only 40% of all identified proteins in the experiment). Independent tests have proved that such a label- free approach tends to be very accurate for unknown samples, and may reflect just how difficult it is to optimize label methods to unknown samples (http://www.abrf.org/index.cfm/group.show/Proteomics.34.html; see also Chen and Yates, 2007).

A variety of free and commercial software exists for label-free quantifications, following slightly different strategies. The most ambitious approaches compare the MS signals over time in a way that is similar to the comparison of 2D gel images (e.g. by comparing contour plots). With hundreds of thousands of features to be compared between two different experiments consisting of several joint RPLC runs, there is an awful lot of data analysis by any standard. You may ask yourself why bother to compare LC MS data when the peptides are not identified by MS/MS anyway? In approaches measuring the elution time and a precise peptide mass (AMT, see Figure 4.9) peptide ion data can be compared to other runs, where MS/MS data were used to identify the peptide. Consistency in elution time and mass may then be enough to align the peptide and the MS signal strength can be used to quantify it. These approaches resemble 2D gel database building and rely heavily on reproducible LC conditions and high resolution mass analyzers.

One of the most accurate methods for shotgun quantifications is MRM using triple quadrupoles or quadrupole-ion trap combinations (q-trap). We have seen that the mass of a peptide combined with its elution time is not informative and specific enough to identify a peptide, or that it can be a complex task to achieve this in

AMT approaches (Figure 4.9). In multiple reaction monitoring (MRM), a further level of specificity is added. The peptide(s) to be quantified have to be known, as has their MS/MS fragmentation pattern. Instead of monitoring the intensity at the mass of a peptide, its mass is chosen to perform MS/MS experiments and the intensity of a couple of the most characteristic/specific MS/MS fragments is used for quantification. Hundreds of instances of such single reaction monitoring can be performed sequentially, forming an MRM experiment. Hundreds of transitions from one precursor ion/fragment pair to another can be performed in cycle times under 10 seconds, so that more than one measurement per eluted peak is performed. The sensitivity and dynamic range of triple quads/q-traps is used by this method to the maximum, as there is no scanning involved; all ions are measured all the time for a precursor/fragment pair and total cycle time is only reduced by the total number of transitions measured. MRM methods need to be developed from prior shotgun MS/MS experiments, to chose reliable transitions (pairs of precursor ion and fragment), and despite increased specificity, manual data validation is still needed (Figure 4.26). MRM offers the highest robustness, reliability accuracy and dynamic range available to date and can also be used for PTM monitoring.

4.5 BIOINFORMATIC APPROACHES FOR MASS SPECTROMETRIC PROTEOME DATA ANALYSIS

Strictly speaking, the whole chapter and much of Chapter 3 deal with bioinformatic approaches in proteomics, and the whole field of proteomics would simply not exist without the massive bioinformatic contributions from data acquisition control to database searching or protein identification and quantification. Figure 4.27 shows a typical data flow for a quantitative LC MS based proteomic study. The data flow for qualitative studies is a lot simpler. What becomes immediately obvious is the modular character of the set-up; data from one stage are exported into another stage, a variety of software tools manipulate the data and integrate data from databases with experimental data to produce results including statistics and quality indicators. What is not shown is the manual intervention that also goes into such an analysis, including optimizing parameters and validation of data or results of manipulations. For a single study the flow of data is mainly one way; database information is used

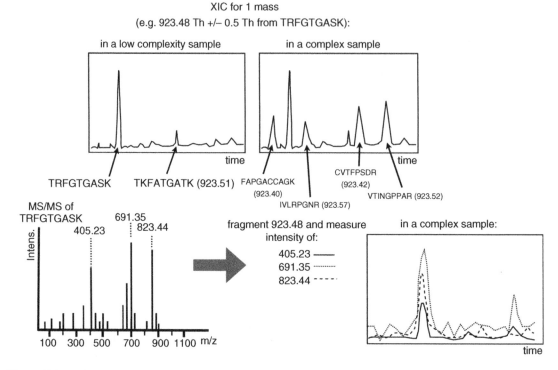

Figure 4.26 MRM for the quantification of shotgun data. To enhance specificity of MS data for quantification triple quadrupole like analyzers can be used to measure the intensity of precursor/fragment combinations. Such coupled pairs are called transitions. It is possible to get highly specific signals for a given peptide even in complex mixtures by measuring about three transitions. Temporal coincidence of the maxima of the transitions shows the specificity. However, the transitions have to be carefully chosen (based on fragmentation data in the given mass analyzer under the same conditions) and evaluated to avoid non-specific interference. The peak volumes of several transitions allow highly accurate quantifications. In the example in the figure the coincidence of all three transition maxima defines a peptide very specifically. If only 2 transitions had been chosen (405.23 Th and 691.35 Th) a second peak, eluting later in the run, could have been erroneously annotated (lower right panel). Up to 200 transitions can be measured continuously during an RPLC elution, allowing the quantification of up to 50–70 peptides.

to enable and support the proteomic study. At a higher level, integrating the data from several proteomic studies, the flow of data goes both ways; databases are used to enable single studies and validated results of these studies are used to complement or extend the databases (Figure 4.28). For the whole process to be possible at all the data have to be generated and stored in a standardized way, otherwise it becomes impossible to integrate the data in a meaningful way. The proteomics community has embraced this task and several organizations have developed standards for submission of data and metadata (e.g. which instruments were used, settings, information on sample preparation, number of biological/technical replicate experiments). Tools have been developed to convert raw and manipulated data so the different

software modules can interact with one another (e.g. Eisenacher *et al.*, 2009); even the commercial software (and mass spectrometry) providers have widely adopted interchangeable standards, or at least there are ways to convert/export the most important features of the data sets into a generic format.

So what are the benefits for the researcher from international, large scale data integrations? Many new tools are generated using the deposited data, and basically your research has not be performed in a way as if it was the first proteomic study on this type of sample. PeptideAtlas, for example, is a peptide oriented database containing spectra of peptides from over 350 data sets (experiments) from all over the world. The actual MS/MS spectra are deposited alongside assignment information, how often

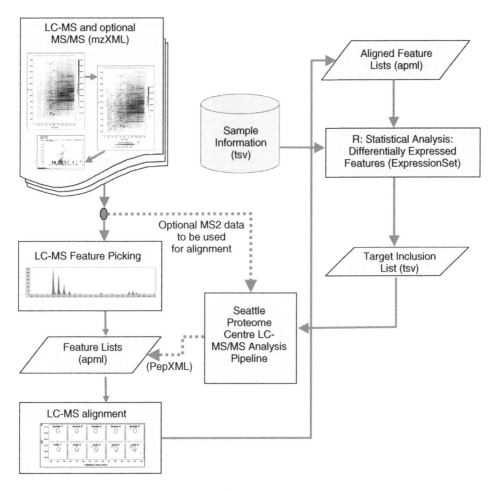

Figure 4.27   Data flow chart for quantitative shotgun proteomics. Corra (http://tools.proteomecenter.org/Corra/corra.html) is a free bioinformatics tool enabling the identification and quantification of proteins from shotgun data in combination with other software tools. As a first step the raw data from mass spectrometers, which have to be able to resolve the charge states/isotopic peaks, have to be converted into the mz/XML format. Corra can than pick the features (peptides) with their elution time, mass, charge and intensities. In tests with 105 LC MS runs and 3 GB of mzXML data per run, this extraction took 25 hours on a cluster of six dual core PCs. The features are then aligned and compared between runs. This can result in a list of peptides that show statistically relevant expression differences, suggesting inclusion lists of hundreds of peptides for repeat MS/MS runs for identification. Reprinted from Brusniak *et al.* (2008), with permission from BioMed Central.

and in which samples the peptide has been found, which instrument was used to generate the spectra and many more. These data can be used to validate experiments with limited data quality, choose transitions for MRM analyses, and many other applications. The data can also be used to validate new or improved data analysis software or strategies, as well as to estimate the quality of your own data or results. The database also contains several tens of thousands of phosphorylation site specific spectra from several species and helps in the planning and targeting of proteomic experiments. Moreover, the database can be searched with access codes, sequences or the spectra of your own experiment. This can give a much higher specificity than a search against a 'theoretical' in silico generated database, as actually observed fragments are searched, and not just predicted ones which may not occur due to physicochemical parameters of the sequence. Peptide data are linked to chromosomes and certain biological

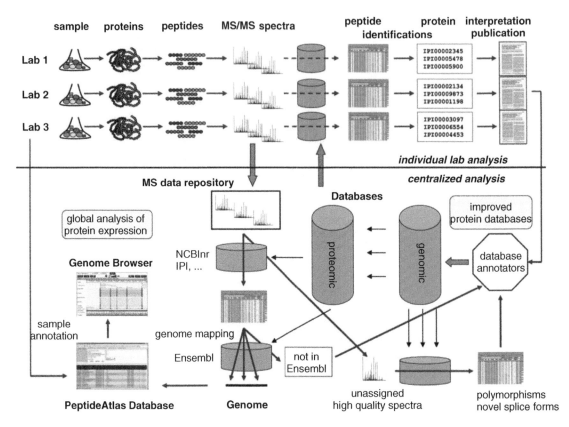

Figure 4.28 Systematic data collection and integration for proteomics. PeptideAtlas (http://www.peptideatlas.org/) contains a collection of peptide shotgun data from various species and different laboratories. Its ultimate aim is to annotate eukaryotic genome databases with proteomic information. Hundreds of thousands of spectra can be retrieved, with all MS/MS information displayed. Such a database has several useful functions, even in its building-up phase. If you wish to know if your peptide has previously been found and which fragments were found (e.g. for MRM monitoring), you query the databases and get original MS/MS spectra from different species, instruments and laboratories. It also contains thousands of PTMs with original data. It is highly desirable that all proteomic data of sufficient quality be published and available in databases/environments such as PeptideAtlas, so that the proteome need not be analyzed from scratch for every new experiment and so as to gain more biological information from proteomics studies. Reproduced from Nesvizhskii & Aebersold, (2005) by permission of the American Society for Biochemistry and Molecular Biology.

functions or diseases, which can be retrieved in linked databases, and can thus help to transform peptide/protein identification data into biological hypotheses.

The bioinformatic resources available for proteomics can be roughly divided into tools and databases, although often they comprise tools used for and with databases using a web based GUI. Table 4.1 shows just some examples of the resources available. Further information can be found on the website of this book.

One of the main problems in proteomics is that the experimental phase often results in a (long) list of interesting proteins (e.g. up or down regulated, phosphorylated,

etc.), but not directly in new biological knowledge. How do we make biological sense of this often huge amount of data? The first major help from bioinformatic sources is the gene ontology (GO) field in the database entries of most proteins/genes. GO information can give a first hint as to which processes are affected in a certain state from which the proteomic sample was derived; if hundreds of proteins are targeted, often an overview is as far as it gets. Finding all the original literature about each of these proteins is not possible for the busy proteomic researcher. GO may offer only limited help here, as the information provided is not very specific. The next step is to see

Table 4.1   Bioinformatic resources for proteomic applications.

Resource	Comments	Source
Databases Uni-Prot/Swiss-Prot Uni-Prot/TrEMBL Ensembl Entrez-Protein-(NCBInr)	These are traditional type databases, containing protein or DNA or RNA sequences, with varying levels of annotation and redundancy/overlap between sequences. They are used for search engines to match MS and MS/MS data to databases entries. Expect monthly updates, some interconnections in unique entry names for each database. Swiss-Prot is the only database that is fully annotated manually. All others are annotated automatically and curated manually. All have some sort of web based GUI and tools like PBLAST to find similar/identical sequences.	http://www.uniprot.org http://www.ensembl./index.html http://www.ebi.ac.uk/pdbe http://www.ncbi.nlm.nih.gov/Database/index.html
GOA	The Gene Ontology Annotation project aims to get high quality data for every protein function, localization. Feeds data into Uni-Prot, NCBI.	http://www.ebi.ac.uk/GOA
PDBe	European version of database for protein 3D structures.	http://www.ebi.ac.uk/pdbe
Search engines MASCOT SEQUEST X!tandem ProtFound Sonar GPM ProteinProspector PepMapper	Most of these search engines are either freely available or offer some free services, and offer more (better) functionality when licensed. Some have versions that work on MS data alone or with MS and MS/MS data together. These are even some open source projects that can cope with huge amounts of data. It is often recommended to install a local version, as one can then also install databases that fit the specific (species) requirements.	http://www.matrixscience.com/ http://www.fields.scripps.edu/sequest/ http://www.thegpm.org/tandem/index.html http://hs2.proteome.ca/prowl/ http://prowl.rockefeller.edu/tandem/thegpm_tandem.html http://prospector.ucsf.edu/prospector/mshome.htm http://www.nwsr.manchester.ac.uk/mapper
Shotgun data tools PeptideAtlas PRIDE Transproteomic Pipeline MSQuant	Various tools for complex shotgun proteomic data handling challenges and also proteomic databases where raw data can be retrieved in a number of ways. These databases also contain elaborate tools to visualize their content. Some of these tools have been introduced in figure legends in this chapter. Some databases contain millions of unique peptides, cross-referenced to a variety of species and tens of thousands of individual experiments. It is highly recommended to submit all proteomic data to such databases.	http://www.peptideatlas.org/ http://www.ebi.ac.uk/pride.init.do http://msquant.alwaysdata.net/

Table 4.1   (*continued*)

Resource	Comments	Source
Interactomics and pathway data REACTOME KEGG Biocarta Pathway Commons IntAct homo Mint Panther	These resources can visualize interactions, show to which pathways proteins of interest belong and what putative functions in a cell are switched on based on pathways that interact for certain functions. Most of the resources are free to use. Visualization tools often have to be installed locally to get intuitive data output. These databases each contain tens of thousands of proteins and typically hundreds of thousands of interactions in thousands of pathways. It is very complex to analyze/visualize a multitude of interconnected pathways and interactions; this is best done with interactive tools in intuitive GUIs.	http://reactome.org/genome.jp/kegg http://biocarta.com http://www.pathwayscommons.org/pc/ http://www.ebi.ac.uk/intact/main.xhtml http://mint.bioinforma2.it/HomoMint/Welcome.do http://www.pantherdb.org/pathway/
General tools ExPASy Systems Biology toolkit	One-stop shops for most proteomic needs, also for gel-based and single analysis. Lots of helps to plan and analyze your data, predict fragments or detect PTMs. Gateways to many databases.	http://www.expasy.ch/ http://db.systemsbiology.net:8080/proteomicsToolkit/

in which pathways and/or protein/protein or protein/gene interactions a target protein has been involved before. Some of the resources in Table 4.1 can help to create and browse networks of interactions or pathways (compare Figures 5.5 and 5.16d). Often several target proteins fit into the same or into related networks, indicating that the biological response in a sample belongs to certain biological pathways. The GO and pathway data are generated and curated in a mixture of automatic and manual annotation, which is a huge amount of work. However, there is still no guarantee that all relevant functions and connections for your proteins in your experiment are completely listed or that the listed ones are not miss-leading, as most proteins have many functions. Following a single lead from a proteomic study can take months or years, but a good study, even a small one, will produce dozens of leads and the pathway software and the GOs can help to streamline and optimize the evaluation efforts.

Bioinformatics is constantly improving the results of proteomic studies on endless 'battlegrounds', often unnoticed. Just as an example, a recent study (Cappadona et al., 2010) was able to show that applying improved noise filters (as opposed to intuitive or instrument derived parameters) to raw MS and MS/MS data can help to achieve a 30% higher success rate in peptide identifications. As most MS/MS spectra in any experiment are towards the low end of the intensity range, this can mean 300 new proteins per analysis or 200 more quantified proteins per shotgun experiment. While these contributions are not available for all experimentalists immediately, they sooner or later filter down into software design of manufacturers or publicly available tools.

REFERENCES

Bitton, D.A., Smith, D.L., Connolly, Y., Scutt, P.J. and Miller, C.J. (2010) An integrated mass spectrometry pipeline identifies novel protein coding-regions in the human genome. *Plos one*, **5** (1), e8849.

Baumgartner, C., Rejtar, T., Kullolli, M., Akella, L.M. and Karger, B.L. (2008) SeMoP: a new computational strategy for the unrestricted search for modified peptides using LC-MS/MS data. *J Proteomics Res*, **7** (9), 4199–4208.

Berndt, P., Hobhom, U. and Langen, H. (1999) Reliable automatic protein identification from matrix assisted laser desorption/ionisation mass spectrometric peptide fingerprints. *Electrophoresis*, **20**, 3521–3526.

Brusniak, M.Y., Bodenmiller, B., Campbell, D. *et al.* (2008) Corra: computational framework and tools for LC-MS discovery and targeted mass spectrometry based proteomics. *BMC Bioinform*, **9**, 542.

Campbell, A.M. and Heyer, L.J. (2003), Discovering Genomics, Proteomics & Bioinformatics, CSHL Press.

Cappadona, S., Nanni, P., Benevento, M. *et al.* (2010) Improved label-free LC-MS analysis by wavelet-based noise reduction. *J Biomed Biotech*, article ID 131505.

Chen, E.I. and Yates, J.R. III (2007) Cancer proteomics by quantitative shotgun proteomics. *Mol Oncol*, **1** (2), 144–159.

Deutsch, E.W., Mendoza, L., Shteynberg, D. *et al.* (2010) A guided tour of the Trans-Proteomic Pipeline. *Proteomics*, **10**, 1150–1159.

Eisenacher, M. *et al.* (2009) Getting a grip on proteomics data – proteomics data collection (PrDaC). *Proteomics*, **9**, 3928–3933.

Gharahdaghi, F., Weinberg, C.R., Meagher, D.A., Imai, B.S. and Mische, S.M. (1999) Mass spectrometric identification of proteins from silver-stained polyacrylamide gels: a method for the removal of silver ions to enhance sensitivity. *Electrophoresis*, **20**, 601–605.

Gloriam, D.E., Orchard, S., Bertinetti, D. *et al.* (2010) A community standard format for the representation of protein affinity reagents. *Mol Cell Proteomics*, **9**, 1–10.

Issaq, H.J., Conrads, T.P., Priete, D.A., Tirumalai, R. and Veenstra, T.D. (2003) SELDI-ToF MS for diagnostic proteomics. *Anal Chem*, **75** (7), 148A–155A.

Keil, B. (1992) *Specificity of Proteolysis*, Springer-Verlag, Berlin, Heidelberg, NewYork, p. 335.

Kussmann, M., Nordhoff, E., Rahbek-Nielsen, H., Haebel, S., Rossel-Larsen, M., Jakobsen, L., Gobom, J., Mirgorodskaya, E., Kroll-Kristensen, A., Palm, L. and Roepstorff, P. (1997) Matrix-assisted laser desorption/ionization mass spectrometry sample preparation techniques designed for various peptide and protein analytes. *J Mass Spectrom*, **32**, 593–601.

Lau, K.W., Jones, A.R., Swainston, N. *et al.* (2007) Capture and analysis of quantitative proteomic data. *Proteomics*, **7** (16), 2787–2799.

Lim, H. *et al.* (2003) Identification of 2D-gel proteins; a comparison of MALDI/ToF peptide mass mapping to micro LC ESI tandem Mass spectrometry. *J Am Soc Mass Spec*, **14**, 957–970.

Lohaus, C., Nolte, A., Bluggel, M., Scheer, C., Klose, J., Gobom, J., Schuler, A., Wiebringhaus, T., Meyer, H.E., and Marcus, K. (2007) Multidimensional Chromatography; a powerful tool for the analysis of membrane proteins in mouse brain. *J Proteome Res*, **6**, 105–113.

Marshall, A.G., Hendrickson, C.L. and Shi, S.D.H. (2002) Scaling MS plateaus with high resolution FT-ICRMS. *Anal Chem* **74** (9), 253A–260A.

Nesvizhskii, A.I. and Aebersold, R. (2005) Interpretation of shotgun proteomic data. *Mol Cell Proteomics*, **4** (10), 1419–1440.

Nucleic Acids Res. (2010) Jan; 38(Database issue), or 2008 Jan; 37, (Database issue).

Ong, S.E. and Mann, M. (2005) Mass spectrometry-based proteomics turns quantitative. *Nature Chem Biol*, **1** (5), 252–262.

Old, W.M., Shabb, J.B., Housel, S. *et al.* (2009) Functional proteomics identifies targets of phosphorylation by B-Raf signaling in melanoma. *Mol Cell*, **34** (1), 115–131.

Quackenbush, J. (2007) Extracting biology from high-dimensional biological data. *J Exp Biol*, **210**, 1507–1517.

Rodriguez, J., Gupta, N., Smith, R.D., and Pevzner, P.A. (2008) Does trypsin cur before proline? *J Proteomics Res*, **7**, 300–305.

Shevchenko, A., Wilm, M., Vorm, O., and Mann, M. (1996) Mass spectrometric sequencing of proteins from silver-stained polyacrylamide gels. *Anal Chem*, **68**, 850–858.

Siepen, J.A., Keevil, E.J., Knight, D. and Hubbard, S.J. (2007) Prediction of missed cleavage site in tryptic peptides aids protein identification in proteomics. *J Proteomics Res*, **6** (1), 399–408.

Suckau, D., Resemann, A., Schuerenberg, M., Hufnagel, P., Franzen, J. and Holle, A. (2003) A novel MALDI LIFT-TOF/TOF mass spectrometer for proteomics. *Anal Bioanal Chem*, **376**, 952–965.

Taylor, C.F., Paton, N.W., Lilley, K.S. *et al.* (2007) The minimum information about a proteomics experiment (MIAPE). *Nature Biotech*, **25** (8), 887–893.

Umar, A., Malgorazata, J., Burges, P.C., Luider, T.M., Foekens, J.A. and Pasa-Tolic, L. (2008) High-throughput proteomics of breast carcinoma cells: a focus on FTICR-MS. *Exp Rev Proteomics*, **5** (3), 445–455.

Wong, J.W.H., Matthew, J.S. and Cagney, G. (2007) Computational methods for the comparative quantification of proteins in label-free LC^{n}-MS experiments. *Brief Bioinform*, **9** (2), 156–165.

Young, N.L., Plazas-Mayorca, M.D., DiMaggio, P.A. *et al.* (2010) Collective mass spectrometry approaches reveal broad and combinatorial modification of high mobility group protein A1a. *J Am Soc Mass Spectrom*, **21**, 960–970.

Zhu, W., Smith, J.W. and Huang, C.M. (2010) Mass spectrometry-based label-free quantitative proteomics. *J Biomed Biotech*, article ID 2010:840518.

5

Strategies in Proteomics

The multitude of strategies employed in proteomics reflects the technical challenges and limited capabilities of the technologies, but also the great variety in proteins, their functions and the resulting questions we wish to answer. There are no strict classifications for all these various strategies, since, for example, gel-free/gel-based technologies are often combined (e.g. GeLC), just as top-down and bottom-up technologies, which can also be combined with gel-free and gel-based separation approaches. The final MS measurement in either of these approaches might be an MS or MS/MS analysis from purified proteins/peptides or complex mixtures. Nevertheless, these classifications are still helpful in evaluating the strength and limitations of certain approaches, and to understand how the data were gathered in the first place.

There is no obvious best approach in using proteomics for biological questions, but there are technologies that work together better than others in tackling a set of specific questions. In this chapter we will see several approaches in the real world. This chapter will present recent original work with some comments on the technologies applied.

5.1 IMAGING MASS SPECTROMETRY

Imaging mass spectrometry (IMS) has seen strong development in recent years. In theory IMS is a perfect proteomic tool; imagine taking a look at a tissue section and determining at the same time all the proteins present in the structures that are visible (Figure 5.1). One can find the locations of proteins at certain morphologically well defined structures and monitor drug enrichments or neuropeptide maturation/secretion. IMS is extremely helpful in pharmacokinetic studies, as even whole body

sections of, for example, mice can be analyzed not only for the original administered peptides, for example, but also for all known/unknown metabolites (Rohner *et al.*, 2004; Stoeckli *et al.*, 2007). Changes in the localization of certain ions, for example after drug treatment or in diseased animals, allow conclusions about the mode of action of drugs or the pathology of diseases (or both, as in Alzheimer's). IMS involving proteins has hitherto mainly used MALDI ToF, ToF/ToF q-ToF or FT MS, and as such the lateral resolution of the IMS 'pictures' is limited to the focus of the LASER, typically 10–300 μm in diameter. This focal size is in turn ultimately limited by the wavelength of the LASER source. The matrix application has to be performed very carefully, so as not to drench the sample in matrix, which would then allow lateral diffusion. To avoid this, special nebulizers or piezo-electric deposition of nano-droplets are used in a repeated fashion, applying several layers of matrix. Latest improvements of the technology allow nano-litre droplets of trypsin solution to be spotted without too much loss of lateral resolution prior to the addition of matrix. This allows in turn MALDI ToF/ToF measurements of tryptic peptides to be taken for the identification of proteins detected on serial tissue slices (not treated with trypsin) by MALDI linear ToF.

One of the limitations of the technology is that at the moment it mainly works with abundant proteins. Although about 500 peptides might be detectable in the tryptic image (compared to about 300–450 features in the range of 2 000–20 000 Da without digestion), proteins of low abundance will not be detected. As historically it was not possible to digest the proteins without loss of lateral resolution (e.g. due to 'leaking' of the peptides in surrounding areas), many applications of IMS have

Introducing Proteomics: from Concepts to Sample Separation, Mass Spectrometry and Data Analysis, Josip Lovrić
© 2011 John Wiley & Sons Ltd

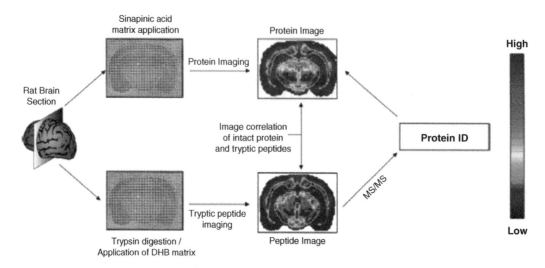

Figure 5.1 Principles of proteomic IMS. Tissues of interest are sliced cryogenically into 10–20 μm sections and mounted onto MALDI target plates. These are carefully coated with either matrix or trypsin followed by matrix for MALDI ToF MS. Sinapinic acid is a better matrix for full-length proteins, while dihydrobenzoic acid (DHB) works better on peptides. A raster of target points is taken and MALDI ToF or ToF/ToF data are measured. From the spatial information and the intensity at a single m/z value, pictures are constructed, with colour coding for the intensity (scale on the left). In the examples in this figure (labelled 'protein image' and 'peptide image') the images were generated using the intensity at 14.2 kDa and at 1 460.90 Da, respectively. In the peptide image ToF/ToF analysis confirmed every single amino acid of the tryptic peptide TQDENPVVHFFK, and together with several other peptides identified the ion at 14.2 kDa as myelin basic protein. Identical distributions of the protein and peptide intensity patterns prove that the peptide is derived from the 14.2 kDa protein. Several thousand images can be generated from one data set and are stored to build a local database, which allows interrogation of the distribution patterns of any mass change, for example after drug treatment. Most IMS studies to date omit tryptic digests. High resolution MS (e.g. FT) can still identify many components or they can be identified by a combination of methods after sample enrichment from the tissue. Reproduced by permission from Groseclose *et al.* (2007).

been on brain sections. These contain large localized concentrations of neuropeptides, which give very good signals in MALDI ToF MS. Lateral image resolution with piezo electrical deposit of trypsin (similar to ink-jet printheads) is typically about 50–200 μm, and data collection can take several hours for each tissue slice. For tissue slices of 20 × 30 mm, resolutions of 50–200 μm would result in 15 000–240 000 image points which need about 1 000 LASER shots each. Even ultra-fast MALDI ToF analyzers can spend long hours and days generating the data. Once the MS data are generated, hundreds of images (each for a different mass) can be created (similar to taking astronomical pictures at different wavelengths). The amount of data is immense and special software (e.g. http://www.maldi-msi.org/) is needed to allow manual image analysis by operators experienced with the morphology. As consecutive slices are all of the same thickness, even 3D distributions of ions across tissues and whole bodies (e.g. from mice) can be reconstructed.

This is not a trivial task, given the huge amount of data involved, and a single set of experiments, even under moderate resolution, would take weeks of measuring time.

Subcellular resolution at around 500 nm is possible in IMS using secondary ion mass spectrometry (SIMS). Here the sample is bombarded with narrowly focused ion beams (e.g. indium, gold (Au3+) or buckminster-fullerene) instead of light. However, biomolecules do not survive this ordeal in one piece and the largest biomolecule-derived fragments surviving the bombardment are around 200 Da (usually stable conjugated rings) and ions of around 80 Da are more typical. This fragmentation during the ionization process allows only limited interpretation of the molecular compositions of the sample (Kurczyk *et al.*, 2010) and protein identification is not possible. However, although the current technology is most sensitive for lipids, it is possible to detect amino acids and to even distinguish closely related cell types based on their fragment pattern (Baker *et al.*, 2008).

Besides MALDI ToF/ToF analyzers, q-ToFs and even FT MS can be used for IMS, usually at the lower end of the lateral resolution scale and on smaller sections, due to the long measurement times involved. Latest developments include DESI imaging MS (see Section 3.2.2) with similar spatial resolution to MALDI IMS; however, mainly metabolites are detected, due to their prevalent ionization under ESI conditions (Wu *et al.*, 2010). Using the specific matrices in MALDI allows for a very efficient ionization of peptides/proteins although they are not the prevalent molecules in the sample.

Complete commercial solutions for sample matrix coating are available and at the time of writing several commercial 'printers' allowing trypsin deposition are available. Thus all components for IMS have become commercially available to allow MS/MS imaging at resolutions below 100 μm. This allows reproducible results, which are needed in clinical studies using IMS (McDonnell *et al.*, 2009) or for biomarker discovery (Figure 5.2). With regard to clinical studies, it is important to note that unlike the classical immunohisto-chemistry analyses using antibodies, IMS will give very poor results from formalin fixed tissues.

5.2 QUALITATIVE PROTEOMICS

Many proteomic studies in the past can be described as 'stamp collection' proteomics. These studies were mainly concerned with identifying as many proteins as possible in certain samples such as micro-organisms, cells, body fluids or tissues. This kind of study is very important to establish the proteins present in a given sample. This qualitative proteomics allows conclusions about which part of the genome is expressed, what expression levels can be expected and how the genomic information is related to the proteome composition. The more challenging task of differential proteomics is often taken on after qualitative studies, as these produce data and experiences (with the sample and the technologies) needed for the quantifications.

One important incentive for many proteomic studies is to see which proteins are expressed in a certain organism, cell type tissue or body fluid. The results of such studies either are important in their own right or may be the basis for more detailed research. However, what can be expected from such analyses is predicted to a great extent by the experimental approaches.

Tears are an interesting fluid from more than just a romantic point of view. They are important in keeping the eyes functional, easily accessible, albeit in small amounts, and might contain biomarkers for eye-related as well as other conditions. Green-Church *et al.* (2008) performed a proteomic study to identify components of tear fluids as well as to compare different tear collection methods with a view to clinical applications. Using a linear ion trap (LIT), in this case an instrument called LTQ, online coupled to nano-RPLC and a two salt step MudPIT approach, 28 proteins were identified. The same samples were separated by SDS PAGE, the gels were cut into 11 slices and proteins from each slice were digested by trypsin and analyzed by nano-RPLC MS with the same mass spectrometer (Figure 5.3). This GeLC approach identified 97 proteins. The analysis of the same tear samples by 2D SDS PAGE and subsequent extraction of spots from the gel and analysis by LC MS/MS led to the identification of 32 proteins, although many more spots were observed, and could even be quantified using the difference gel electrophoresis (DIGE) approach. Some valuable conclusions can be drawn from this study. It could be shown that different sample collection methods will not only have an influence on the proteins detected in the sample; based on gene ontology (GO) information, it was shown that one of the sample collection methods (a standard method used for many years) is more 'invasive' and that the tears contain systematically different proteins compared to other methods, likely to come from an interaction of the collection method with the ocular surface epithelium. The study also shows how in samples that suffer from the predominance of a few proteins, the MudPIT approach can easily end up measuring the peptides from the same, abundant, proteins over and over again. This is particularly true for body fluids or cell extracts where a few very abundant proteins represent more than 80% of the total protein content. Excluding these abundant proteins from most fractions (in this case by SDS PAGE and slicing of the gel) significantly increases the number of different proteins detected. Although several approaches were combined in this study, the number of proteins identified remains modest, for such a complex sample as tears.

Another study aiming to identify the proteomic composition of tears was performed earlier. Using a very similar approach in terms of sample collection and separation, 491 proteins were identified (de Souza *et al.*, 2006, Figure 5.3). In concordance with the Green-Church

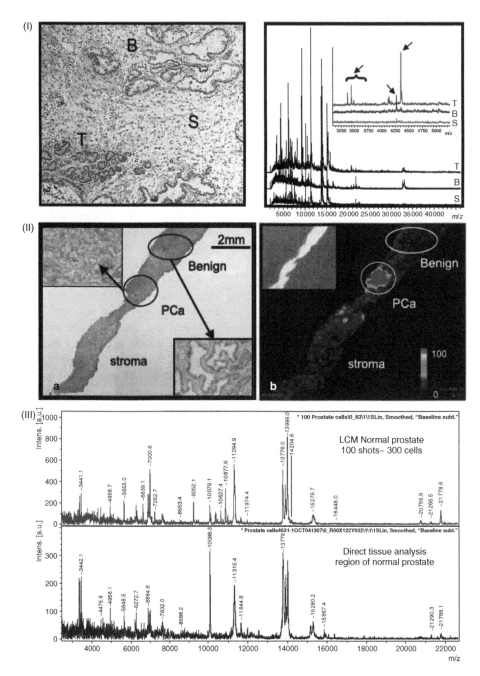

Figure 5.2   IMS for the discovery of biomarkers. (I) A histological sample (left) of a prostate with cancerous (tumour, T) benign (B) and normal tissue (stroma, S) and typical MALDI ToF MS spectra from the same regions (right). Several ions are specifically measured in cancerous tissue. Panel (II) shows serial tissue section stained by haematoxylin and eosin (a) and the IMS for the mass 4 355-Da, (b). A cancerous section (PCa) is clearly stained by IMS. Panel (III) shows that data from cancerous cells collected by LASER capture microscopy looks very similar to the IMS spectrum at a single point of a tissue section. The ion at 4 355 Da was identified from an IMS preparation by MALDI ToF/ToF data as a fragment of MEKK2 (MAP/ERK kinase kinase 2), a signalling molecule. Additional work, including MS/MS of tryptic digests from cancer lysate and Western blots from tissues, confirmed that MEKK2 is indeed over-expressed in prostate cancer samples. Note that light capture microscopy (LCM, compare Figure 4.9) was performed on 300 cells only. Cazares *et al.* (2009), Figure 1a–c and Supplement Fig 1.

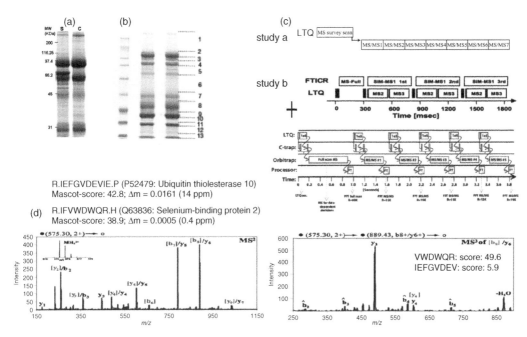

R.IEFGVDEVIE.P (P52479: Ubiquitin thiolesterase 10)
Mascot-score: 42.8; Δm = 0.0161 (14 ppm)

(d)   R.IFVWDWQR.H (Q63836: Selenium-binding protein 2)
Mascot-score: 38.9; Δm = 0.0005 (0.4 ppm)

Figure 5.3 Analyzing the tear proteome I. Comparing two studies that analyzed the tear proteome (see text for details), it is clear that both used similar starting material (study in (a) used 11 slices of the gel, study in (b) 13 slices) for GeLC analysis. However the mass spectrometers used were very different; study 'a' used a quadrupole IT, study 'b' two different hybrid FT mass spectrometers, an LIT/FTICR mass spectrometer and an LIT/orbitrap instrument. Both FT instruments are composed of two mass analyzers that can produce data in parallel and were also used in this way (c), whereas the LIT in study 'a' used a top ten approach: an MS scan followed by MS/MS scans of up to 10 of the strongest ions. Study 'a' identified 97 proteins (including the usage of 2D gels), study 'b' nearly 500 (see text for details). (c) Combining the high accuracy of an FTICR MS measurement and the MS³ capability of the LIT in the same instrument can improve specificity without sacrificing in throughput. While the FTICR is measuring the peptide mass with high accuracy, the IT can perform an MS/MS and MS³ experiment. The results in (d) show how even with high accuracy and good fragment spectra, the wrong peptide (R.IEFGVDEVIE.P) can have the higher score (panel on the left); only the combination of very accurate MS (0.4 ppm) and the MS³ fragments (panel on the right) identifies the correct peptide (R.IFVWDWQR.H). Study 'a' is de Souza *et al.* (2006); study 'b' is Green-Church *et al.* (2008). (a) Reprinted from Green-Church (2008). (b) Reprinted from de Souza *et al.* (2006), with permission from BioMed Central. (c) (part) Modified with permission from Olsen & Mann, (2004). © (2004) National Academy of Sciences, U.S.A. (part) Modified from Olsen *et al.* (2005), by permission of the American Society for Biochemistry and Molecular Biology. (d) Reproduced with permission from Olsen & Mann, (2004). © (2004) National Academy of Sciences, U.S.A.

study, far fewer proteins were identified using a pure LC-MS approach (59 proteins), a number that could not be increased by using four times more starting material (62 proteins). The biggest obvious conceptual difference between the studies is the use of high performance mass spectrometers in the de Souza study; instead of an LIT, not one but two different high performance hybrid FT mass spectrometers were used, generating 386 and 320 protein identifications each, combining to 491 different proteins in total, which means the overlap between the two instruments was 208 proteins. In other words, each of the high performance mass spectrometers in a very

elaborate strategy produced between 35% and 46% of identifications unique to this mass analyzer (Figure 5.3).

The data acquisition and data analysis strategy using these different instruments is worth looking at in more detail, as it is crucial to the success of the studies. The data acquisition and data analysis strategy in the Green-Church study using an LIT was the so-called 'top ten' approach; a survey scan over the entire mass range used for the analysis (e.g. 300–1 500 m/z) followed by MS/MS of the 10 most intense peaks. Such a measurement cycle takes about 3–8 seconds, depending on the resolution and sensitivity settings used. An LIT under these conditions will work with mass accuracies of several hundred parts per million for MS and MS/MS data. In the de Souza study different data acquisition strategies were used for each instrument, based on the different performance characteristics of the hybrid FT MS used. Taking advantage of the ability to run the two mass analyzers in the linear trap quadrupole (LTQ) FT MS independently, a strategy to maximize measurement time and accuracy was employed, similar to the one shown in Figure 5.3. In principle, the data acquisition cycles consisted of a short survey scan (at a resolution of 25 000, which is low for the FT MS), based on which the three strongest ions were chosen for consecutive selected ion monitoring (SIM) high resolution scans for these ions. In this scan type, a small m/z region is analyzed at very high resolution by the FT analyzer, leading to high accuracy MS data for the tryptic fragment in question (around 1 ppm at a resolution of 50 000). In the meantime, the same tryptic peptide is analyzed by the LIT analyzer in MS/MS mode at relatively low resolution, ensuring speed and sensitivity. While the original SIM measurement is still under way, an additional MS^3 fragmentation on the strongest ion from the MS/MS analysis is performed on the LIT. After MS/MS and MS^3 scans of the first ion chosen from the survey scan are performed, the next two chosen ions get the SIM, MS/MS and MS^3 treatments, before a new survey scan is performed. Once chosen for further analysis the ions are put on an exclusion list for 30 seconds (dynamic exclusion list); that is, even if they are still the strongest ions in the next survey scan they are ignored and the next strongest ions are chosen instead. Scan times vary, but are in the region of 2–3 seconds for a complete cycle, allowing about 10 complete scans over the elution time of a typical peptide, and thus the analysis of around a dozen co-eluting peptides at any given time. The LTQ orbitrap used the same LC set-up, but with a straightforward 'top five' data acquisition strategy with dynamic exclusion (Figure 5.3 c). The survey scan was taken at a resolution of 60 000 (taking about 0.9 seconds to measure by FT), which is not the highest possible resolution (as this would take 1.5 seconds to measure). MS/MS data were taken at a resolution of 15 000, or 0.25 s analysis time. Including filling, activation and FT measurement, a complete cycle with five MS/MS data sets lasts around 4 seconds. A very high accuracy of all data was achieved by internal calibrations (lock mass mode) on an environmental

contaminant (Olsen *et al.*, 2005), thus increasing the accuracy from around 4 ppm to better than 1 ppm.

The stringency settings for database searching have a major impact on the scoring and thus on the significance of any particular database hits. With the above detailed strategies, the data for both instruments (three different mass analyzers) were searched with an accuracy setting about twice lower than the experimentally determined accuracy. The criteria for a positive match were also very specific; apart from matching certain accuracies for the MS data (better than 3 ppm for FTICR and 5 ppm for orbitrap), a protein hit was determined as significant when it comprised at least two peptides with a MASCOT score higher than 21 (significance level 0.01) for the orbitrap data, or two peptides with a score higher than 27 (significance level 0.01) for the LTQ FT, or one peptide alone with additional MS^3 data, raising the score to above 54 (significance level 0.0001). Testing the complete data set against a reverse database did produce some false positive peptide hits, but not a single protein hit with the above criteria.

Given these two high performance mass spectrometers, why is there such a low level of overlap in the proteins identified? Most of the non-overlapping proteins were identified by one (FTICR with MS^3 data) or two peptides only. In the reciprocal data sets from the two mass spectrometers, single peptides for all the proteins exclusively identified by the corresponding alternative mass analyzer were found, but deemed not significant on their own. Interestingly the same is true for the Green-Church study, where many single peptides of proteins identified with the high performance mass analyzer where detected, but could not contribute to a significant protein score. The proteins identified by one or a few peptides are the less abundant proteins. Thus to dig deeper into the proteome, high accuracy and optimized data analysis are inevitable. This is not to say that the LTQ is not a superbly performing mass spectrometer for shotgun proteomics; it is used in many very successful shotgun proteomic studies and outperforms mass spectrometers with clearly superior accuracy, based on the ability to generate very good MS/MS ion series, even of long peptides (Elias *et al.*, 2005). In samples with more balanced protein abundances, for example in cellular lysates, an LTQ can easily identify thousands of peptides resulting in around 1 000 protein identifications. However, challenging mixtures such as tears, urine or plasma pose problems in any proteomic approach. Since most of the proteins identified in any shotgun approach are based on just a few peptides, it is very difficult to quantify these proteins (see Section 5.3) and the number of proteins identified by just a few peptides is very high in samples where 80–90% of the total protein content is composed of a handful of highly abundant proteins (e.g. in erythrocytes, chloroplasts, plasma, urine or tears).

How do gels perform on the tear proteins? Gels show some 500 different spots, many of which exhibit clear indications of PTMs in their migration patterns, with predominantly glycosylation-like trains of spots

being visible (Figure 5.4). The presence of extensive glycosylation can be assumed from gel patterns; direct proof came from another study, which was using LC MS/MS (Li *et al.*, 2005) and concentrated on the analysis of this PTM. However, no direct attempts have been made (or published) to identify all the proteins in tears with gel-based methods. Gel-based proteomics was rather used to identify differences between normal and diseased patients, which we will discuss further below. The reasons for this are obvious; if you compare the differential spots in Figure 5.4, you will see more than

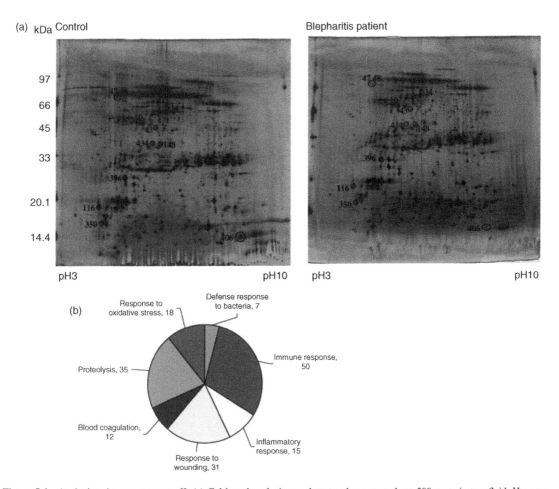

Figure 5.4 Analyzing the tear proteome II. (a) Gel-based analysis can detect and compare about 500 spots in tear fluid. However, most of these cannot be identified – the study from which the gel images are taken only managed to identify 12 proteins (see circled spots), which were differential in patients with blepharitis (causing a chronic eyelid inflammation) in a study with more than 30 clinical samples. The pie chart (b) shows the GO classifications of the proteins identified in tear fluid from study 'b' in Figure 5.3. See text for more details. (a) Reprinted with permission from Koo *et al.* (2005). © 2005, American Chemical Society. (b) Reprinted from de Souza *et al.* (2006), *Genome Biology*, 7:R72, with permission from BioMed Central.

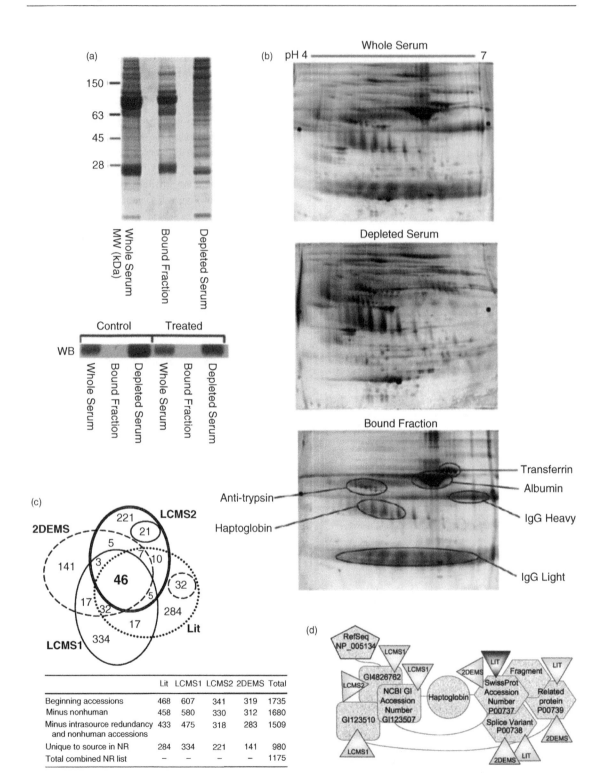

(a)

MW (kDa)
150
63
45
28

Whole Serum
Bound Fraction
Depleted Serum

Control   Treated

WB

Whole Serum
Bound Fraction
Depleted Serum
Whole Serum
Bound Fraction
Depleted Serum

(b)   pH 4 ——————————— 7

Whole Serum

Depleted Serum

Bound Fraction

Anti-trypsin
Haptoglobin
Transferrin
Albumin
IgG Heavy
IgG Light

(c)

2DEMS
LCMS2
221
21
5
141
7
3
10
46
32
17
5
32
284
17
LCMS1
334
Lit

	Lit	LCMS1	LCMS2	2DEMS	Total
Beginning accessions	468	607	341	319	1735
Minus nonhuman	458	580	330	312	1680
Minus intrasource redundancy and nonhuman accessions	433	475	318	283	1509
Unique to source in NR	284	334	221	141	980
Total combined NR list	–	–	–	–	1175

(d)

RefSeq NP_005134
LCMS1
LCMS1
GI4826762
LCMS2
NCBI GI Accession Number GI123507
2DEMS
LIT
Fragment
LIT
SwissProt Accession Number P00737
Related protein P00739
Haptoglobin
GI123510
Splice Variant P00738
LCMS1
2DEMS
LIT
2DEMS

30 easily, with the unaided eye, yet 'only' 12 differential spots could actually be identified in this study, due to the low sensitivity (in terms of proteins identification) of gel-based proteomic approaches. The 2D gels show clearly how four to five groups of spots (presumably each group is one protein with PTMs) make up almost the complete total protein amount on the gel, and all these pieces of information (excluding protein identifications) can be gained within a week or two of having the samples, by simply running some repeat gels and overlaying their images.

Which brings us to the question of the benefits of performing quantitative proteomics on tears. One of the most interesting direct results of knowing about 500 proteins contained in tears is their (possible) functions, as derived by GO notations in databases. Some of the biological processes in which the proteins from tear are presumably involved were of course expected, such as proteolysis and immune response (Figure 5.4). The presence of about 60 protease/hydroxylase inhibitors underlines the importance of a complex balance of protease activity and inhibition of this activity for a healthy eye. Moreover, 18 proteins involved in oxidative stress response were found, 16 of which had never been described as being present in tears before this proteomic study. These proteins shed a light on the generation of tear fluid and its function and, most importantly, show the wide array of proteins one has to analyze in order to explain the function of tear fluid. Of course, all these proteins also open up new avenues for research – their function, how they are regulated, where they come from, how they affect the health of the eye, what role they play in diseases/ageing and which of them are potential biomarkers for eye-related conditions, or

even for unrelated conditions in other parts of the body (as most proteins in body fluids are shared by two or three fluids; Li *et al.*, 2008). Of course, for all these analyses proteomics might not be the best approach and classical biochemistry might come in handy, leading to an explosion of workload and intellectual involvement. In essence, 'stamp collection' proteomic studies are very valuable and can give us a wealth of information on a variety of biological processes.

You may wonder why we look at a relative obscure body fluid like tears. Using this example it is relatively easy to explain the major concepts, which also apply to other body fluids (or indeed any other proteomic samples). Sample preparation and analysis methods have to be adopted (e.g. urine turns out to be difficult to analyze as it contains a mixture of proteins with special properties and stabilities), and for more classical body fluids the amount of data is overwhelming. Plasma is one of the best analyzed body fluids. However, its composition is still far from being known or even understood. Long before the onset of proteomics, plasma and serum were analyzed for diagnostic markers (biomarkers), from metabolites to salts and proteins ranging from the ubiquitous haemoglobin to the more elusive peptide hormones. The proteins in plasma span one of the biggest dynamic ranges encountered in proteomics. How can proteomics cope with this most complex body fluid? An obvious approach is to remove the most abundant proteins, and promising in this respect are depletion columns, for example with antibodies against the most common proteins (Figure 5.5), removing 83% of the total protein (even though the abundant proteins are

Figure 5.5 Analyzing the plasma/serum proteome. (a/b) Serum depletion allows more sensitive detection of minor protein components. Serum was depleted using a MARS column containing immobilized polyclonal antibodies against albumin, transferrin, IgG, IgA, -a1-antitrypsin and haptoglobin (see results of SDS PAGE (a) and (b) of 2D gels). In both gel systems many more spots/bands of minor protein components can be observed after the depletion. Also note the multitude of protein isoforms resulting from PTMs visible in the 2D gels. The Western blot (lower panel in 'a') against APOA1/A2 (apolipoproteins are part of lipoprotein complexes in the blood) shows an increased signal intensity after depletion. Identical protein amounts were loaded in all gels/blots. (c) The Venn diagram and the table show the results of a study aimed to identify as many human plasma proteins as possible, using 2D gel electrophoresis in conjunction with MALDI ToF MS based PMF and LC MS/MS (2DEMS), a shotgun proteomics approach (LCMS1), a shotgun proteomics approach with all proteins larger than 30 kDa removed by filtration (LCMS2) and a literature search on thousands of previous publications (Lit). The identified sequences were cleared of non-human and redundant entries (within one source and across sources, see table). See text for further details. (d) shows how all different sequences associated to haptoglobin from the various approaches were collapsed into one entry from SwissProt (SP) to avoid redundancies in the final resulting database. (a/b) Reprinted from Freeman *et al.* (2006), courtesy of Elsevier. © 2006 Elsevier. (c/d) Reproduced from Leigh Anderson *et al.* (2004), by permission of the American Society for Biochemistry and Molecular Biology.

not removed completely), and enhancing, for example, Western blot signals of weaker proteins nearly sixfold and the number of spots visible in 2D gels more than twofold (Figure 5.5). Even with sophisticated methods, each approach alone can only detect a limited number of proteins, for various technological reasons. For example, a strategy which removed the most abundant proteins, created dozens of fractions by sequential anion exchange and size exclusion chromatography (separating proteins by charge and size) followed by 2D SDS PAGE and MALDI ToF PMF combined with RPLC ESI IT MS/MS analysis of all coomassie stained spots 'only' yielded 283 different proteins (Figure 5.5). Equally sophisticated LC MS approaches, for example targeted at small proteins (such as peptide hormones) by including the removal of all proteins larger than 30 kDa in the sample clean-up, resulted in comparable (if slightly larger) numbers of proteins for each individual method. In fact, the most efficient method of collecting proteins for a plasma protein database was a literature research (over 450 original publications contained relevant data). About half of all proteins found were unique to each method, which means only half of the proteins showed some kind of overlap between different methods and only 46 of over 1 000 non-redundant proteins were detected by all methods! The take-home message here is that many different methods can be used to analyze complex proteomes, and most of them are to a great extent complementary. This shows us how far away we are from 'truly' proteomic technologies, which should be able to detect several thousands to tens of thousands of different proteins in serum/plasma. Establishing (part of) the qualitative proteome of a certain sample can and should ultimately lead to the addition of the results into a database. The proteomes of tears and plasma are represented in several databases, for example MAPU, Sys-BodyFluid or Plasma Proteome Database (PPD), which are available to the public. Depending on the quality and depth of the stored data, these databases can be used to design quantitative experiments (telling us, for example, which peptides have been identified) or screen with, for example, antibody chips for potential biomarkers. In fact, the biggest proteomic databases for plasma contain about 10 000 entries. Although these databases are much better curated (i.e. manually controlled) for redundancies and other data is attached to the entries (e.g. PTMs, single-nucleotide polymorphisms, isoforms, splice variants, signalling

peptides, GO information), which is much more curation than the 'average' DNA database, it is impossible to remove all redundancies, even if researchers go to great lengths to do so (Figure 5.5(d)). These problems are also based on the complexity of the underlying gene expression, and the nature of the generation of entries in DNA/RNA databases, and no level of manual or automated annotation can bridge the current gaps in, for example, PTM coverage (Sections 4.5 and 5.4). Proteomic studies often end up characterizing proteins that are not contained in any database or are in conflict with some of the database entries, even when the analysis is aimed at very abundant and well researched proteins (see, for example, John *et al.*, 2006).

The thought that there is no genomic equivalent for body fluids is an interesting one; while we can predict (often correctly) if a protein will be extracellular, there is no way to predict if it will end up in a body fluid. We even know less about which cells actually produce each body fluid. Only proteomics can deliver these answers, and studies have so far also shown us that many of the proteins found in body fluids were not expected to be there. Many of the body fluid proteins are of intercellular origin; and this is after all efforts are undertaken to eliminate the body fluid's contamination by any cellular particles.

Other examples of samples where 'stamp collection' (or quantitative) proteomics is extremely useful for biological questions are lysates of cells and tissues or poorly characterized organisms. These organisms are typically micro-organisms, where the genome (i.e. the DNA sequence) is 'known' but it is rather unclear which proteins are expressed at all, or under what conditions which parts of the genome are expressed. With such organisms the sample preparation can be an issue, as bacteria, yeast fungi and plants can have rigid cell walls that might pose problems for sample preparation. However, even in relative easily solubilized samples (e.g. mammalian cells), the problem remains that particular classes of proteins might be underrepresented by specific proteomic approaches. Typical examples of poorly represented proteins in proteomic studies are small proteins (below about 30 kDa). For the analysis of small proteins either specific techniques can be applied to enrich them (see Leigh Anderson *et al.*, 2004 and Figure 5.5, using size exclusion filtration) or CE can be used instead of RPLC, again preferring small proteins (see Mischak *et al.*, (2009), and Section 5.3.3).

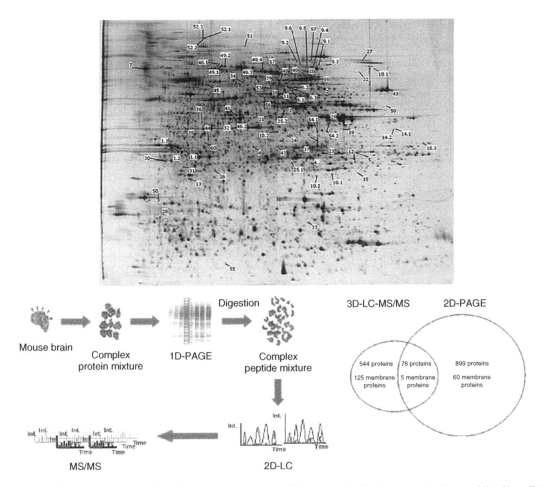

Figure 5.6 Comparison of gel-based and hyphenated proteomics of brain samples. In the same study large mobile pH gradient gels (40 × 30 cm, see Section 'Mobile pH gradients') followed by protein ID using MALDI ToF MS PMF was compared to GeLC MS/MS approaches with 2D LC of gel slices and nano-HPLC ESI-LC IT MS/MS. Although the 2D gel based approach was able to identify more proteins in total, the GeLC approach identified more membrane proteins. For the 2D gel-based approach 11 000 spots were picked form three gels (automatic robot), 1 900 different proteins in total were identified and condensed to 899 unique proteins after PBLAST similarity searches. See text for more details. Reproduced from Lohaus *et al.* (2007), by permission of the American Society for Biochemistry and Molecular Biology.

Next to small proteins, membrane proteins are another class of proteins known to be underrepresented in some proteomic studies. While there are buffers to keep every membrane protein in solution, these are not universal buffers and not compatible with, for example, gel separation of proteins. Another hurdle for many proteins is the transition between the first and second dimension of 2D gel electrophoresis. As a result many membrane proteins, particularly the integral ones with several transmembrane domains, are lost. In the example of brain samples (Figure 5.6), a carefully designed study

allowed the identification of nearly 900 brain proteins by 2D SDS PAGE followed by MALDI ToF PMF and ToF/ToF MS/MS analysis. The same study tried very hard to identify as many proteins as possible using a 3D LC MS/MS approach and ion trap MS/MS. Despite using several different proteases and repeat experiments, the 3D LC MS/MS approach 'only' identified 544 proteins. The prize for the highest number of identified proteins went to the gel-based separation process. The excellent results from 2D gels were surely helped by the fact that Joachim Klose was one of the co-authors; he invented (independently and at the same time as O'Farrel) high resolution 2D SDS PAGE as we know it and Figure 5.6 shows a high quality gel, resolving more than 4 000 spots. The gel-based approach also benefitted greatly from the expert utilization of robots for cutting and protein digestion. However, no expertise in the world could help the fact that the gel-based approach only identified half as many membrane proteins as the LC MS approach. It is interesting to note that the authors still concluded that the LC approaches were not feasible beyond this study; the manual validation of the automatically generated database search results alone took several person-months of work. Also noteworthy is the discussion about the stringency of the identifications: the authors did not feel comfortable with the settings suggested by the stringency criteria of the software providers; using these less stringent settings, some 11 000 proteins would have been detected form nearly half a million MS/MS spectra, albeit with unacceptably high rates (whatever that means) of false positive identifications. The gel-based approach also consumed far less protein/sample than the combined LC MS approaches. Apart from the very strong take-home messages mentioned above, this example also teaches us that proteomics results often depend on the experience and preferences of the labs performing the experiments – there is more than one way to generate long lists of proteins. While the software situation for the analysis of shotgun data has improved since the study was performed, analyzing complex shotgun data is still not for the faint-hearted.

A constant problem in proteomic studies is the bias against proteins of low abundance. It is difficult of course to prove that rare proteins are underrepresented and not just 'not expressed', as we do not know how many of these proteins to expect before we perform the proteomic study. In this regard, the analysis of yeast has been very beneficial in showing the limitations of different proteomic approaches. In yeast and other fast growing species, there is a distinct codon bias for proteins expressed in different amounts. Proteins with high expression levels prefer to use certain codons of the degenerate triple code for amino acids, allowing an efficient translation of the sequence. One can calculate a 'codon adaptation index' (CAI) for each gene, ranging from about 0.05 to about 1. If only the most commonly used codons are used in a gene, the index will be near 1; if almost all codons are the rarely used ones, the CAI will be near 0. While CAI values are not a strict prediction for the absolute level of protein expression, there is a strong correlation between CAI value and protein abundance. However, proteins with the same CAI value can differ as much as 100-fold in their expression level.

Figure 5.7a shows the CAI distribution for all potential yeast proteins inferred from the genomic data. While several 2D gel-based studies, identifying hundred of proteins by MS/MS, only ever found proteins with CAI values larger than 0.2, shotgun proteomics with 2D LC MS/MS identified about 25% of all predicted proteins, with most proteins having CAI values of 0.11–0.3. Although the data of shotgun proteomics are biased towards higher CAI values (and thus higher protein expression levels) than the genomic distribution, shotgun proteomics is more powerful in detecting low abundance proteins than 2D gel-based approaches. Figure 5.7 also shows that the shotgun proteomic approach in a particular study showed a bias towards proteins with high pI, large size and peripheral membrane proteins (presumably due to sample preparation), and that small and acidic proteins were underrepresented, as well as those with low CAI values (and thus inferred low expression levels). Impressively transmembrane proteins were well represented, something that would be impossible with gel-based approaches (see Figure 5.6 and further below).

These results show an impressive performance for shotgun proteomic approaches, at least in this model organism, but what about the 75% of the proteins predicted by the genome and not detected by proteomics? Are they not expressed under the growth conditions of the experiment? Taking a very different proteomic approach, some interesting conclusions can be derived (Ghaemmaghami *et al.*, 2003), but let us start at the beginning. The advantages of a model system, like yeast, are that a number of technologies as well as a huge amount of data are available to test some fundamental (proteomic) concepts. Is it possible to measure the expression of every single gene in a certain condition (such as log-phase growth)? We know it is not possible (at the moment) with MS based proteomic approaches; we can measure 25% of the proteins, biased towards high CAI values or expression levels. A more radical approach is to clone a (Western blot) tag into every single gene and then measure the expression of this recombinant protein from its native environment in the genome (Figure 5.7). From thousands of controlled Western blots (from an equal number of yeast clones) the expression of over

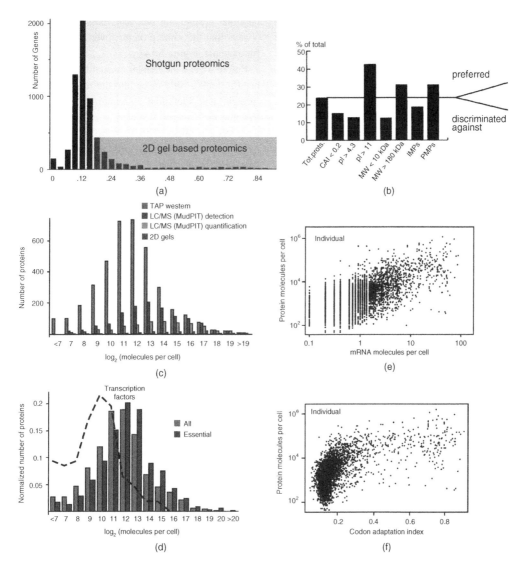

Figure 5.7   Estimating the depth of qualitative proteomics. (a) Codon adaptation index distribution in yeast (see text for details). The shaded areas indicate the level of CAI generally covered by shotgun proteomic and 2D gel-based proteomics. (b) Sensitivity of shotgun proteomics to certain protein classes. For example, 22% of all expected genes were covered in the study, but only 15% and 12% of all expected proteins with a CAI smaller than 0.2 or a pI > 4.3, respectively, thus these classes of proteins are underrepresented in shotgun approaches (IMP/PMP=integral/peripheral membrane proteins). (c) Comparison of the results from quantitative TAP Western blot analysis with other proteomic methods in yeast. Note the $\log_2$ scale for molecules per cell: 7 represents 128 molecules per cell and 19 represents 524 288. The study had a dynamic range for protein quantification over 10 000. (d) Expression contribution of all proteins, essential proteins and transcription factors in molecules per cell. Comparison with (c) shows that around 40% of transcription factors cannot be covered by shotgun proteomics in yeast. (e) Pairwise comparison of 4 000 proteins in yeast between protein and mRNA level per cell. While the general trend correlates, the protein abundance varies within 2 orders of magnitude for any given mRNA abundance. There are about 2 000 proteins for every single molecule of mRNA per cell. (f) Protein abundance against CAI, showing that the CAI allows a reliable prediction of protein abundance. Protein abundance is very low for small CAIs and shows a variation of about 2 orders of magnitude for any given CAI. (a) After Futcher *et al.* (1999). (b) Reproduced from Washburn *et al.* (2001), courtesy of Nature Publishing Group. © 2001 Nature Publishing Group. (c)–(f) Reproduced from Ghaemmaghami *et al.* (2003), courtesy of Nature Publishing Group. © 2003 Nature Publishing Group.

4 000 proteins could be established (see tandem affinity purification (TAP) Western blot analysis in Figure 5.7c,d), and using green fluorescent protein fusion under microscopic control another 270 proteins could be added, leaving only 1 700 expected or potential proteins undetected. Many of these 1 700 proteins were presumably not expressed under the experimental conditions.

Comparing the results of this study to other proteomic studies, it becomes clear how much is left to be desired in proteomics. This study clearly shows that none of the MS based proteomics studies had a chance of detecting (let alone quantifying) the vast majority of proteins; below an expression level of about 1 000 proteins per cell, shotgun proteomics can no longer cope, which incidentally means that some (but not even most) of the transcription factors can be analyzed. It is important to bear in mind that these studies were performed several years ago; MS based proteomics is now more sensitive, but not by several orders of magnitude. Also, these studies were performed on one of the best analyzed model organisms we have, with amazing genetic tools and an unlimited potential to produce samples. None of this is possible with more complex organisms, although mice are being targeted for the construction of extensive knockout libraries and about 8 000 antibodies to human proteins have been collected by now with the aim of obtaining one for every gene/protein. So what these experiments show is that even our best efforts in proteomics can only reveal a partial picture. Even with more sensitive and targeted approaches we cannot assume to be able to analyze more than 40% of all proteins of any given complex proteome, even under ideal conditions and with well characterized model organisms. The proteins we can detect with current proteomic methods will preferentially be of the more abundant type; we can expect to detect something like 90% of the most abundant proteins and about 10–20% of the less abundant proteins. To even detect the central biological 'switches' we need better tools and better ways to analyze our data. MS based proteomics is still the best and most versatile technology we have, with a lot of scope for improvements, as we will be made even more blissfully aware in Section 5.3 on quantification.

But even with all these limitations discussed thus far, pure qualitative proteomics is still able to detect new genes and can deliver information to improve the presumed structure of previously described genes of the human genome, using approaches mainly measuring high abundance proteins (Bitton *et al.*, 2010). This is still

possible, despite the fact that most proteomic methods always find the same proteins, over and over again (see, for example, Figure 5.14).

Before we end the chapter on qualitative proteomics, we need to look a little more into the meaning of the terms 'protein' and 'identified' in relation to proteomic studies. Imagine the scenario depicted in Figure 5.8, where proteins share peptides. This scenario is not uncommon, as based on their evolutionary origin many protein contain at least some highly homologous/identical domains resulting from duplications, which evolved to have different functions but kept large parts of the original structure (and sequence!). If proteins A and B are 'identified' by shared peptides but only one is exclusive to B, we have no easy way to tell whether A is really present in the sample! There are several other possible combinations of related peptides between proteins. In a list with several hundred or thousand of proteins identified from shotgun approaches, the chances of a 'false positive' identification are usually set to below 1 in 100 or 1 in 1 000 for protein identifications. However, this statistic does not consider the chances of missing the slight differences between related proteins, for example family members with different functions or a splice variant, which might easily be missed as it differs in one peptide only from an already identified protein. Often the same peptides are allocated to different proteins/sequences, and only manual evaluation can make us aware of the problem. In many (but not all) cases manual evaluation might also be able to solve the problems and decide which sequence to use, that is, to decide which protein has actually been identified by the peptide based data. Gel-based approaches, on the other hand, are more limited in the fraction of proteins that can be analyzed, particular regarding proteins of low abundance.

5.3 DIFFERENTIAL AND QUANTITATIVE PROTEOMICS

Differential proteomics has taken over from qualitative proteomics as one of the main strategies in use today. The application, development and evaluation of quantitative proteomics are the aims of many ongoing projects. You may wonder why there is a needs for a distinction between differential and quantitative proteomics, but in practice there are considerable differences at the technological level as well as in the potential outcomes of a study and the ways in which the data can be used.

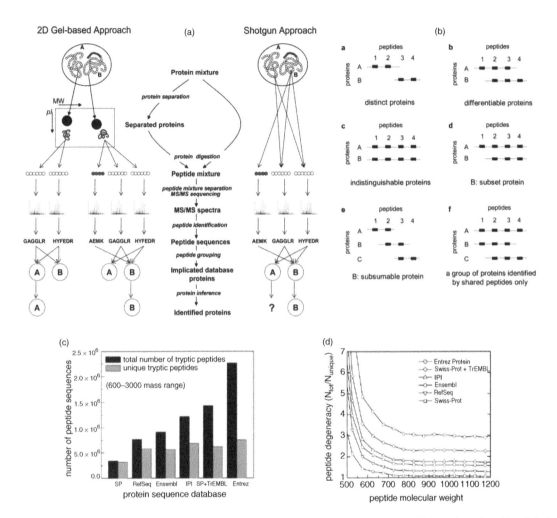

Figure 5.8   Principles of protein identification in gel-based and shotgun proteomics. The loss of information of peptide origin in shotgun proteomics can lead to loss of information on quantitative and qualitative data on the protein level (a). In the example the peptides GAGGLR and HYFEDR are present in proteins A and B, and AEMK only in B. With the help of gel position it is clear in the gel-based approach that both proteins are present. In the shotgun approach there is no way of knowing if A was expressed at all, as the presence of all measured peptide can be explained by the presence of protein B alone. Homologies among proteins are common (see text) and (b) explains how proteins can be related to each other with regard to shared peptides. This presents a substantial problem when dealing with complex mixtures and thousands of shared peptides. Taking only the human part of several databases (c) shows how many (sequence) identical peptides exist in the databases (SP = SwissProt). Increased database size increases the amount of redundant peptide out of proportion. Up to a length of 7–8 amino acids the degeneracy of peptides decreases and then stays nearly constant (d). Reproduced from Nesvizhskii & Aebersold (2005), by permission of the American Society for Biochemistry and Molecular Biology.

Differential proteomic approaches are often planned and executed in a 'semi'-quantitative fashion. This means that researchers are not interested in the quantification of all proteins/peptides between certain samples, rather they only compare features between different samples and regard them as differential when the measured intensities vary by more than 20–40% (typical cut-off values for gel-based and gel-free approaches). This is not to say that the real expression levels vary by these amounts; often no attempts are made to calibrate any of the measurements. It is obvious that finer differences are lost in these analyses, but equally obvious that they are very powerful approaches to identify biomarkers, as we will see further below. We will also go into a little more detail as to what 'identify' might mean in proteomic terms, as well as what might constitute a 'biomarker' in proteomic studies.

In 'proper' quantitative proteomic approaches the aim is to quantify as many features (proteins or peptides) as possible. Here we light upon a big difference between gel-based approaches and shotgun proteomics based approaches; gel-based approaches can deliver differential proteins (more precisely, differential 'spots') from a semi-quantitative or purely comparative approach. Shotgun proteomics based approaches need careful quantification to identify differential peptides and proteins (Figure 5.8). This approach to peptide based proteomics is much more complex and software to analyze the data 'semi-automatically' is only just becoming available. However, stable isotopic labelling in peptide quantification approaches allows a much simpler comparative approach for shotgun proteomics as well. Just as with gel-based proteomics, this allows for 'comparatively' rapid identification of biomarkers. What, then, is the advantage of 'proper' quantitative approaches? Quantitative data analysis will deliver an output in the useful unit of 'proteins per cell' or per millilitre of body fluid or per milligram of sample. Data sets of this type can easily be compared between a variety of labs, with samples of different origin, including different species or even different

kingdoms (Figure 5.9). This cannot be done with purely comparative proteomic data. From the point of view of understanding and modelling an organism or a pathway within an organism, rate constants and absolute protein concentrations are needed; thus proper quantification is essential (see Hood *et al.*, 2004, and Sections 4.5 and 5.6).

However, most quantitative proteomics does not have such far-reaching goals. Thus comparative, semi-quantitative approaches are deemed sufficient, and proteomics is challenging enough even without absolute quantifications. Nevertheless, one should encourage (as indeed several organizations and journals do) all researchers to at least supply the data in a format that allows comprehensive analysis by others. This is important if researchers are not going to analyze the same proteomes over and over again; if experiments become comparable, new experiments and biological insights can be based on incorporating previous data and thus have a better foundation and allow significant progress.

However, in the literature there is often no such distinction between comparative and quantitative proteomics, and we will discuss some examples of research projects that cover both strategies.

When it comes to quantitative proteomic approaches (semi- or otherwise), there are a large number of possible strategies and Figure 5.10 can only show an overview

(a)

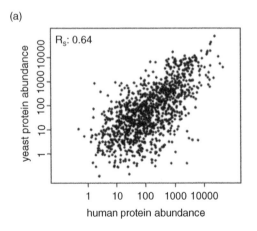

(b)

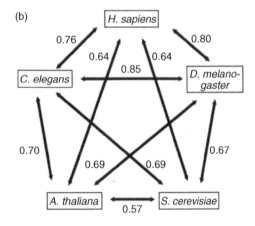

Figure 5.9   Absolute quantifications for proteome comparison of different species. (a) Pairwise comparison of protein abundances per cell from orthologues between yeast and humans. The variation is up to 3 orders of magnitude, but better for most proteins. (b) From this and other comparisons the correlation of protein expression between different species can be deduced. The analysis was based on over 1 500 protein families, spanning more than 4 orders of magnitude in expression. Best correlations are seen for data from spectral counting, indicating the surprisingly good quantifications possible with this type of analysis. See text for further details. Reprinted from Weiss *et al.* (2010), courtesy of Elsevier. © 2010 Elsevier.

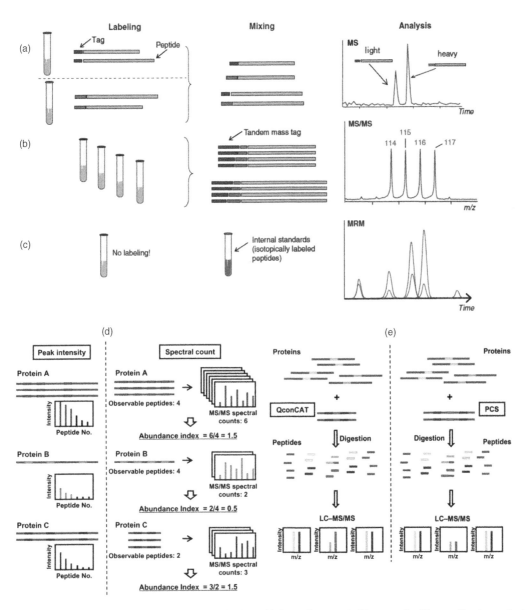

Figure 5.10 Quantitative approaches in proteomics. Labelling with heavy isotopes (a, b) can work with quantification at the MS or MS/MS level. These labelling approaches can be used for shotgun or 2D gel-based approaches. Label-free approaches encompass MRM (c), which is targeted to specific proteins (Figure 4.26) or the comparisons of peak intensities or spectral counting (d). The label-free methods work best when several peptides per protein are considered and outliers are removed. Spectral counting has to be normalized on protein length. For absolute quantification internal (stable isotope labelled) standards can be used for all methods. Instead of generating and purifying several separate internal standard proteins, concatenated peptides can be used. These can be just carefully chosen tryptic peptides (QconCAT) or tryptic peptides with spacers mimicking the natural sequences surrounding the tryptic sites to account for cutting efficiencies (PCS = protein concatenated standard). Concatenated standards can be expressed in and isolated form labelled bacteria. By estimating the amount of many proteins from a few standards of known concentration ('bootstrapping') many proteins can be quantified absolutely, assuming that identical intensities in MS and MS/MS signals originate from identical amounts of different peptides/fragments. For gel-based quantification approaches see Figures 2.13, 2.14 and 5.14. Isotopic labelling approaches also work for the absolute or relative quantification of proteins from 2D gels. (a)–(c) Domon & Aebersold (2006), © 2006 by the American Association for the Advancement of Science. (d), (e) Kito and Ito (2008). © 2008 Bentham Science Publishers Ltd.

of those most commonly used (see also Sections 2.3.2 and 4.4). As always when huge numbers of different technologies compete, this indicates that none of these approaches is perfect and none is without merit. As on prior occasions when such a variety was encountered (e.g. proteome display or separation technologies, mass spectrometers), it comes down to preferences, experience and suitability for the task at hand when a certain quantitative method needs to be chosen. A broad division can be made between methods that quantify proteins and those that quantitify peptides. Both have their advantages and disadvantages, many of which we have touched upon already as their characteristics follow though from protein detection and characterization; the quantification adds several new layers of challenges. The result of this is that often only a subset (the most abundant) of the detected/identified proteins can actually also be reliably quantified, usually about 30–60%. The protein/peptide divide also divides gel-based from gel-free approaches. The gel-based approaches have the usual problem of not reaching very deep into the proteome, while they are excellent at dealing with various family members/isoforms and PTMs, as we will see in an example later. Also, if identification is not necessary, the gel-based approaches can be very sensitive in comparing spots.

Another divide is that between labelling approaches and label-free approaches; this divide works for gel-based approaches (e.g. comparing different gels of unlabelled samples vs. DIGE) just as well as for gel-free approaches (e.g. spectral counting vs. SILAC). Unfortunately there are only a few easy rules (e.g. it is impossible to label humans by SILAC) when we need a straightforward answer to the question which technology would work best. Many factors need to be considered, among them experience, sample suitability, available separation equipment, software, time available in person-hours for each task, the mass spectrometers available, financial constraints and the ultimate aim of the study, to name just a few.

After all this theory, let us see how some quantification methods work in practice. In a comparison of three popular proteomic quantification methods, Wu *et al.* (2005) analyzed how each method would quantify a mixture of six proteins, and all three methods (DIGE, ICAT and isobaric tag for relative and absolute quantitation (iTRAQ)) performed nearly equally well with accuracies better than 85%, regardless of whether the sample mix

was analyzed diluted in buffer or plasma (Figure 5.11a). However, for the gel-based analysis about 10 times more protein was used. When cellular lysates from cell lines either positive or negative for the tumour suppressor p53 were analyzed, there were interesting differences in the performance of the three methods: while the approach was by no means tweaked to achieve the largest amount of differential proteins, iTRAQ was the most sensitive, identifying 19 differentially expressed proteins, followed by ICAT and the gel-based DIGE (Figure 5.11a). While every method had its own problems (ICAT is biased towards cysteine rich proteins, iTRAQ struggles with low precursor ion selection resolution, gel-based methods with co-migrating proteins), there was very little overlap between the three methods; each method has its own bias, which in the end leads to the selection of a subset of different proteins that are amenable to analysis. While it appears that 'investment' in further improving each of the three approaches would have resulted in a higher number of differential proteins, the combination of the three methods was also very effective.

Another comparative study shows the strength and weaknesses of gel-based and gel-free quantification approaches from another angle. Schmidt *et al.* (2004) compared 2D SDS PAGE with silver/coomassie staining and ICAT approaches for the comparative quantification of two different mycobacterial strains (Figure 5.11b–d). ICAT could reliably quantify nearly three times as many proteins compared to the 2D gel-based approach, but only 10% of these proteins were quantified by both methods! This small overlap can be partially explained by ICAT being biased against small proteins (below 30 kDa) and towards larger ones, while 2D gels worked best for small proteins and not so well for proteins larger than 60 kDa, or membrane proteins. Other reasons for this small overlap are suspected in a combination of other preferences/peculiarities of the technologies. It is interesting to note that the ICAT analysis started with some 60 000 MS/MS spectra, 2 000 of which could be assigned clearly as ICAT labelled tryptic peptides by IT MS and the automated software. However, after manual (!) inspection of the data, only 800 peptides were further used, quantifying the resulting in 280 proteins with about three peptides per protein. The 2D gel analysis started with 1 800 spots, of which 270 could be identified by MALDI ToF MS based PMF. After removal of the least intense spots (to get reliable quantifications) some 180

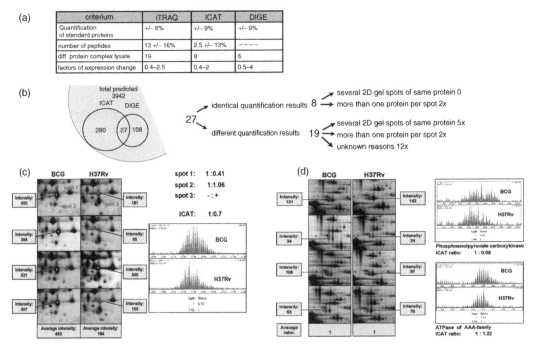

Figure 5.11 Comparison of quantitative proteomics technologies. (a) Summary of the results from a systematic comparison of three quantification methods using MALDI ToF ToF for protein identification and/ or quantification. All three methods worked equally well for the detection/quantification of standard proteins in simple mixtures or diluted in complex cell lysates with similar deviations from the expected values (first row). Note that iTRAQ results were based on a lot more peptides per protein than ICAT (second row). When used to find differentially expressed proteins, in p53 positive/negative mammalian cells, iTRAQ was more successful (third row). It is remarkable that only one protein with differential expression was found by all three methods. The factors for the differential expression of this one protein between the three methods ranged from 1.5 to 1.68, showing how different experimental approaches lead to comparable results. The range of differential expressions found by the MS based quantifications was narrower than that of the gel based quantification (fourth row) (Wu *et al.*, 2005). (b) Summary of the results from a comparison of DIGE and ICAT with IT MS for the quantification of differentially expressed proteins between two strains of *Mycobacterium tuberculosis*. Both methods are highly complementary, quantifying only 27 overlapping proteins from a total of 388 quantified proteins (about 10% of the genome containing 3 942 predicted genes). From the 27 proteins quantified by both methods, 19 showed significantly different results, some of them even reversed changes. This could only be partly explained by more than one protein being in one spot of the 2D gels used for quantification or by proteins delivering several spots on a gel (PTMs?). (c) Results for a protein that produced three spots in 2D gels, with two of them strongly differential. As ICAT cannot distinguish between these different protein forms, the protein is not differential according to ICAT results. The ICAT lane shows the time axis (X) with the intensities for the light (BCG strain) and heavy (H37Rv strain) ICAT peptides (Y axis). (d) Example of two proteins found in one spot. Although both are differentially expressed (ICAT results) the co-migration in the gel means DIGE cannot detect any differences. Note that the heavy labelled ICAT peptides elute consistently earlier than the light version from RPLC and that the y axis have a different scale in every graph, a typical setting in MS, where the strongest peak in the filed of view determines the scale. The scales in the examples range from 2.3 to 6.5 $\times$ 10^7 (arbitrary instrument units). (c) and (d) Reproduced from Schmidt *et al.* (2004), by permission of the American Society for Biochemistry and Molecular Biology.

spots remained, representing the 108 unique proteins further analyzed in the study. The gel-based approach used about 1.5 times the amount of proteins used for the gel-free approach and altogether a modest total amount of about 2 mg of protein from each strain was used throughout the study.

The Schmidt *et al.* study clearly shows the problems inherent in comparing results from quantifications achieved at the level of proteins (sometimes also called top-down quantifications) as opposed to quantifications performed at the level of peptides (sometimes called bottom-up quantifications): gels can suffer from several proteins co-migrating in one spot, and thus miss changes in abundances, peptide quantifications can miss changes in different forms (whatever that means) of proteins and come up with average figures, which might miss PTMs or other modifications. Although not an extensive study, it is clear that out of only 27 overlapping quantified proteins 19 (70%) deliver method-specific results. The discrepancies can also only be explained for a minority of these results, and are likely to be found in the complexities of the technologies.

In another example, the comparison of iTRAQ to 2D gel-based proteome quantification of *E. coli* lysates after induction of rhsA (which plays a role in polysaccharide metabolism) showed a higher consistency of results (Figure 5.12). Comparing four different cellular

conditions with four biological repeats from 2D gel electrophoresis and only one shotgun experiment, the authors conclude that both methods deliver reproducible results. The coefficient of variation (CV) for the 2D gel derived quantification is on average a little worse than the CV of the iTRAQ results (based on different peptides for the same protein from a single experiment, compared to 4 gel based experiments) but, more to the point, only about 20% of the proteins show a deviation larger than 100% between the two methods in any of the four compared cellular conditions. Around 50% of proteins show differences in quantification results of below 50% between the two approaches. However, it was possible to quantify many more proteins using iTRAQ than by using the gel-based approach. Despite the higher number of

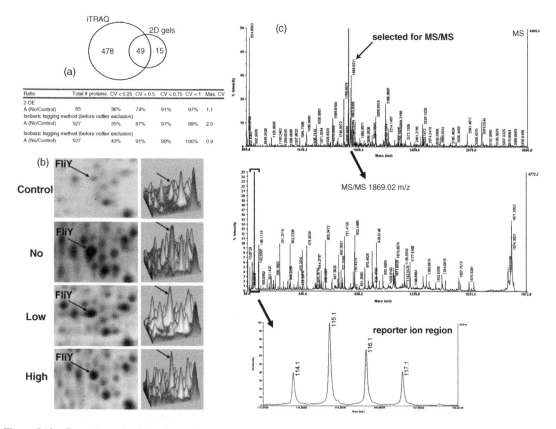

Figure 5.12 Comparison of gel-based quantification with iTRAQ. (a) Summary of overall results and some analyses of the spread of standard deviation of the mean. Discarding of outlier peptides in iTRAQ results in more consistent data (i.e. higher fraction of peptides with lower coefficient of variation (CV)) (b). Gel-based results for the FLiY protein in four different biological situations (control cultures, no indiction of rhsA by IPTG, indiction of rhsA by a low amount of IPTG and induction of rhsA by a high amount of IPTG). (c) Results from one iTRAQ labelled peptide from FLiY. The relations of the peak intensities in the lowest part of the MS/MS spectrum are used for quantification (control 114 Da, no 115 Da, low 116 Da, high 117 Da). Choe *et al.* (2005).

proteins quantified by iTRAQ, not all proteins quantified by gels could be quantified by iTRAQ (Figure 5.12). The better correlation of gel-free and gel-based results in this study might well be due to technical differences in the studies (e.g. different variability in the proteomes) but it is likely to reflect the fact that the average number of peptides considered for quantification per protein was around 17, and thus a lot higher for iTRAQ than for ICAT (see previous comparisons, Figure 5.11). The use of MALDI ToF/ToF with the averaging of 2 500 LASER shots per MS/MS spectrum for the iTRAQ quantification is a huge sacrifice of measuring time, but also makes the data highly reproducible. The example of a single protein, in Figure 5.12 b and c, shows what variability is still to be expected, even with very consistent data. The quantification of the FLiY protein gave an increase of a factor 4.27 with a CV of 0.55 after induction of rhsA from the gel-based results. The results of iTRAQ gave an increase of a factor of 2.7 with a CV of 0.24. These results indicate that the 'real' factor of increase (after many measurements, not just four) is somewhere in between 2.35 and 6.62 (gel-based) or 2.1 and 3.35 (iTRAQ) and thus, although the factors measured by the different methods seem very different, they are actually in good agreement.

These results cannot of course be taken without a pinch of salt and transferred to other (your own) proteomic approaches; it appears obvious that even highly controlled experiments can only deliver trends in protein expression. The best consistency is useless if it is a consistently wrong result (as it might be when the quantification is based on peptides alone), and screening methods such as comparative proteomics with high throughput are not necessarily of the highest reliability in quantitative terms.

5.3.1 Quantitative analyses using gel-based approaches

Two-dimensional gel electrophoresis based approaches are still the mainstay of quantitative proteomics. There are some four or five times more studies on quantitative proteomics using gel-based approaches than gel-free approaches at the time of writing this book. We have observed that gel-free approaches can cover a larger proportion of the expected proteomes – why, then, are gels still so popular? The answer is simply that they work well. If there is a situation that requires quantitative proteome analysis, if the samples are available in

high microgram to low milligram amounts, if the 2D procedure is established in the lab, then it will deliver results very fast, within weeks. At every stage of the analysis the results are tangible; even before one can see the end result, it is obvious after staining several gels whether the sample preparation is appropriate, whether the samples are of comparable quality and comparable overall protein content, and just how similar or different the samples are. Modern software allows for relatively easy comparison of the results no matter how 'shallow' the depth of proteomic penetration (Millioni *et al.*, 2010), and software for gel comparison and analysis can even be had for free (e.g. http://web.mpiib-berlin.mpg.de/pdbs/2d-page/downloads.html). Also on the positive side, 2D gels can be compared between different labs and from different experiments of the same lab, with years between the actual experiments. The better (more standardized) the procedures, the better the results of this comparison will be. All these are points with regard to which gel-free approaches have not (yet) delivered. Using gels, no one needs to be a proteomics specialist to get 40–50% of the number of differential proteins the best labs in the world using these technologies would obtain. As quantification is performed on the gels, if only differential proteins need to be identified by MS the workload is not immense, and collaborations to set up the identification of 20–30 gel spots are straightforward. As a rule of thumb, here is what to expect from 2D gel electrophoresis based approaches. For a single experiment, about 50–200 µg of protein of a complex mixture or between 4×10^5 and 1×10^7 cells are enough. Expect to see some 200–2 000 spots, and some 10–50 differential proteins between closely related samples (e.g. induction of one gene). Within a reasonable time frame a fraction of these interesting spots will be identified (typically less than half). This is the stage at which proteomics has done its job for many people; it may by then have generated 5–30 proteins of interest and follow-up experiments to try and figure out why these proteins are changed and shed more light on the biology behind the changes may take years. Of course, gel-based proteomics in this form can only scratch the surface of the proteome and often barely quantify 5% of all expected proteins. However, it is the method of choice to get a relatively quick comparative overview. With investment in ultra-high resolution gels, biological pre-fractionation, DIGE and other fluorescence detection

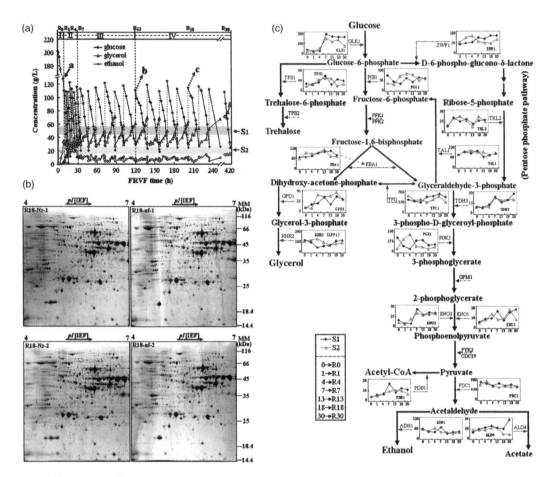

Figure 5.13 Gel-based quantitative proteomic approaches in industrial application. 2D gel-based proteomics for the monitoring of a complex, industrial scale, repeated vacuum fermentation by yeast (a). Samples were taken during the 20-day fermentation run and analyzed by 2D gel electrophoresis ((b) just shows a few of the gel images). Among other proteins, all enzymes of the main fermentation pathway could be quantitatively monitored throughout the fermentation process (c). See text for more details. Cheng *et al.*, (2009). Reprinted with kind permission of the Spinger Verlag. © Springer Verlag 2009.

technology or even isotope labelling of proteins prior to the separation, LC MS/MS for protein identification, and so on, it becomes feasible to analyze a larger proportion of proteins that are expressed at relatively low levels, perhaps around 10 000 proteins per cell. However, some proteins will forever elude quantitative analysis by 2D gels, be it because they are hard to detect (see Section 5.2) or because they are highly dynamic in, for example, PTM content. Let us see how modern gel-based quantitative proteomics performs in practice.

In an attempt to rationalize industrial fermentation conditions for yeast during ethanol production, 900 μg samples were taken at various point in a complex fermentation cycle and subjected to 2D gel electrophoresis (18 × 20 cm, pH 4–7, 12% SDS PAGE; see Figure 5.13). The gels were coomassie stained and all spots were quantified as percentages of the total protein of all spots on the gel and set in relation to the first time point. On every gel of the analysis (performed in biological duplicate) some 600 spots were detected and quantified per gel. From all the variable spots, 106 in total were identified by MALDI TOF MS PMF analyses, resulting in 68 different proteins. In a study this size it is particularly difficult to follow many spots through all experimental conditions. Among other targets a comprehensive quantification of enzymes of the glucose metabolism was achieved. This study was very successful and allows a detailed analysis aimed at further optimization of the industrial process. From a proteomic point of view there are immediate points for improvement; only a limited pH/size range of proteins was analyzed (utilization of different pH gradients, SDS PAGE compositions would extend this range) and not the most sensitive stain was used (use fluorescence/silver stains) and a good third of the gels was not usable, probably due to a combination of sample contamination and suboptimal conditions. A better sample clean-up, loading less sample and using a more sensitive stain might perhaps have helped in this respect. Presumably more proteins would have been detected if the PMF approach had been supplemented with LC MS/MS analyses. However, as it is the authors have more avenues of further research and optimization to pursue than they can possibly carry out in a short time, and the chosen

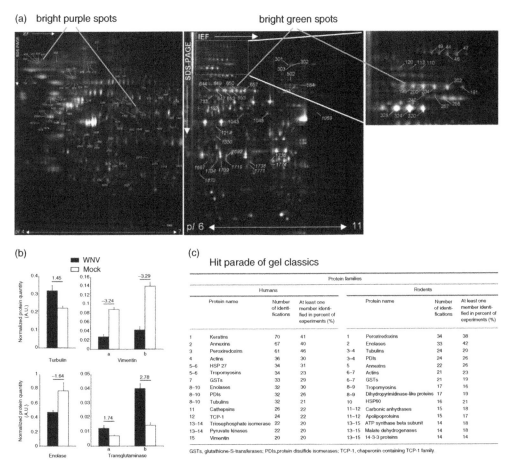

Figure 5.14 Gel-based quantitative proteomic approach for the analysis of viral infection. (a) Overview of DIGE gels used to identify differential proteins in mammalian cell lysates after infection with a flavovirus. The original gel images are colourful, with differential spots ranging from bright purple to bright green (depending on which way round the changes are), while unchanged spots are white against a black background. Combining gels with two pH ranges allows to analyze a wider pH range in high resolution. The detailed view shows the spot trains, characteristic for glycosylations. (b) The bar charts show the expression changes in proteins often used as 'controls' or 'housekeeping' proteins in other studies: vimentin ($-2.5 \times$ change), tubulin ($+2.2 \times$ change), enolase ($-2.5 \times$ change), transglutaminase ($+2.2 \times$ change). (c) The table shows the results of comparing 170 gel-based proteomics studies; the indicated proteins were identified as being differential in many of the 180 independent experiments from all sorts of cells and biological situations. More details are given in the text (a) and (b). Pastorino *et al.* (2009). (c) reprinted from Petrak *et al.* (2008). © 2008 Elsevier.

approach is very efficient; all the improvements would also result in a larger workload. Here is the simplicity of gel-based quantitative proteomics at its best – the best value for money with a simple workable approach.

Slightly more sophisticated 2D gel-based technology, implementing some of the suggestions made above, was used in the research project presented in Figure 5.14. The study employed staining of proteins with different fluorescent dyes for different samples and running the samples as mixtures on one gel (2D fluorescence DIGE), to analyze changes in the cellular/viral proteome after viral infection. A standard cell line for virological research was used as host, velo cells from African green monkeys. The cells were infected with the West Nile virus, a flavivirus. Using pH 4–7 gels (18 cm, 10% SDS PAGE) 1 950 spots (79 differential) were detected. Adding gels with pH from 6 to 11 added another 1 674 spots (48 differential). Using the DIGE system and two biological replicates meant that for this assessment only four gels in total were needed. However, the DIGE system needs optimizing to ensure that no unreproducible changes are induced during the labelling of thousands of proteins in a procedure that changes weight, hydrophobicity and charge of the labelled proteins. From the total of 127 differential spots (cut-off criteria: change greater than factor 2, *p* value for reliability of observed change less than 0.05) an impressive 93 spots (representing 69 unique proteins, 14 of them viral) or 73% were identified by MS. This is testimonial to the very sensitive and specific approach of digesting the proteins with trypsin from the gel pieces and analyzing the resulting fragments by nano-RPLC MS/MS using an ESI q-ToF. The searches were made highly specific by only accepting proteins identified by at least two peptides with high scores and high accuracy (error smaller than 0.05 Da). It is interesting to note that the same gels that were used for DIGE were stained again with SYPRO Ruby and used for protein identification by MS and overall a 100 µg of protein was loaded per gel. This low amount of protein load also explains the very high resolution achieved on the gels.

A small part of the protein changes was confirmed by Western blots. The 2D gels showed several isoforms that were not resolved by 1D SDS PAGE Western blots, but the Western blots still confirmed the initial DIGE results. This study was very efficient from the proteomic point of view; using DIGE, not many gels had to be run, the gel

comparison is rather easy and the combination of a high sensitivity LC MS/MS protein identification strategy allowed the identification of most of the differential proteins. Overall this study has supported earlier notions about rearrangements of the cytoskeleton, induction of apoptosis and stress response in infected cells. Among known or suspected mediators of RNA trafficking, several RNA binding proteins never before implicated in viral infections were found to be up-regulated, just like several factors regulating translation, proteases and other aspects of host-virus interactions. Many of these proteins now can be analyzed as new potential targets for anti-viral medical research. Nevertheless, the depth at which the proteome can be analyzed by such a study is limited. Latest (qualitative) proteomic studies of comparable human cells aimed at complete proteome coverage and reaching down to low expression levels have identified about 7 000 unique proteins, albeit without any quantification. In this study shotgun approaches were combined with cell pre-fractionation into cellular compartments and an improved digestion protocol (Wiśniewski *et al.*, 2009). Seven thousand different proteins would result in over 10 000 spots on 2D gels, while 'only' about 3 600 spots were observed in the study on Figure 5.14. Similar high numbers of spots have been repeatedly reported for approaches based on giant gels (not commercially available, and with carrier ampholyte based isoelectric focusing), but again without quantification of all the spots and without identification of the vast majority of spots/proteins. Using pre-fractionation and commercial immobilized pH gradient isoelectric focusing gels about 5 000 spots have been resolved for bacteria and up to 8 000 for mammalian cells (Richardson *et al.*, 2008; Lee and Pi, 2010) but again without quantification. The reason why single studies using gels do not get much deeper into the proteome is that the workload increases immensely and the returns diminish, the deeper one tries to dig in the proteome. Although the number of spots can be increased several-fold, the workload, the accuracy, the amount of sample needed and the requirements for quality control for quantitative analysis increase much more. Some research also shows that it is questionable whether the additional spots represent more, low abundant proteins, or mostly new isoforms of proteins identified at lower gel resolutions (Vasudev *et al.*, 2008). In essence, to go deeper into the proteome and analyze more than the top 2 000–4 000 spots (representing about

1 000–1 500 proteins) requires expert knowledge and transforms the straightforward gel-based quantitative approach into a very complex analysis. Although new technologies (e.g. for the analysis of membrane proteins; Braun *et al.*, 2007) are constantly being developed, for the time being 2D gels are the only technology able to display and quantify full proteins in significant numbers, including their different isoforms and PTMs. Figure 5.14 also contains a list of proteins that are repeatedly found in quantitative studies using 2D gels. These represent a mixture of potentially genuine changes (e.g. stress induced or cytoskeletal proteins) which might be affected by many differential conditions. However, finding any of the proteins on the list should trigger a 'red alert' as they might represent artificially introduced changes, and careful evaluation of the results is needed. This 'hit parade' also highlights the fact that some of the proteins often used as internal standards (e.g. tubulins), are not very likely to be 'unchanged', and are thus poor choices for internal standards.

5.3.2 Quantitative analysis using gel-free approaches

Quantitative analysis of peptides is currently the most buoyant field in proteomics. Given the limitations of gel-based approaches, it is tempting to extend the much increased and nearly unbiased proteome coverage of shotgun proteomics to the quantitative analysis of proteomes. As indicated by the huge variety of approaches currently in use or evaluation, this is fraught with problems. A lot of these problems result from the enormous data explosion when dealing with peptides; in extensive studies it is not uncommon to deal with MS/MS data on hundreds of thousands of peptides. Recognizing the same peptides, identifying them in databases and putting them together into 'real' proteins all pose immense technological and conceptual problems and the correct reconstruction of proteins from peptide mixtures is simply not always possible. To add quantification to this is asking for trouble. Reliable quantification needs reproducible data, and as we have seen, in Section 4.4 'bottom-down' quantification is able to deliver this kind of data, just not for all peptides that can be measured or identified. As a rule of thumb, proteins that are detected by four, five or more peptides are very likely to be quantifiable by a variety of methods. Some methods will even deliver reproducible results with two to three peptides per

protein (see ICAT example in Figure 5.11). By the same rule of thumb only about 30–60% of the proteins that can be reliably identified by shotgun approaches can also be reliably quantified within the constraints of the method (see Section 4.4). However, not all the problems are of a fundamental nature; many problems can be solved with better bioinformatic tools, developments in mass spectrometry and database curation. As the field is relatively new (compared to gel-based technologies for quantitative proteomics) and has immense potential, developments can be expected to lead to 'off the shelf' and commercial solutions that will in the very near future match what specialist laboratories can achieve right now. Latest results of MRM based studies allow researchers to reliably quantify (label-free and with modest amount of sample) low abundance proteins down to a level of 50 protein copies per cell, a feature previously only achieved by (in most organisms unworkable) fluorescent fusion protein detection (see Figure 5.7). However, MRM approaches also only work for 50–200 proteins per single experiment. So all the different methods have their own limitations, and at the moment there is no single method combining all the sensitivity, accuracy and comprehensiveness of the single methods. Furthermore, not all methods are applicable to all types of samples, as some require in vivo labelling or prior knowledge of the proteome.

Given the rapid development of a variety of quantitative shotgun proteomic technologies, it will only be possible to give examples of some of the most often encountered or most inspirational approaches.

Isotope coded protein labelling (ICPL) is a relatively new, affordable and versatile quantification method for complex proteomic samples. Compared to ICAT (see Figures 5.10 and 5.11), it should label more peptides per protein. The labelling chemistry is directed towards free amino groups as found in lysine or the amino terminus of proteins/peptides (Lys is over three times more common than Cys, which is the second rarest amino acid). Unlike the iTRAQ method, all mass spectrometers capable of MS/MS can be used for ICPL, including the versatile ion traps. However, the label used in ICPL (nicotinylation) interferes negatively with the peptide fragmentation by electron transfer dissociation, but not colision induced dissociation, which is used in most proteomic experiments. Unlike iTRAQ, but like in SILAC, ICPL is sensitive to 'crowding' of the spectra

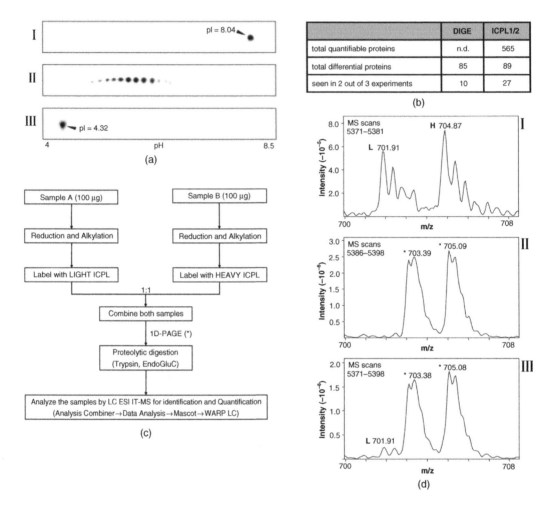

Figure 5.15   ICPL and DIGE for quantitative proteomics. ICPL can be used for gel-based and shotgun proteomics alike. (a) Position of myoglobin on 2D gels (I) before, (II) after 10 seconds, and (III) after 2 hours of labelling with the ICPL reagent. (b) Summary of the results for the Colomé *et al.* (2010) study (see text for details). All changes had to be at least a factor of 1.5 up or down to be considered differential, and also had to be observed in two of the three experiments (1 × DIGE, 2 × ICPL). (c) A typical ICPL strategy and (d) one of the problems of quantifying MS data from complex samples. The ICPL pair of peptides eluted during scan 5371–5381 (I). However a substantially stronger ICPL peptide pair elutes during scans 5386–5398 (II). The strong peaks at 703 Th/2 109 Da distorted the measurement of the weaker peaks that were measured at only two Th difference (701.9 Th/1 402.8 Da) and vice versa (III). The IT MS used in the study is set to run at the lowest useful resolution, and thus the highest scan speed and sensitivity, so that it barely shows the charge state of the +3 peptides. For more details see text. (a) Reprinted from Schmidt *et al.* (2005), courtesy of Elsevier. © 2005 Elsevier. (c)–(d) Reprinted from Paradela *et al.* (2010), courtesy of Elsevier. © 2010 Elsevier.

and co-eluting peaks can influence the quantification of ICPL peptides (Figure 5.15 d). Usually complete proteins are nicotinylated before any separation and proteolytic digest (Figure 5.15). ICPL can also be used to quantify proteins on 2D gels, but it changes the migration behavior of the proteins due to the charged nicotine label, similar to the changes induced by fluorescent labels in DIGE.

While incomplete labelling can lead to complicated analyses, the acidic shift allows the gel-based analysis of notoriously difficult-to-analyze proteins with high pI, for example histones. ICPL results show an extensive overlap with differential proteins identified by gel-based DIGE. The proteome coverage achieved by ICPL ranges from being comparable to gel-based approaches to about twice or three times greater than gel-based approaches. Despite the fact that MALDI normally preferably detects arginine-containing tryptic peptides, the lysine labelling based ICPL works very well with MALDI ToF/ToF detection and quantification as well as MS/MS identification, just as iTRAQ. The offline MALDI ToF/ToF analysis of complex RPLC runs from ICPL labelled lysates allows the detection (software based, supported by manual input and evaluation) of peptides with a mass difference of 6 Da and a different intensity. Only these pairs of unlabelled and ICPL labelled peptides of differential intensity need to be analyzed by MS/MS. As the ICPL label adds additional specificity to the MS/MS analysis (e.g. fragments that differ by the weight of Lys + 6 Da, two identical peptides per sequence), many researchers feel one peptide identification of proteins and even quantifications of proteins based on a single peptide analysis is reliable enough. Overall, ICPL is a versatile and affordable technology, a real alternative to gel-based quantifications in many situations, with comparable or better sensitivity than gel-based approaches. ICPL can be multiplexed at the moment with four different reagents, so that four different experimental conditions can be compared. Although the analysis of PTMs is in principle possible, given the large theoretical peptide coverage, none have been reported and one major problem of the approach is getting all principally accessible proteins/peptides labelled. Many peptides remain unlabelled in a reproducible fashion, which is presumably due to the fact that the label is usually introduced at the level of complete proteins, which may display some steric hindrances to the labeling of all peptides. The labelling at the peptide level could improve the labelling efficiency, but would also increase the number of manipulation steps before different samples are mixed. In a recent study by Colomé *et al.* (2010) two independent ICPL experiments were combined with a DIGE experiment. In mammalian cells transfected with an auto-immune regulatory protein some 1 588 unlabelled/labelled peptide pairs from a total of 565 different proteins could be quantified. Interestingly the authors chose as an additional criterion that the identical change had to be seen in at least two of the three experiments to be classed as significant. This criterion led to a significant drop in differential proteins, and only just above 10% of the gel identified differential proteins 'survived' this criterion (Figure 5.15b). The authors

were able to establish a new function for auto-immune regulatory protein – the induction of apoptosis which was supported by additional tests. Interestingly the gel-based proteins did not show the proteins involved in apoptosis as being differential, and thus ICPL was deemed more sensitive. ICPL was performed using 200 µg of sample and a GeLC MS approach with a Pauli ion trap.

Another very popular labelling approach is SILAC (Figure 5.16). Although the labelling has to be performed 'in vivo' and is mainly performed on cultured cells, even whole animals up to the size of mice are being used (at a cost). However, human or clinical samples cannot be analyzed directly by SILAC. While this is of course a drawback, SILAC labelled samples can be used to deliver internal standards (thousands of them!) for clinical proteomic studies and thus allow very reliable quantifications of different samples compared to the SILAC standard. To achieve a near 100% labelling even for proteins with a slow turnover time, the labelling period has to last several generation times of the cells in question. This predisposes SILAC to use with fast growing micro-organisms, but mammalian cells are also often used. The big advantages of SILAC are that all steps after the labelling are performed with mixed samples (labelled/unlabelled) and thus no heterogeneities are introduced and that all proteins are labelled to nearly 100%, allowing reliable quantifications. Using 'heavy' lysine and arginine, all tryptic peptides are labelled at least once, so the method is perfectly amenable to the analysis of all PTMs, or even interaction studies (see Sections 5.4 and 5.5). Using essential amino acids (for the organism in question) can be another way to ensure high labelling efficiency. All mass spectrometers can be used, and most software for shotgun quantification has a way (fully automated or manually assisted) to deal with SILAC data. However, it can be costly to obtain special media and labelled amino acids in the quantities needed and the heavy peaks can crowd the spectra, just as with ICPL (see, for example, Figure 5.15). In most SILAC studies the fraction of the proteins that can be quantified out of all the identifiable proteins is closer to 100% than in any of the other shotgun quantitation approaches. This can be attributed (among other characteristics) to the high confidence that even 'one-hit wonder' identifications/quantifications deliver; even a single peptide is present twice and delivers in one case label-specific fragments. In addition,

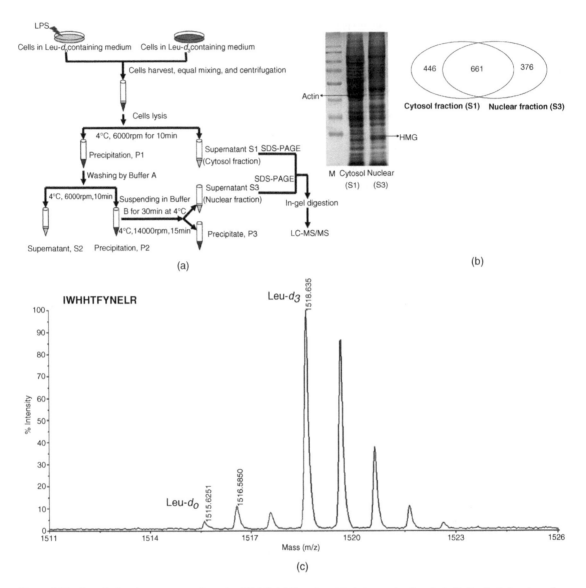

(a)

(b)

(c)

Figure 5.16   Quantitative shotgun proteomics using SILAC. (a) Experimental strategy using Leu d₃ (with 3× deuterium to replace hydrogen) for SILAC and isolation of nuclear and cytoplasmic fraction to be able to see translocations after LPS stimulation of macrophages. (b) Enrichment of cytosolic and nuclear proteins as well as the overall distribution of the 1 483 proteins that were quantified. Over 500 proteins showed elevated levels in the cytosol, over 600 were elevated in the nucleus and another 80 proteins were reduced in the nucleus after 10 minutes of LPS stimulation. (c) Nearly complete label of actin after labelling the cells for seven generations (about a week); only very small levels of unlabelled peptide (Leu-d$_o$) are found in peptides containing one, two and three Leu in their sequence. In the peptide with 3× Leu, only very minute amounts of 2× labelled peptide are found (labelled with 'X'), allowing highly accurate quantifications. Also note the high resolution signal in the example spectra from the LIT used in the study. (d) Detail of some results from the PANTHER analysis of pathways and functions influenced by LPS activations. Reprinted with permission from Du *et al.* (2010). © 2010, American Chemical Society.

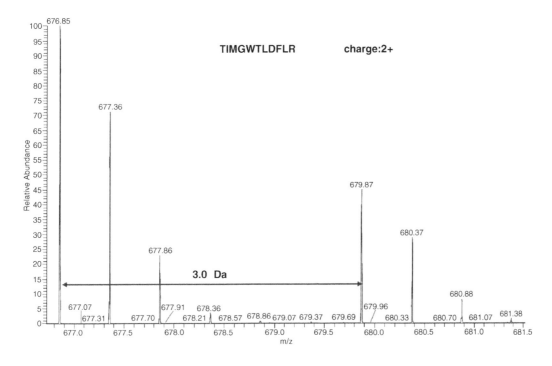

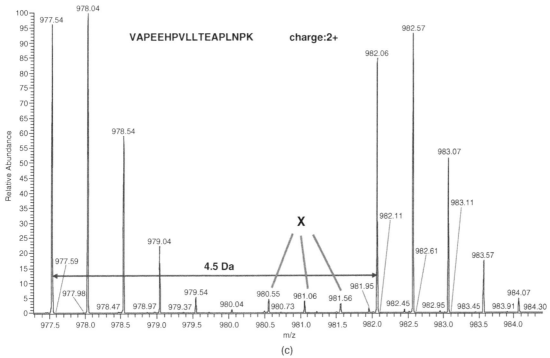

Figure 5.16   (*continued*)

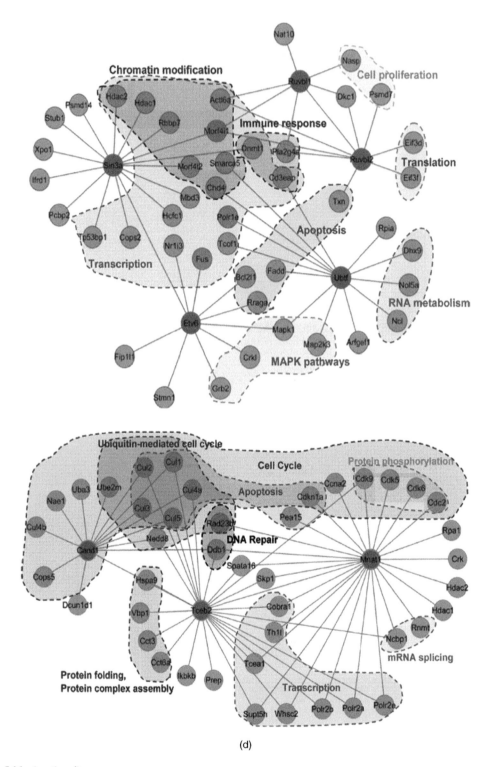

(d)

Figure 5.16   (*continued*)

as opposed to ICPL, all peptides are labelled equally well if the label period is long enough.

A study by Du *et al.* (2010) shows how powerful SILAC approaches can be. Analyzing the reaction of macrophages to lipopolysaccharide (LPS), they used the SILAC approach in combination with a (relatively) simple enrichment for nuclei and cytoplasm, and could identify over 1 000 proteins in each fraction (Figure 5.16). An incredible number of over 600 proteins changed their abundance in the cytosol and over 700 had changed abundances in the nucleus. The sensitivity was such that kinases of known low abundance (among them c-RAF and IKB kinase) or transcription factors (among them Rb1, Sp1) were found with changed ratios in the cytoplasm or nucleus. This amazing sensitivity does not mean that all of the low expressed proteins (like kinases and transcription factors) are quantified – the largest qualitative proteome analyses of mammalian cells show over 8 000 proteins and are still not 'complete') – but it certainly means that these methods have the potential to deliver new insights at the level of regulation of biological processes, rather than their execution by more abundant proteins. The superb sensitivity in the study by Du *et al.* is achieved because of its very sensitive quantification by SILAC, reaching down to changes of ±20%, a very sensitive and accurate mass spectrometer (LIT-orbitrap), a high resolution separation approach (GeLCMS with 20 gel sections and peptide separation by RPLC) as well as the 'biologically' intelligent sample complexity reduction into cytoplasm and nuclei. The samples were compared after 10 minutes of LPS stimulation, which means that most abundance changes reflect translocation between nucleus/cytoplasm and potentially other cellular compartments. If whole cell lysates had been analyzed, only a few changes would have been detected. Based on the considerable coverage of low abundance signalling and transcriptional regulators, the authors could reconstruct the interactions and cross-talk between several signal transduction and transcription control networks. This showed, for the first time for LPS induction, several connections (cross-talk) directly.

Another study using SILAC (Graumann *et al.*, 2008) was able to quantify over 5 000 proteins in murine embryonic stem cells, again using a combination of high resolution mass spectrometry, cell compartimentalization (cytoplasm, nuclei and membranes, which added over 2 000 compartment-specific proteins) and high resolution

protein separation by GeLC MS. In addition, this study used an OFFGEL approach on the complete cell lysate, which led to the quantification of nearly 4 000 proteins with less than half the workload involved in the subcellular compartment analyses. SILAC is surprisingly versatile, it can be multiplexed up to three samples (light, medium and heavy), and it is not limited to cell culture material; mice and other animals have been analyzed by SILAC as well. In a very interesting approach, analyzing the enormous regeneration potential of amphibians, newts where fed livers of SILAC labelled mice (Looso *et al.*, 2010). After three weeks half of the newts had their tail amputated. It regenerated to a great extent in the following 40 days (still being fed SILAC labelled murine livers) and the regenerated tails were than compared to the normal tails. Over 3 000 proteins could be quantified, which is all the more impressive considering that the genome and proteome of newts are not very well characterized. In fact the SILAC approach added specificity to the extensive, homology search based peptide/protein identification approach; by the shifting masses the numbers of labelled amino acids per peptide and thus the numbers of lysines in the sequence were known. The quantification of over 3 000 proteins now allows a much better understanding of the regeneration process than hitherto possible, and shows how proteomic methods can be creatively adapted for individual research purposes.

The single most popular and successful quantification approach for LC MS based proteomics technologies is iTRAQ. As the name suggests, iTRAQ can be used for absolute protein quantifications. For this, a labelled standard of peptides has to be used, for which the total, absolute amount used in the experiment is known. However, most researchers are not very interested in absolute quantities of proteins (see Section 4.4) and avoid the additional experimental complications. Thus iTRAQ is mainly used for relative quantifications. There are up to eight different tags available (i.e. eight different samples can be compared in one experiment) and the labelling is performed at the peptide stage, after tryptic digestion and before LC separations (see Figures 5.10 b). Unlike the other labelling approaches, iTRAQ uses isobaric tags; that is, peptides are labelled with different tags of identical mass. Thus no difference between the various samples can be detected in MS analyses. This avoids the 'crowding' of spectra with new (labelled) peaks and the 'dilution' of peptides needed for MS/MS based identification; all the identical peptides

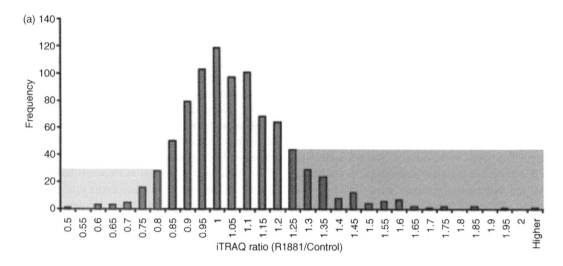

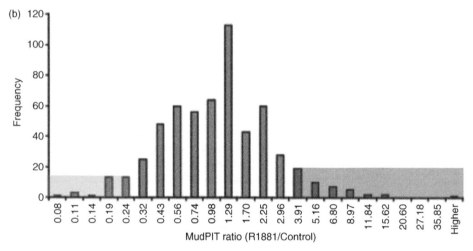

Figure 5.17 Quantitative shotgun proteomics using iTRAQ. Prostate cancer cells were either left untreated or were treated with the androgen R1881, and proteins were analyzed by 2D HPLC. Runs were analyzed either by iTRAQ using a MALDI ToF/ToF MS or by were quantified by spectral counting, using an online ESI quadrupole IT MS/MS. (For details on iTRAQ and spectral counting see Figures 5.10 and 5.12.) (a) Ratio for iTRAQ signals; (b) ratio for normalized spectral counts. Based on statistical considerations, cut-off levels of 1.2 standard deviations either side of the mean for up-regulation and 0.83 standard deviations for down-regulation upon androgen treatment were set. The shaded areas show the proteins with changes deemed relevant. Note that due to the greater variation of the means for spectral counting the cut-off ratios were larger for the spectral counting method. (c) In the iTRAQ data 875 proteins passed quality controls and resulted in 39 down- and 70 up-regulated proteins (top Venn diagram). In the MudPIT approach 502 out of 844 peptides showed no significant changes and were detected in both lysates. Ninety-six proteins only detected in control cells showed not enough spectral counts to be considered for significant statements. Together with 30 proteins that were detected in androgen treated cells at significantly lower levels this left a total of $27 + 30 = 57$ proteins with down regulated expression levels (middle Venn diagram). Similar calculations lead to 65 up-regulated proteins identified by MudPIT and spectral counting. The bottom Venn diagram shows that the MALDI based iTRAQ and ESI based spectral counting approaches (denoted with 'MudPIT') are very complimentary and perform at a similar level, albeit with a small overlap in the proteins identified and quantified. Vellaichamy *et al.* (2009), reprinted with permission from PLoSone.

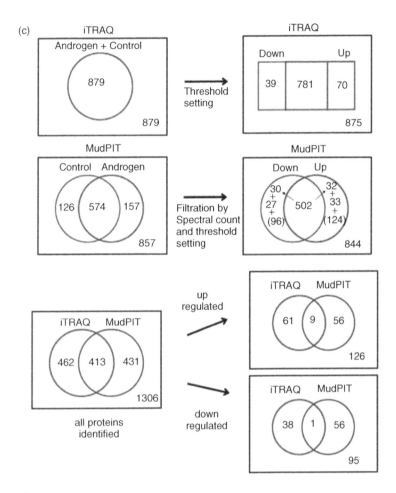

Figure 5.17   (*continued*)

from the various sample are united and contribute to the MS/MS fragments needed for identification of the peptide. Not so the reporter part of the tag; during MS/MS each type of tag generates a specific fragment that can be measured in a low, but usually empty region of the MS/MS spectrum (around 117 Da). It is the differences in ratio of these reporter fragments that are used for quantification in iTRAQ. The tags also produce a specific balancer fragment each (in the region of 29 Da), which is only needed to allow the tags to be isobaric in MS. While the idea is ingenious, not all mass spectrometers can be used for iTRAQ, as they cannot retain/measure the intensity of the relatively small reporter fragments. The otherwise widely used Pauli ion traps cannot be used, while LITs can be 'persuaded' to measure iTRAQ quite well. No conclusions can be drawn from the MS spectrum alone, so every peptide needs

to be fragmented in order to see the usually small fraction of differential peptides. Thus, fast mass spectrometers with high transmission and sensitivities for all fragment ions are needed, making triple quadrupoles or Q-Traps not very popular for iTRAQ. LIT-orbitraps have only recently been used for iTRAQ, leaving ToF/ToFs and q-ToFs as the instruments of choice for iTRAQ analyses. On the positive side, all sorts of samples can be used, no in vivo labelling is necessary, the multiplexing allows internal controls and the system really works well for a large number of users. Several software packages are available to help cope with the massive amount of MS/MS data produced. Usually tens of thousands of MS/MS spectra have to be evaluated, many of which will be rejected for not delivering sufficient fragmentation data for peptide identification or reporter fragment quantification. Here, MALDI ToF/ToF instruments have advantages; they deliver 'harder' colision induced dissociation fragmentation at higher energy with stronger signals for the reporter fragment and have high resolution and good sensitivity. However, most iTRAQ experiments involve LC MS separations. These are measured a lot more rapidly online. The offline coupling to MALDI introduces additional steps and takes a long time to measure after the LC separation, as millions of LASER shots are needed for the whole experiment (typically around a thousand LASER shots per MS and more than a thousand for MS/MS, or 3–11 hours for the complete measurement of one RPLC run). Since quantifications rely on MS/MS data, a high resolution parent ion selection is essential to prevent neighbouring peptides from contributing to reporter ion signals, and this can cause problems during crowded LC MS runs. As with many quantifications in proteomics, for all these technical reasons the resulting relative ratios would not stand a rigorous clinical test; but most researchers are very happy to find hundreds of differential proteins, even if the accuracy of the relative quantification is 'only' 40% (i.e. a 10-fold change might be 6- or 14-fold), as long as the tendency is right (particularly since the accuracy is best around the all important 1 : 1 ratio). The sensitivity of iTRAQ depends on the sensitivity of the underlying MudPIT approach; typically close to a thousand proteins are quantified in a single experiment, often with hundreds of differential proteins (typically at accuracies of ±30% and better between different peptides).

Interestingly some researchers find that different LC MS quantification approaches deliver mutually exclusive results (see Figure 5.17). Although the explanations for these findings are not always straightforward, using MALDI based MS/MS quantification compared to ESI based spectral counting (i.e. counting how many MS/MS spectra are generated per peptide) can deliver highly complementary results. The question arises as to how reliable these results are, if they can be so different. If we barely scratch at the surface of a proteome (e.g. by quantifying 'only' about 900 proteins out of a presumed 10 000 or more). Which proteins ones we actually see can differ dramatically from one experiment to the another within one method (which is why it may be advisable to repeat LC MS quantification until no more new proteins are found, i.e. about three to nine times) and will differ even more so when different methods are used (see above). As long as no one thinks they analyzed a complete proteome, the results can complement each other and

should be interpreted as such. Independent verifications are also very important to substantiate at least some of the findings. The Association of Biomolecular Resource Facilities (ABRF) poses an annual challenge asking participants to, for example, quantify an unknown sample in a couple of days. The results are very interesting and indicate that experience with a certain method is more important than the choice of technology in order to deliver reliable quantitative information.

These competitions also show that label-free shotgun quantification methods can be very reliable tools (e.g. beating all other methods in the PRG 2007 study of the ABRF, but not in other years). Label-free LC MS quantifications based on peptide intensity can be extremely powerful (see Figure 5.17), allowing the quantification of over 5 000 proteins, down to the level of low expressed proteins, for example kinases. This in turns allows complex signalling events or immunological reactions to be analyzed with unprecendented accuracy. However, comparing peptide intensities without labelling needs an exceptional computational effort. High sensitivity studies with significant coverage of the proteome are only possible with high resolution/accuracy mass spectrometers; the added specificity of labelling approaches (e.g. the second 'heavy' peptide, heavy fragments, knowing the number of labelled amino acids per peptide) is missing, and peptides have to be compared across different LC runs, rather than in the same run. For these reasons, one limiting factor for label-free quantifications based on extracted peptide intensities is the data analysis software, and for deep proteomics penetration 'in house' developed software has been used up to now. Very efficient software became available only recently (e.g. Zhu, *et al.*, 2010). Still, it is a good idea to have IT specialists at hand and allow much more time for data analysis than for data generation, which is also true for large-scale labelling approaches. Spectral counting (see Section 4.4 and Figures 5.10 d and 5.17) is semi-quantitative, and needs several peptides of the same protein for good quantification, so although the standard deviations are a little higher than for other methods, it is also a very valuable method with the added advantage of being a lot easier on the software/data analysis side of the project as opposed to using the peptide intensity for quantification.

All the quantification approaches presented hitherto have been 'discovery' approaches, unbiased by knowledge of the proteome (and sometimes even the genome). Targeted approaches such as MRM can be much more accurate and sensitive than these discovery type analyses.

In fact, MRM allows a much better quantification and determination of absolute protein abundances than any of the other methods presented so far. The drawback of MRM is of course that it is targeted; for MRM a measurement strategy has to be developed based on prior knowledge of the peptides of the protein of interest that are actually experimentally observed, and several fragment ions that are easily observed for each of these peptides have to be known. Using very fast IT MS, several hundred proteins can be reliably quantified in a single LC MS run. As the sample amount requirements for MRM are low and run-to-run differences in LC elution times are not important (the specificity comes from the parent mass and the fragments alone), the method is very reliable. Figure 5.18 shows an example of a strategy where MRM is used to internally calibrate ion extraction

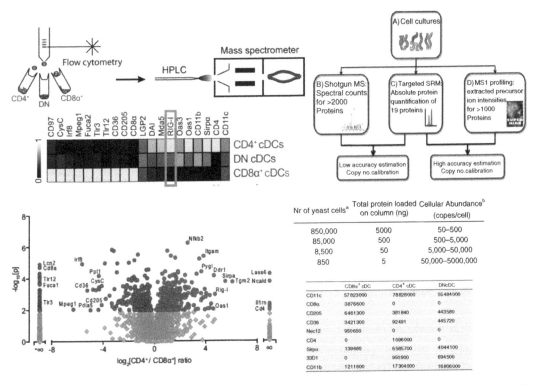

Figure 5.18 Label-free quantification for shotgun proteomics. Dendritic cells (DCs) were isolated by fluorescence activated cell sorting according to sets of surface markers (top two panels on the left) and separated into three different populations. Using a hybrid LIT/orbitrap FT MS in a shotgun approach, over 1 million peptides were identified in all three DC populations. These corresponded to nearly 100 000 independent peptides, and around 5 000 proteins for each DC population. The peptide intensities were compared and differential proteins identified by statistical analysis (volcano plot, bottom left). In a volcano plot the log of the difference in expression (on the x-axis) is plotted against the significance of the observed differences (y-axis), with results outside the area in grey in the lower part of the y-scale being deemed significant. The results were not obviously biased against rare proteins (e.g. 3.6% of quantified proteins were kinases, and kinases represent 4.2% of the genome). The table on the lower right shows how known markers were clearly differentiated in the DC populations. The units are ion counts (instrument units), as no absolute quantification was performed. These very impressive results show how sensitive label-free shotgun quantifications can be. The workfloor schema in the upper right panel shows the strategy to achieve absolute quantification of low copy proteins. This mixture of absolute quantification of 19 proteins and bootstrapping of the rest allows the absolute quantification of proteins down to a couple of proteins per cell (middle panel on the right). Moreover, the quantification by targeted single-reaction monitoring is highly reliable, allowing integration of results in system biology approaches. Panels on the left and lower right panel reprinted from Luber *et al.* (2010), courtesy of Elsevier. © 2010 Elsevier. Top panel on the right reproduced from Malmstrom *et al.* (2009, Suppl.), courtesy of Nature Publishing Group. © 2009 Nature Publishing Group. Middle panel on the right reprinted from Picotti *et al.* (2009), courtesy of Elsevier. © 2009 Elsevier.

and spectral counting, allowing the quantification of more than 50% of the predicted open reading frames of *Leptospira interrogans* (which causes yellow fever in humans) under exposure to antibiotics. The quantification ranges from less than 50 to more than 40 000 proteins per cells and the proteome is presumably analyzed to more than 80% coverage (as not all proteins are expressed under all culture conditions), thus allowing a unique insight into the defence mechanisms of *Leptospira*. Using the 'right' mass spectrometer (triple quadrupole or Q-Trap), MRM allows a much better and more accurate quantification of data than discovery approaches, with reproducibilities requested for clinical settings or meaningful 'systems biology' type analyses, as we will see in Section 5.6. Recent single-cell proteomic analyses also show that more noise is introduced at this very sensitive level of proteome analysis, as every cell is in a slightly different 'status', for example for translation from specific rare mRNAs (one mRNA molecule often correlates to hundreds of protein copies per cell). However, except for yeast (with optical proteomics), the accuracy and sensitivity level of MRM is superior to any other approach. This shows what will be relevant for future proteomics analyses (e.g. what pushes a single cell over the edge to become apoptotic or oncogenic?). Only at the single-cell level will we be able to understand why only a very small but relevant proportion of cells decide to, for example, change their fate.

5.3.3 Quantitative proteomics for biomarker discovery

Biomarker discovery is the main application for proteomics in terms of sheer volume. The term biomarker can be defined very loosely, as any protein (for us, of course, also metabolites or even salts can be used as biomarkers) that might change under certain conditions can be used a marker for a certain cell cycle state, nutritional situation, immunological state or any disease state imaginable. In a more stringent definition a biomarker would be a protein that is highly specific for a certain condition or disease and would allow the easy screening or test for this condition or disease, in order to start or adopt an appropriate treatment. We will here only consider the more stringent definition in this chapter, as examples of a variety of loosely defined 'biomarkers' have been given throughout the book.

Serious biomarker research must include several non-proteomic considerations, such as availability of sample material, evaluation, and cheap and robust detection possibilities, in order to be successful (see Mischak *et al.*, (2009), for an excellent summary).

Taking SELDI as an example, it is obvious why it is so popular in biomarker screening, in particular with clinical researchers. SELDI offers simple sample handling, apparently robust measurement of features, and a manageable amount of data as an output. This makes it a feasible technology for the busy researcher. On the down side, it can be difficult to identify the features (as differential proteins or as ubiquitous proteins with new minor modifications). Due to 'hidden' complexities (e.g. many components that cannot usually be seen), potential markers often cannot be substantiated, as without knowing the identity of a feature no alternative evaluation approach is possible (e.g. immuno-histochemistry or enzyme-linked inmmunosorbent assay). While the high rate of attrition is sadly true for most approaches generating potential biomarkers, the deeper the understanding of the biomarker from its discovery process, the higher the chance of survival. Biomarkers can often be combined for enhanced specificity, which is difficult if there are only a limited amount of features to start with.

More complex studies, such as label-free quantification of pre-fractionated (e.g. by OFFGEL) samples, which can reach very deep into the proteome, generate many more features to compare. However, the robustness and workload using complex sample preparation methods are such that a clinical study has little chance of being started or finished. This is also due to the extreme variability in clinical sample material (you may have noticed that most fancy 'deep' proteomic approaches were performed on up to four different situations under very well controlled conditions, not with 400 clinical samples from individuals of different age, sex and with other heterogeneities).

Most biomarker studies are thus compromises, with technologies that work reasonably well and are easy to implement and reproducible. This all explains to a great extent the prevalence of gel-based studies for biomarker discovery. Often, however, it is difficult to find new biomarkers, due to the preferential nature of every method; there is a bias towards certain proteins (see Figure 5.14), and a higher resolution of any single method often only results in the identification of more peptides or isoforms of the same proteins. The combination of proteomic approaches that are tailored to a specific sample and disease is the best way forward to a more efficient detection of biomarkers (Figure 5.19).

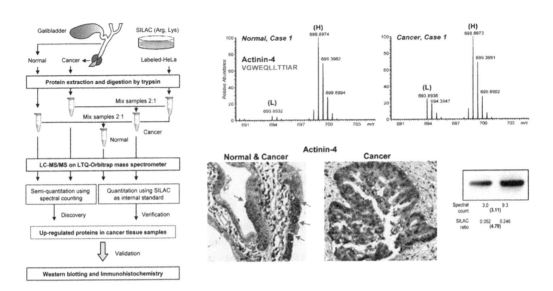

Figure 5.19   Combination of proteomic approaches for the discovery of biomarkers. Surgical samples cannot be labelled by SILAC, but SILAC can still be used for quantification of surgical samples. In a clever strategy (left), cholangiocarcinomas were compared to normal tissue by shotgun proteomics and spectral counting to discover any cancer-specific changes. These changes were verified and quantified even further by mixing them with lysates from cultured cancer cells, which were SILAC treated. The cultured cells are likely to express most if not all the proteins of the cancers in clinical samples, but the expression levels will be changed due to culturing the cells. The SILAC labelled peptides were used as 'internal standards' against which both normal and cancer cells were compared. The spectra in the top right panel show one peptide of one of the newly identified biomarkers, actinin-4 from a control and a cancer sample next to the SILAC labeled (H) internal standard. The new biomarker candidates, among them actinin-4, were validated by immunohistochemistry and Western blots (middle and right lower panel, respectively), with good correlation in the case of actinin-4 for all methods. Actinin-4 was expressed strongest in the zone with mixed normal/cancer cells (arrows) and may thus be an early biomarker for this cancer. Reprinted with permission from Kawase *et al.* (2009). © 2009, American Chemical Society.

At the moment proteomic tools deliver a variety of potential biomarkers from body fluids (in particular, serum) and a variety of tissues. However, to make proteomics applications more successful for biomarkers that survive the clinical evaluation process several suggestions have been made. Robust pre-fractionation and cell/protein enrichment protocols (such as OFFGEL, depletion of abundant proteins, fluorescence activated cell sorting or LASER capture microdissection, to name but a few) allow us to go 'deeper' into the proteome. Sample handling needs to be much more standardized, potentially with the help of robotic stages, as standard operating procedures have to be tightly controlled due to the complexity of the effects they may have on proteins. In general, ways have to be found so that the proteome is not analyzed 'as if for the first time' in every new study. Depositing data in a useful format is only possible with reliable (absolute) quantifications. Several approaches have been undertaken over the years to achieve this

goal, with gel images lending themselves to such a data exchange. A useful data set for the shotgun proteomics age in this respect would be the unique (empirically validated) identifiers for each protein; these could be either characteristic peptides with good correlation to the amount of a protein or specific internal fragments to distinguish them from similar peptides and allow specific MRM. Another route towards unique protein identification could be chips with antibodies for each of the estimated 25 000 human proteins. Both of these routes (and some others) are followed by international proteomic consortia, trying to integrate individual efforts for a common goal of more efficient biomarker discovery.

5.4 ANALYSIS OF POSTTRANSLATIONAL MODIFICATIONS

PTM analysis is a discipline unique to proteomics. While transcriptomics can do a reasonable job of detecting many (though never all) proteins with altered expressions (for a fraction of the cost and effort compared to proteomics), there is no alternative to proteomics for detecting PTMs. There are basically two ways to approach the problem: detect and/or quantify as many PTMs as possible from a complex mixture of proteins, or find all PTMs in a single protein or small group of proteins. While strategies have recently been developed for the former approach, results from the latter tell us that it is impossible to detect all PTMs – although thousands might be detected in a single experiment, those are the ones that 'play ball'; they are compatible with the selected enrichment and detection/quantification approach. As these approaches are even more complex than pure protein/peptide detection approaches and for every protein there are many PTMs, the 'depth' of PTM analyses regarding the proteome coverage is lower than that of the protein identifications. However, clever biological strategies can increase this depth considerably. At the moment we do not know the functional significance of many of the PTMs, but it is widely assumed that many might not be significant under all cellular circumstances. Comparing closely related cellular situations thus allows enrichment for potentially significant PTMs that might be significant in defining or regulating a certain cellular state. Another problem with PTMs is their diversity (see Figures 1.4, 4.21 and 5.20). Even for the best analyzed of PTMs, phosphorylations, there is a very complex set of strategies for their detection. Given the diverse chemical nature of PTMs, trying to analyze different ones simultaneously involves even more highly specialized strategies. It is thus impossible in a book like this to introduce a significant proportion of these strategies in detail. However, there are commonalities in the successful proteomic strategies dealing with PTMs, which we will look at in some examples.

To analyze PTMs the approaches have to be targeted; in untargeted approaches PTMs represent something of a nuisance, although some might be detected anecdotically. Because PTMs are so complex, accommodating all of them in the database search strategy of untargeted approaches would normally destroy any chance of identifying a significant proportion of the proteins/peptides in the study. If all potential PTMs were considered, every single peptide sequence would produce a large number of new potential masses in the search engine. Often (e.g. glycosylation) the exact weight of the modification cannot be

predicted and is thus impossible to integrate in a screening approach (Figure 5.20).

The same (or another) PTM can occur several times on one peptide, and on a protein. In combination with the fact that even highly significant PTMs occur in sub-stoichiometric fractions (e.g. 5% of one site, 40% of another) The problems of combinatorial PTM can only be solved by the analysis of complete proteins or full top-down proteomics. However, the throughput and feasibility of top-down approaches is severely limited, so that combinations with bottom-up analyses are performed or pure bottom-up data are used, fully accepting the limitations of these approaches.

While gel-based separation can often separate many different combinatorial PTM states, the sensitivity of subsequent MS analyses is not sufficient for a complete elucidation of the PTMs. This is where top-down proteomics is very powerful. However, it is only applicable to relatively small, abundant proteins that are easy to purify. It usually uses more than 1 µg of purified proteins per measurement, often also separated into differentially modified (charged or size separated) isoforms, with robust stoichiometry of modification. Typical examples are histones (Figure 5.21), HMG proteins and troponin. However, even with these small, highly expressed proteins and using FTICR, a complete deciphering of PTMs is complex and often not possible. Modifications that account for less than 5% of the total protein amount cannot be detected unless specific isoforms can be highly enriched using a number of charge-sensitive methods.

To gain more detailed insights into complex PTM patterns, or to handle significantly larger or less abundant proteins, top-down and bottom-up approaches can be combined, often with 'middle-up' proteomics (producing large fragment peptides by rare cutting proteases). In fact, most of the analyses of PTMs are of the bottom-up variety, as they are more versatile (but deliver less precise information on the protein level). All the peptide based quantification methods are available for modification analysis, and Figure 5.21 shows an example of the screening for glycoproteins in brain tissues relating to Alzheimer's. To analyze the often complex structure of glycosylation 'trees', more complex strategies have to be employed, which currently do not allow high throughput. As glycosylations are very important for functional properties of proteins (activity, half life, membrane orientation, to name a few) and slightly altered

PTM	amino acids	weight shift (MS)
phosphorylation	Ser / Thr / Tyr	+ 80
sulfonation	Ser / Thr / Tyr / Trp	+ 80
methylation (also 2–3 x)	Ser / Thr / Lys / Asn	+ 14 (28/42)
acetylation	Lys / n-term	+ 42
lipidation (various molecules eg.palmotylation myristoylation)	n-term, c-term, Cys	various,palm +238, myr +206/208/210)
glycosylation (great variety)	Asn / Ser / Thr / Lys	more than +162, up to several kDa
ubiquitination sumoylation	Lys	+8.5/ 12 kDa polym. to 70 kDa

e.g. Histone H3.3 n-terminal tail

ARTKQTARKSTGGKAPRKQLATKAARKSAP STGGVKKPHR YRPGTVALRE

peptide SAPSTGGV; possible further PTMs in over 30 combinations, adding 80 Da-13 kDa

● Monomethylation
♦ Dimethylation
▲ Trimethylation
★ Phosphorylation
▲ Acetylation

strategy	methods	application
labelling of PTM	chemical / tags stable isotopic label radioactive isotopic label	various PTMs; phosph., glyc., lipid. Inductively coupled plasma MS, (phosph./sulf.),MuDPIT SILAC/ICPL various (^{32}P, 33P, ^{35}S etc.) also for lipid. and glyc.
enrichment of PTM	enrich modified proteins/ peptides PTM specific separations Charge specific protein isoform separation	metal affinity (phosph.), anti P-Tyr antibidies, lectine bdg. (glyc.) mP-AGE (glyc.) Top Down proteomics for many PTMs,FFEIEF,OFFGEL, ion exchange chromatography
enzymatic/ chemical treatments of PTM	reversal of modification conversation into another molecule	many PTMs, e.g. phosphatases, glycosidases, homoserine formation
Mass spectrometric setting / strategy	precursor ion scans changed polarity (–) Inductively coupled plasma MS manual/ automated data dependent MS/ MS or MS^3 neutral loss scans special fragmentation methods high accuracy/ increased confidence	e.g. 98(pThr.pSer) 80(pTyr, sSer/sTyr), 204(HexNac) e.g. for precursor -79(phosph./sulp.) phosph./sulf.detection/ quantification e.g. in ion traps or hybrid orbitraps or FT ICR analyzers phosph./sulf./glysos. ECD/ETD all PTMs, counteracting increase in searchspace
MS data analysis optimization	search special databases targeted approaches (known PTMs, MRM) exclusion lists for unmodified peptides	subset-search with all PTMs for all PTMs for all PTMs

Figure 5.20 Proteomic strategies for the analysis of PTMs. Histone example after Jung *et al.* (2010). For inductively coupled plasma MS in proteomics, see Zinn *et al.* (2009).

glycosylations often represent, for example, biomarkers for cancer, this complex field of proteomics will need to be developed further towards higher throughput.

The main PTM analyzed by proteomics has for decades been (and still is) phosphorylation. Here, nearly as many strategies as independent proteomic studies exist. The most successful strategies have now TiO_2 as affinity matrix for phospho amino acids on the peptide level in common. A variety of matrices were used in the past and showed scope for improvement (too unspecific to isolate complete phosphoproteins with high specificity) and all matrices show more or less bias towards different subtypes of phosphopeptides (e.g. depending on the surrounding amino acid sequences). Using these modern peptide based methods, it is possible to detect hundreds to thousands of phosphorylation sites,

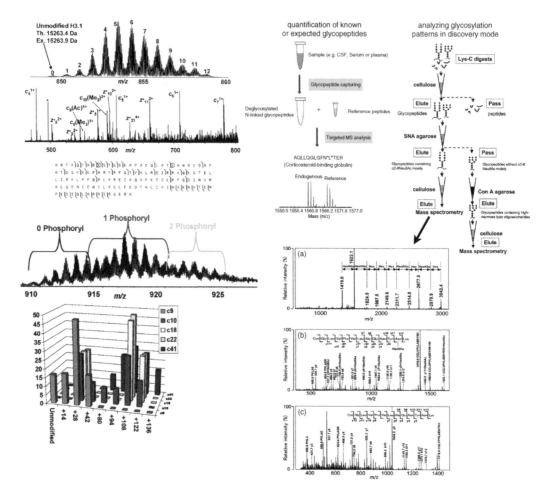

Figure 5.21 Proteomic analysis of PTMs I. Histones are highly modified with a variety of PTMs. Due to their high expression level and relatively small size they can be purified and analyzed by top-down proteomics (panels on the left). The top panel shows the various acetylated forms of histone H3.1 from HeLa cells, isolated by various LC chromatography steps. The numbers on the peaks indicate the acetylation status (shift of 14 Da per acetylation), shown at the 18+ charge state of an ESI quadrupole FTICR MS hybrid analyzer measurement. The spectrum below shows a small area of the ECD spectrum from the above peaks, and the sequence below indicates the fragments and acetylation sites that could be assigned with 20 ppm accuracy. Arresting the cells in prometaphase with colchicine results in altered PTMs on histone H3.1, with apparent mono- and diphosphorylaton, shown at the +17 charge state (middle panel on the left). The combination of phosphorylation at positions 9, 10, 18, 22 and 41 with the acetylation status (as deduced from ECD fragments) is shown in a 3D graphic (lower left panel). Similar complexities are to be expected for many proteins (e.g. signalling or tumour suppressor proteins), but cannot be analyzed with current methods, including bottom-up approaches (Thomas *et al.*, 2006). The middle panel in the top row shows a successful strategy for the quantification of known or expected glycopeptides/proteins. The sensitivity of the method is high enough to deliver results from small amounts of clinical material such as brain or cerebrospinal fluid (CSF) samples. The glycopeptides are isolated by hydrazide chemistry based solid phase extraction and then de-glycosylated. The MS never actually takes an MS/MS measurement of the glycosylation, but allows quantification by comparison with standards (Hwang *et al.*, 2010). The discovery mode introduced in the top panel on (the right) makes full use of the advantages of offline MALDI quadrupole IT ToF MS/MS for structural analysis of glycopeptides. Differential enrichment methods are used systematically to achieve some pre-selection before MS/MS (a) shows the glycosylation specific fragments (4 × HeXNAc and 5 × Hex were identified in the example). (b and c) MS³ of different fragments, chosen manually for the assumed information content, allows further fragments matching the exact glycosylation site to be found and the amino acid sequence to be confirmed (Kubota *et al.*, 2008). All panels on the left reprinted with permission from Thomas *et al.* (2006). © 2006, American Chemical Society. Top panel middle, reproduced with permission from Hwang *et al.* (2010). All other panels, reprinted with permission from Kubota *et al.* (2008). © 2008, American Chemical Society.

or even to quantify similar numbers of sites. It is a very recent development that phosphoproteomics allows similar absolute numbers of identifications/quantifications as shotgun approaches allow for unmodified peptide identification/quantification. However, about 5–10 times more material is needed, as the phosphopeptides need to be enriched first. Also, since there are potentially rather more different phosphopeptides than unmodified peptides, the depth of proteome coverage of the analyses is lower. The important issue of combinatorial modifications cannot be resolved by peptide based approaches. This is significant as one phosphorylation in vivo is often the prerequisite for another, and biological significance follows the complex phosphorylation patterns. Much more light (but with low sensitivity/throughput) can be cast on these problems with gel-based approaches (Figure 5.22). However, in proteomics strength seems to be in numbers and it is hard argue with several hundred (and up to 5 000–10 000) quantified phosphorylation sites in a single shotgun based SILAC study for PTMs compared to several dozens of proteins from individual gel-based studies (Figure 5.22). With the new peptide based technologies being commonplace, surely future generations of researches will sort out the exact stoichiometries of phosphorylations on complete proteins. Furthermore, while the gel-based proteomic approaches can show clearly that a protein is phosphorylated, with a sequence coverage of 20–40% on peptide identifications (strictly biased against phosphopeptides) most sites of phosphorylation can only be identified by further complex analyses. However, there are always a 'lucky few' modification sites that can be identified straight from analytical 2-D gels.

The phosphoproteomic data that can now be generated is well beyond what was possible in terms of either scale or throughput just a few years ago. The developments will continue as the perfect is the enemy of the good, and current approaches are far from perfect. The possible new qualitative insights arising from more data will be mentioned in Section 5.6; suffice it to say here that systems biology (for example) needs much higher qualitative data for its purposes than most proteomic studies can deliver right now. While it is amazing to be able to get so much information on PTMs, the data are most beneficial if seen in the context of differential association or subcellular translocation.

Special approaches have been developed to identify substrates of enzymes responsible for PTMs, for example kinases. The large scale analysis of substrates (as KESTREL assay or micro-array based) is typical for this line of inquiry and similar assays have been developed for proteases or phosphates. Next to the classical approaches based on the expression of inducible kinases (or the conditional expression of kinases), the approach of mutating a kinase so that it is able to use a special form of ATP are now delivering good results as well. This approach gives very good sensitivity and specificity, as the background of other kinases' activity is drastically reduced. This assay also eliminates a typical problem in signalling cascades: if we activate kinase x, it will activate kinase y which will activate kinase z (e.g. MEKK(RAF) MEK/ERK cascades). Since the signal is amplified at every level, how do we know that a substrate we identified is a direct substrate and not a substrate of a kinase further down the cascade? Mutant kinases with ATP analogues allow a clear-cut answer, as the other kinases cannot use the ATP analogue.

5.5 INTERACTION PROTEOMICS

No protein functions as single protein but always in complexes with other molecules. Yet we do not even have a half-decent understanding of all these interactions. The best data are available for yeast. In a way the yeast (*Saccharomyces cerevisiae*) community has shown the rest of the (proteomics) world what needs to be done to really understand protein interactions and functions.

However, for many reasons it will be impossible for some time yet to catch up with the pioneering work in yeast interactomics. Yeast allows genetic manipulations on a grand scale; its genome was first published in 1997 (and its number of known proteins has since grown by 100%), nearly complete cloning of knockout strains for every gene has been achieved, and all essential genes are known. Nearly every single protein has been cloned with tags or fluorescence markers, allowing isolation and life imaging, down to the single-cell level (although there is one such study in human cancer cells as well). All these technologies, together with size and easy availability of high numbers of cells, allow an exceptionally high proteome coverage with high curation level of the data. Last but not least, yeast is 'home' to the groundbreaking yeast two hybrid (Y2H) system for the detection of protein/protein interactions. Y2H screens are the source of most proteome-wide interaction studies to date (Figure 5.23).

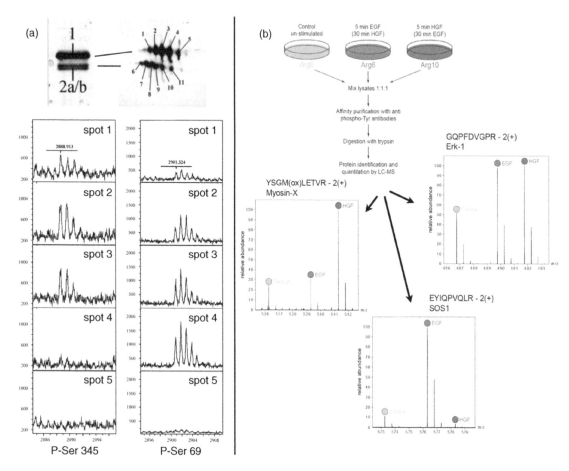

Figure 5.22 Proteomic analysis of PTMs II (a). The analysis of ovalbumin by SDS PAGE (left) shows two isoforms, which show several charge (and size) isoforms in 2D PAGE (right), typical of combinatorial phosphorylations. MALDI ToF MS analysis was able to show differential phosphorylations at two serines (lower panels) in the five charge isoforms. It can be assumed that even more phosphorylation sites are involved in creating the 2D pattern, but they could not be easily analyzed by MALDI ToF MS. The clearly changed phosphorylation pattern of the different spots would be lost in analyses from SDS PAGE bands or even shotgun proteomic peptides. A strategy to distinguish between phosphorylations induced by epidermal (EGF) and hepatocyte growth factor (HGF) is shown in (b). SILAC with two different Arg forms (+6 Da and +10 Da) was used in combination with a hybrid LIT/orbitrap FT MS analyzer to quantify peptides highly enriched by phospho-tyrosine specific antibodies. It is notoriously difficult to analyze the specificity of signalling pathways that use overlapping kinase cascades. In the study presented in (b) 274 different phosphoproteins were analyzed. Furthermore, targets common to EGF and HGF signalling (e.g. ERK-1) could be distinguished from EGF- or HGF-specific targets (SOS-1 or Myosin-X, respectively). This is essential to decipher interdependent signalling networks. The final example shows another double SILAC strategy, that was used to establish time courses for phosphorylation of more than 6 600 phophorylation sites on a total of 2 244 proteins after EGF stimulation of mammalian cells (c). The phosphorylated peptides were enriched and the phosphorylations detected by MS/MS on a hybrid LIT/FTICR MS analyzer. The MS^3 capabilities were used for the amino acid identifying fragmentations and single ion monitoring (= zoom scan) of the FT MS for exact determination of peptide mass. These and other details allowed the huge number of phosphorylation events to be analyzed. More than half of the phosphorylations came from proteins with more than one phosphorylation event and each phosphorylation site had its own kinetics, often one phosphorylation going up while another phosphorylation on the same peptide/protein goes down in intensity (Olsen *et al.*, 2006). (a) Reproduced from Hunzinger *et al.* (2006), courtesy of Elsevier. © 2006 Elsevier. (b) Reprinted with permission from Hammond *et al.* (2010). © 2010 American Chemical Society. (c) Reproduced from Olsen *et al.* (2006), courtesy of Elsevier. © 2006 Elsevier.

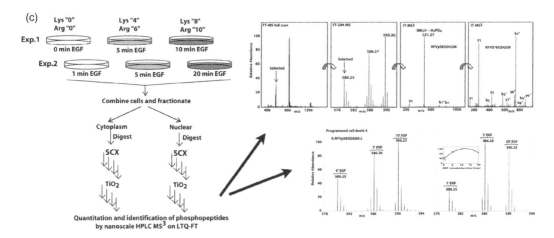

Figure 5.22 (*continued*)

By cloning foreign genes into yeast strains (or whole libraries with thousands of sequences), Y2H screens can also be applied to proteins from other species (measured in yeast). Y2H data have been invaluable in estimating the size of the playing field for interactomics; yeast with its 6 700 proteins is estimated to have some 10–60 000 protein/protein interactions, while humans presumably have around 700 000 interactions, with estimates ranging from 125 000 to 1 500 000. Similar estimates exist for worms and flies, with intermediate numbers. These estimates are based on complex statistical models and one of the most obvious bases for these models is the amount of interactions found overlapping in independent but similar studies. Despite the huge amount of interaction data created by Y2H screens, Y2H also has serious limitations: interactions are always 'one on one', always occur *in vivo*, but not the right environment (always in the nucleus, although some Y2H variants allow for cytoplasmic interactions to be measured) and thus produce many false positive or negative results (estimated to range from 75% for membrane proteins down to 25% for transcription factors). Proteins might interact that never come near each other in a cell (e.g. nuclear and mitochondrial proteins). Using appropriate controls, and validation strategies in 'smaller scale' interaction studies with just a couple of 'baits', much better accuracies can be achieved, leaving the problem of 'one on one' interactions and the non-physiological interaction environment. For instance, it is difficult to follow the dynamic of interactions of membrane proteins during

signalling or cell differentiation. Proteome-wide alternatives are interaction studies on micro-arrays (chips) or mass spectrometry based approaches. Chips contain hundreds of immobilized proteins that can then be targeted by hundreds of labelled compounds (Figure 5.23). Chip based approaches work based on the level of domains and peptides, for reasons involving the cost and workload as well as expression/solubility problems. The great advantage of chip based approaches is that they can deliver binding constants, which are of vital importance for systems biology approaches. However, the interactions occur in vitro and are again mainly one on one, often limited to domains and peptides being present only.

In contrast to the other technologies, large-scale mass spectrometry based interactomic approaches are able to detect proteins in multi-protein complexes (Figure 5.24). Although there are several varieties of MS based interactomic approaches, they are all based on expressing and isolating single proteins (which in turn carry over the associated proteins/complexes) and then analyzing the complexes by GeLC MS. The various technologies use either antibodies directed against the native specific proteins used as bait or affinity purification with antibodies against one or two tag epitopes cloned at either end of the bait proteins. Tags allow highly reproducible isolation conditions and use, for example, readily available high affinity antibodies. The very successful TAP tag contains one tag detected by an anti-IgG matrix in the first step. The isolated complexes are then eluted by a protease specifically cutting an amino acid sequence after the first

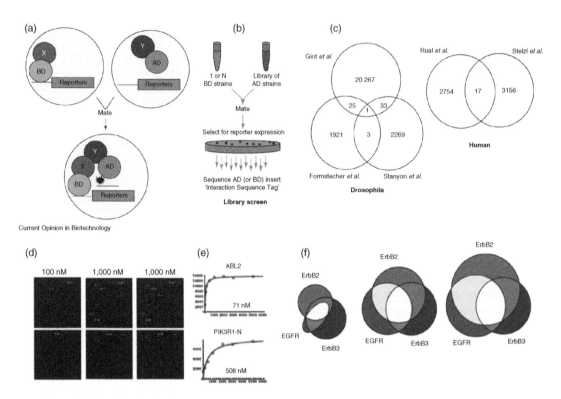

Figure 5.23 Interaction studies on a proteomic scale. Y2H is a standard tool for large-scale interaction proteomics (a). The 'bait' protein is fused with a DNA binding domain (BD), specific for a region in front of a selectable reporter gene. The 'prey' is fused with a transactivation domain (AD). When the prey catches (interacts with) the bait in the yeast nucleus upon mating yeast strains expressing the cloned fusion proteins, the transactivation domain is brought in the vicinity of the reporter gene promotor, thus the reporter gene is activated and the yeast can grow. For high throughput screening one or several baits are usually mated with libraries of prey strains (typically thousands of clones), and the resulting clones are screened for interaction (b). Sequencing the surviving strains establishes the protein interacting with the bait. Although thousands of interactions are detected, the studies must be incomplete, as there is virtually no overlap between very similar individual studies (c). In this example of a chip-based interaction study, 159 specific Src homology region 2 (SH2) domains were bound to microarrays ('chips', (d)). These phosphotyrosine binding domains were then incubated with different concentrations of 66 different fluorescence-labelled phopshotyrosine-peptides, allowing the determination of binding rates by reading out the fluorescence intensity on the chips. From this the binding constants of each phosphopeptide to each SH2 domain can be independently determined (e). With increasing concentrations of phosphopeptides, the SH2 domains get decorated to varying proportions with phosphopeptides from different receptors, which is a model for the recruitment of different proteins towards SH2 domains derived from different receptors ((f), the receptors are EGFR, ErbB2 and ErB3). These analyses can give us a glimpse as to how signalling networks are regulated at different levels of signalling intensity (i.e. Tyr phosphorylation) and how receptors might be decorated with various signalling molecules, allowing more specific responses under varying cellular situations. (a)–(c) Reproduced from Parrish *et al.* (2006), courtesy of Elsevier. © 2006 Elsevier. (d)–(f) Reproduced from Jones *et al.* (2006), courtesy of Nature Publishing Group. © 2006 Nature Publishing Group.

tag. The proteins and protein complexes released by this cut from the first tag are then bound by calmodulin interaction with the second tag, from which they are released after washing by calcium removal. This procedure allows very clean and background free isolation of complexes. The bait can be expressed from its native promoter (to analyze interactions at physiological expression levels) or using high expression promoters for better efficiencies (losing some physiological aspects of the interactions).

Using yeast as a model organism, the TAP MS approach has shown that most proteins are part of one or more ever-changing large protein complexes. These complexes in themselves are made up of core complexes and modules as well as attached proteins (Figure 5.24).

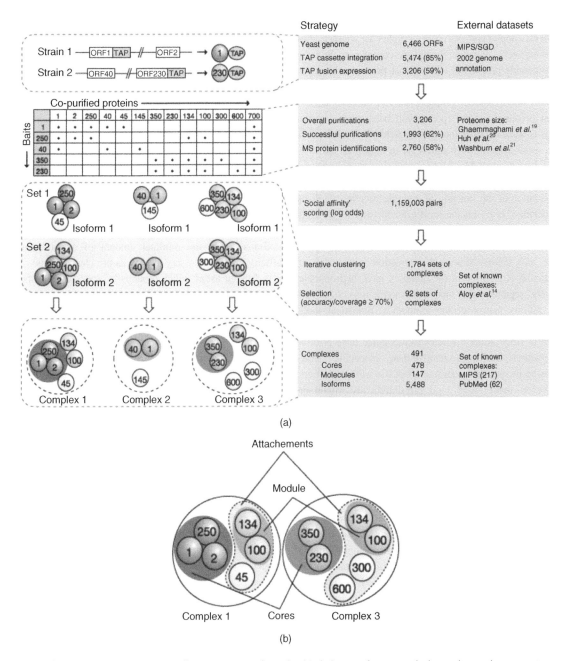

Figure 5.24 MS based interaction studies on a proteomic scale. (a) A large-scale systematic interaction study on yeast was performed starting with the generation of several thousand yeast strains expressing fusion proteins with a TAP tag. For each MS experiment 2 litres of yeast culture were used. The proteins were isolated using the double affinity tag strategy. The highly enriched associated proteins were separated on SDS PAGE gels, which were sliced into 1.25 mm wide pieces. On average 80 proteins were identified per gel, between one and eight per slice, by automated PMF with MALDI ToF MS. This allowed the detection of proteins with an expression level down to 30 proteins per cell to be analyzed in over 52 000 samples, identifying over 36 000 proteins in total (2 750 unique ones). The systematic bioinformatic analysis of data established that proteins spend their time in complexes that consist of stable cores, with attached modules made up of several proteins (b). Reproduced from Gavin *et al.* (2006), courtesy of Nature Publishing Group. © 2006 Nature Publishing Group.

During physiological changes the modules and attached proteins 'hop' from one core complex to another, thus defining very dynamic, extended complexes. Since a considerable proportion of the yeast proteome has been sampled and other lines of evidence support the concept of core and extended complexes, it can be assumed that most proteins in other species will behave in a similar way as well, defining the playing field for interactomics in a more dynamic way than Y2H or chips can do.

However, most MS based interaction studies are not of this high throughput. A more typical scenario is the elucidation of the function of a single protein. 'Guilt by association' is by far the most efficient way to go about this. False positive results are a big problem with these MS based interaction studies, as once the cells are lysed, new and non-physiological interaction can occur, in particular if proteins in the complex have hydrophobic, 'sticky' regions. Other unspecific interactions occur with the affinity matrix or the antibodies. Statistical tools, used in high throughput screens (Figure 5.24), cannot be used in single protein studies to recognize false positive interactions. It is easy to remove many unspecific interactions by increasing the stringency of the washing buffers; however, many weak (and physiologically important) interaction partners will also be lost by this change. The answer to this dilemma is to use high resolution detection methods (GeLC MS of the isolated complexes) under mild washing conditions and SILAC (or in principle also iTRAQ) as in-built control experiment (Figure 5.25). Incubating the complexes isolated from light lysates with heavy label lysates would show both isotopic patterns in MS (or MS/MS with iTRAQ) for unspecific interactions occurring after cell lysis. Only peptides that show up exclusively (or overwhelmingly) in the light version are derived from a specific interaction; the others are formed during and after lysis. It is important to accept the many unspecific bands in order to identify the few specific associations in this approach. Another approach to finding weak or transient interactions is the use of chemical cross linking prior to isolation of specific complexes. After cross-linking, the complexes can be washed very stringently, removing most false positive interactions. Even in mild conditions 'true' interactions can easily be lost; every clean-up will take a certain time (typically 1 hour), and if the proteins in the complex have a fast rate of dissociation even the 'true' interactors will have left the complex, as their

protein concentrations are drastically reduced during the washing steps – only cross-linking can avoid this. On the other hand, cross-linking can add a huge number of unspecific interactions, as everything 'near' the complex can get attached to it, so a combination of cross-linking and SILAC offers higher specificity (Figure 5.25).

Of course, interaction analysis of single proteins can be performed without SILAC, cross-linking or tandem tags, or indeed any tags at all. Such analyses can be very useful in finding a function for a specific protein or analyzing changes in its binding partners in changed signalling or other circumstances. A good antibody to a reasonably expressed protein will deliver many co-immunoprecipitating proteins when procedures are optimized (use the minimal amount of antibody and affinity matrix necessary as well as pre-clear lysates). As it is difficult to say which proteins are specifically associated, good controls will be of paramount importance (e.g. point-mutated bait proteins, isotype matched unrelated control antibody). For approaches such as this, 2D gels are usually not suited to separate the immunoprecipitations; the 2D gels will do more harm than good, as many proteins get 'lost' and diluted. GeLC MS and even simple RPLC MS/MS of the whole immunoprecipitation are more sensitive and can deliver specific interaction partners even if no differences can be seen on, for example, SDS PAGE gels due to a high background of unspecific interactions with antibodies or affinity matrices.

As an alternative to targeting the complexes formed by fishing for single proteins, protein complexes can be analyzed directly. Blue native electrophoresis (see Section 2.2.3) allows the separation of many intact protein complexes. By cutting the blue native gels into slices and analyzing them by GeLC MS, intensities for single proteins can be determined along the gel. Proteins showing concentration distributions with overlapping peaks are part of the same protein complexes.

5.6 PROTEOMICS AS PART OF INTEGRATED APPROACHES

Apart from asking relatively simple questions (simple in theory, that is), how can we maximize the knowledge gained from analyzing proteomics data?

Proteomic studies generate a lot of data, the interpretation of which is more time-consuming and often less technologically advanced than its acquisition. It

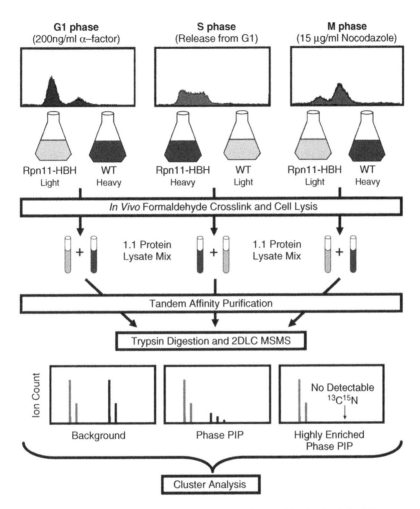

Figure 5.25 Analysis of interactions using cross-linking and SILAC. The Rpn11 subunit of the 26S proteasome in yeast was expressed under its own promoter as a fusion protein with a TAP tag. An especially tailored schema was used to analyze specific proteosome interacting proteins (PIPs), in three stages of the cell cycle (G1, S and M phase). Mixing of heavy/light SILAC labelled yeast lysates allowed unspecific interactions to be excluded and cell cycle phase-specific interactions to be determined; for example in the analysis of the G1 phase, the heavy labelled WT lysate only contains the not TAP-tagged version of Rpn11 and all proteins in this lysate are in the 'heavy' version. All 'heavy' peptides in immunoprecipitations of the 'light' TAP tagged Rpn11 from G1 phase must derive from unspecific background contaminations. Specific interacting proteins would consist exclusively of light peptides (see lower panel). The study established 677 specific interactions, 359 of which were found in all cell cycle stages, 143 in two and 175 in only one cell cycle stage. A similar study by the same laboratory using a q-ToF MS instead of LTQ orbitrap FT MS and without synchronization in different cell cycle stage identified a total of 411 interactions. The proteosome is the driver of the cell cycle, and was in this study also found to be connected to the pheromone pathway. Reprinted with permission from Kaake *et al.* (2010). © 2010 American Chemical Society.

is thus perfectly possible to publish proteomics results without actually performing experiments; if the data are deposited in a standardized way in the public domain (entered into a database), they can be retrieved by others and analyzed beyond the initial intentions. Examples

are the annotations of genomic databases, which have been corrected and extended. Thousands of new genes have been found, and presumed pseudogenes have been proven to be expressed. The usual suspects are the N-termini of proteins; thousands of corrections have been added due to proteomic data. Other examples are database conflicts (e.g. mutations, single-nucleotide polymorphisms), GO information and obviously PTM and association data. This will lead to better databases for many different lines of research, not least proteomics.

The buzzword for multiplying the informations deduced from proteomic studies is data integration. At the very beginning of modern systems biology there was great enthusiasm for the integration of proteomics data with transcriptomics, genomics and metabolomics (or better even metabonomics) data, in order to achieve a full understanding of the regulation of systems and their modelling at every level, even integrating evolutionary and social aspects of behaviour (e.g. underlying the development of cancer). Systems biology has gone on to integrate a great deal of data (although the number of species with system wide modelling is a hundred times lower than the number of species with sequenced genomes), but proteomic data are not included very often in the models. In a typical (experimental) systems biology study, transcriptomic and metabonomic data is integrated. There are only a few studies taking proteomic data on board as well. Why is proteomics in practical terms not included in most system analyses? There are several reasons, one being that the proteomic data are often not of high enough quality. The accuracies even of the few 'real' quantitative proteomic studies are not good enough for a precise integration in systemic applications; while the trend of protein expression changes is right, most studies do not have sufficient biological replicates, and the huge amount of data makes reliable measurements difficult. It is possible to generate better quantitative data in proteomics, but studies have to be designed with this goal in mind (Figure 5.26a).

Another reason why proteomic data are given the cold shoulder in system analyses is the low depth or coverage of most proteomic studies. There are of course technological reasons for this (no standard format, no easy comparison between different approaches, while most transcriptomic data come from one commercial platform), but also fundamental biological reasons; the core proteome of all living beings is estimated to contain some 5 million proteins. This pan-proteome is relatively difficult to compare and integrate (although cross-species comparison approaches have been very successful). Metabolomics has a smaller variety for the core processes in comparison; about 600 metabolites are

estimated for yeast, 1 200 metabolites are presumed in *E. coli*, 3 500 metabolites is an agreed value for humans. Plants are metabolic 'dinosaurs'; they make up for their lack of mobility with diverse metabolic pathways and may have between 5 000 and 25 000 metabolites per species, and all plant species together are estimated to produce 200 000 different metabolites. All members of the tree of life are related so closely in their biochemistry that the core pan-metabolome might be several orders of magnitude smaller than the pan-proteome. The dilemma for these predictions is of course that the metabolome cannot be 'sequenced'; it is based on enzymes and their (often unknown and combinatorial) activities. One enzyme might produce different metabolites, depending on which starting material it is presented with at which concentrations. Nevertheless, it is technologically relatively easier to achieve a high coverage of the metabolome than of the proteome. The transcriptomic coverage (often used in lieu of protein content) can also be high in well-established and studied species (chips are cheaper if a lot of the same ones are produced). Furthermore, there are conceptual difficulties; how can the different traits of proteins (e.g. composition of complexes) be treated in the data analysis? Pathways are often used as a summary for complex protein behaviour.

The take-home message here is as far as possible to design studies in such a way as to make them comparable. This includes absolute quantifications, even if it means 'bootstrapping' of spectral counts or ion intensities to deduce numbers for 'protein copies per cell' using a few standards. Measuring enough biological repeats to be able to estimate standard deviation and mean values is another prerequisite for data integration. Data integration also depends on contributing the data (even of smaller studies) to one of the bigger databases (in a standardized format), so they are in the public domain in a useable format.

Despite the relative lack of proteomic data in system biology, the early predictions about the power of systemic approaches has been vindicated (if only on a small scale) if multidimensional data are integrated, even with limited coverage of the various 'omes' (Figure 5.26b). The combination of metabonomic and proteomic data allows the interpretation of, for example, the connection between protein synthesis and glutathione metabolism and can explain some counterintuitive findings that could not be understood from either data set alone.

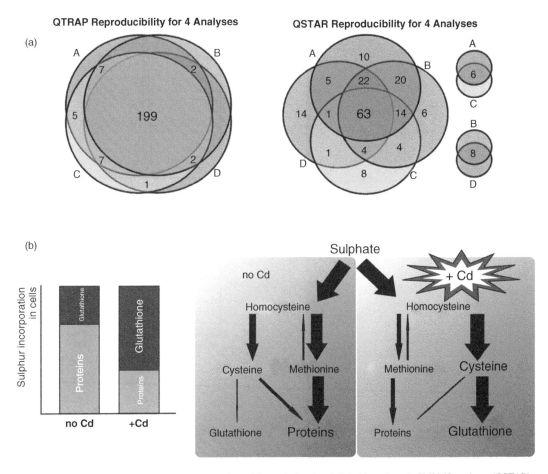

Figure 5.26   Accuracy of proteomic data and integration with metabolomics. A hybrid quadrupole ToF MS analyzer (QSTAR) was used to establish over 200 transitions for a targeted MRM analysis (see Figure 4.26) of phosphorylation events after EGF stimulation of mammalian cells. This initial discovery mode was compared in four assays to the MRM mode using a q-trap analyzer (a). While of the 223 initially established phosphopeptides 199 were identified/ quantified in four technical repeats (A–D) by MRM measurements on a hybrid quadrupole/LIT mass analyzer (q-trap), only 63 of the peptides were identified in all four technical repeats using the QSTAR in a shotgun-like approach. (b) Integrating metabonomic and quantitative shotgun proteomic data in one experiment explains how exposure to cadmium (Cd) redirects the sulfur flux from proteins to glutathione. The panel on the left shows how incorporation of sulfur into cells is shifted from being directed into proteins towards glutathione in the presence of Cd. The panel on the right shows the flux of sulfur under normal conditions and in the presence of Cd. In the presence of Cd all cysteine-containing proteins were expressed at a lower level. This confusing phenomenon is explained by the sulfur flux towards the detoxifying glutathione. (a) Reproduced with permission from Wolf-Yadlin *et al.* (2007). © (2007) National Academy of Sciences, U.S.A. (b) After Lafaye *et al.* (2005).

One unexpected form of data reduction in proteomics is the number of structures available to proteins. Although the rigorous 3D structural analysis of proteins is picking up speed, the number of 3D structures solved is much lower than the number of proteins/genes analyzed. At the moment some 60 000 3D structures are known. Despite the bias towards more stable structures in this number, it seems clear that only a limited number of folds (local structural elements) are used by the dazzling variety of proteins encountered. Often the 3D structure for one protein is the template for a whole group of proteins with similar functions or mechanisms of action. However, the sequences between these proteins can be difficult to compare as there are many ways to

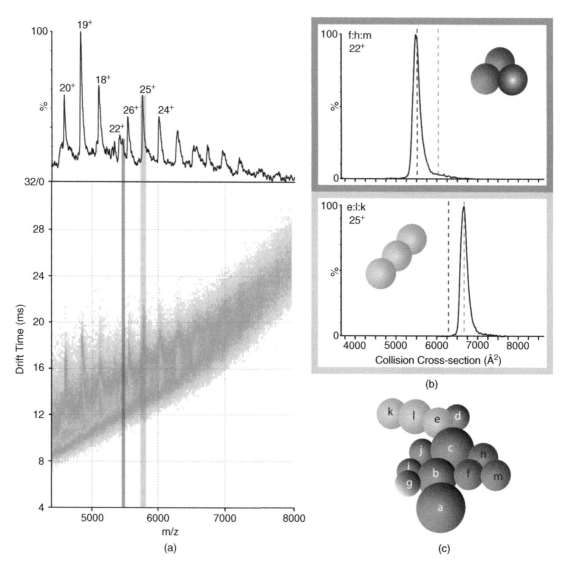

Figure 5.27 Native MS and ion mobility (IM) MS for the elucidation of protein complexes. In IM MS the flight time of ions on the way to the final mass analyzer can be measured. As they have to pass through gas, ions of different shape and size will arrive at different times at the analyzer. Combining the information with the m/z of the ion, measured after the IM range, IM MS can give an indication of the size and shape of molecules in the gas phases. PTMs can make peptides bulkier and thus lead to the detection of putative PTMs in IM MS. In native ESI, molecules retain many of their native characteristics, namely, their interactions and overall folds. (a) IM MS data on the eukaryotic translation initiation factor sIF3, a multiprotein complex. The contour plot of the IM measurement is overlaid with the MS data. The ion at the +22 charge state represents the core complex with the addition of the complex of the proteins f, h and m. The ions at +26, +25 and +24 charge states represent the complex with the addition of the complex of proteins k, l and e. Note the distinctively longer drift time in the IM range for the latter. (b) Measurements of the drift times of the two trimeric complexes alone (in the same charge state as observed in the complex). The complex of f, h, m has a shorter drift time and thus compact packaging, while the complex of e, l, k has a long drift time and must thus have an elongated shape. (c) The complex structure, deduced from the presented IM MS data in conjunction with other data. Reproduced from Pukala *et al.* (2009), courtesy of Elsevier. © 2009 Elsevier.

get the same structure. In particular, there is a 'gap' between the rapidly arising good quality interactomic data and a real understanding of the structures involved. A detailed knowledge of these structures would also allow us to understand which subset of possible PTMs is to be expected (i.e. which of the serines, threonines or tyrosines are really phosphorylated at any given time) and how this can influence structure/function at the level of protein complexes. The relatively young field of native mass spectrometry might be able to bridge part of this gap, particularly when used in combination with ion mobility MS, where the shape of ions can be estimated in the mass spectrometer right before MS data are acquired (Figure 5.27).

REFERENCES

Baker, M.J., Brown, M.D., Gazi, E., Clarke N.W., Vickerman, J.C. and Lockyer, N.P. (2008) Discrimination of prostate cancer cells and non-malignant cells using secondary ion mass spectrometry. *The Analyst*, **133**, 175–179.

Baumeister, W. (2005) From proteomic inventory to architecture. *FEBS Lett*, **579**, 933–937.

Becker, J.S. and Jakubowski, N. (2009) The synergy of elemental and biomolecular mass spectrometry; new analytical strategies in life science. *Chem Soc Rev*, **38**, 1969–1983.

Bitton, D.A., Smith, D.L., Connolly, Y. *et al.* (2010) An integrated mass spectrometry pipeline identifies novel protein coding-regions in the human genome. *PLoS One*, **5** (1), e8849.

Braun, R.J., Kinkl, N., Beer, M. and Ueffing, M. (2007) Two-dimensional electrophoresis of membrane proteins. *Anal Bioanal Chem*, **389** (4), 1033–1045.

Cazares, L.H., Troyer, D., Mendrinos, S. *et al.* (2009) Imaging mass spectrometry of a specific fragment of mitogen activated protein kinase/extracellular signal regulated kinase kinase kinase 2 discriminates cancer from uninvolved prostate tissue. *Clin Cancer Res*, **15** (17), 5541–5551.

Cheng, J.S., Zhou, X., Ding, M.Z. and Yuan, Y.J. (2009) Proteomic insight into adaptive responses of Sacharomyces cerevisiae to the repeated vacuum fermentation. *Appl Microbiol Biotechnol*, **83**, 909–923.

Choe, L.H., Aggawall, K., Franck, Z. and Lee, K.H. (2005) A comparison of the consistency of proteome quantitation using two-dimensional electrophoresis and shotgun isobaric tagging in *Escherichia coli* cells. *Electrophoresis*, **26**, 2437–2449.

Collura, V. and Biossy, G. (2007) From protein-protein complexes to interactomics, Chapter 8 in *Subcellular Proteomics* (Bertrand, E. and Faupel, M., Eds.), Springer, 135–183.

Colomé, N., Collado, J., Bech-Serra, J.J. *et al.* (2010) Increased apoptosis after autoimmune regulator expression in epithelial cells revealed by a combined quantitative proteomics approach. *J Proteome Res*, **9** (5), 2600–2609.

Domon, B. and Aebersold, R. (2006) Mass spectrometry and protein analysis. *Science*, **312**, 212–218.

Du, R., Long, J., Yao, J. *et al.* (2010) Subcellular quantitative proteomics reveals multiple pathway cross-talk that coordinates specific signaling and transcriptional regulation for the early host response to LPS. *J Proteome Res*, **9** (4), 1805–1821.

Elias, J.E., Haas, W., Faherty, B.K. and Gygi, S.P. (2005) Comparative evaluation of mass spectrometry platforms used in large-scale proteomics. *Nature Meth*, **2** (9), 667–675.

Farley, A.R. and Link, A.J. (2009) Identification and quantification of protein posttranslational modifications. *Meth Enzym*, **463**, 725–763.

Figeys, D. (Ed) (2005) *Industrial Proteomics*, Wiley.

Freeman, W.M., Lull, M.E., Guilford, M.T. and Vrana, K.E. (2006) Depletion of abundant proteins from non-human primate serum for biomarker studies. *Proteomics*, **6**, 3109–3113.

Futcher, B., Latter, G.I., Monardo, P. *et al.* (1999) A sampling of the yeast proteome. *Mol Cell Biol*, **19** (11), 7357–7368.

Gavin, A.C., Aloy, P., Grandi, P. *et al.* (2006) Proteome survey reveals modularity of the yeast cell machinery. *Nature*, **440**, 631–636.

Ghaemmaghami, S., Huh, W.K., Bower, K. *et al.* (2003) Global analysis of protein expression in yeast. *Nature*, **425**, 737–741.

Graumann, J., Hubner, N.C., Kim, J.B. *et al.* (2008) Stable isotope labeling by amin acids in cell culture (SILAC) and proteome quantitation of mouse embryo stem cells to a depth of 5,111 proteins. *Mol Cell Proteomics*, **7** (4), 672–683.

Green-Church, K.B., Nichols, K.K., Kleinholz, N.M., Zhang, L. and Nichols, J.J. (2008) Investigation of the human tear film proteome using multiple proteomic approaches. *Mol Vision*, **14**, 456–470.

Grimsrud, P.A., Swaney, D.L., Wenger, C.D., Beauchene, N.A. and Coon, J.J. (2010) Phosphoproteomics for the masses. *ACS Chem Biology*, **5** (1), 105–109.

Groseclose, M.R., Andersson, M., Hardesty, W.M. and Caprioli, R.M. (2007) Identification of proteins directly from tissue: in situ tryptic digestion coupled with imaging mass spectrometry. *J Mass Spectrom*, **42**, 254–262.

Hammond, D.E., Hyde, R., Kratchmarova, I., Beynon, R.J., Blagoev, B. and Clague, M.J. (2010) Quantitative analysis of HGF and EGF-dependent phosphotyrosine signaling networks. *J Proteome Res*, **9** (5), 2734–2742.

Hood, L., Heath, J.R., Phelps, M.E. and Lin, B. (2004) System biology and new technologies enable predictive and preventative medicine. *Science*, **306**, 640–643.

Hunzinger, C., Schrattenholz, A., Poznanovic, S., Schwall, G.P. and Stegmann, W. (2006) Comparison of different separation technologies for proteome analyses: Isoform resolution as a prerequisite for the definition of protein biomarkers on the level of posttranslational modifications. *J Chrom A*, **1123**, 170–181.

Hwang, H., Zhang, J., Chung, K.A. *et al.* (2010) Glycoproteomics in neurodegenerative diseases. *Mass Spectrom Rev*, **29** (1), 79–125.

John, J.P.P., Anrather, D., Pollak, A. and Lubec, G. (2006) Mass spectrometric verification of stomatin-like protein 2 (SLP-2) primary structure. *Proteins*, **64**, 543–551.

Jones, R.B., Gordus, A., Krall, J.A. and Macbeath, G. (2006) A quantitative protein interaction network for the ErbB receptors using protein microarrays. *Nature*, **439**, 168–174.

Jung, H.R., Pasini, D., Helin, K. and Jensen, O.N. (2010) Quantitative mass spectrometry of histone H3.2 and H3.3 in Suz12 deficient mouse ES cells reveals distinct, dynamic post-translational modifications at K27 and K36. *Mol Cel Proteomics*, **9** (5), 838–850.

Kaake, R.M., Milenković, T., Pržulj, N., Kaiser, P. and Huang, L. (2010) Characterization of cell cycle specific protein interaction networks of the yeast 26S proteasome complex by the QTAX strategy. *J Proteome Res*, **9** (4), 2016–2029.

Kawase, H., Fujii, K., Miyamoto, M., *et al.* (2009) Differential LC-MS based proteomics of surgical human cholangiocarcinima tissues. *J Proteome Res*, **8**, 4092–4103.

Kito, K. and Ito, T. (2008) Mass spectrometry based approaches towards absolute quantitative proteomics. *Cur Genomics*, **9** (4), 263–274.

Kolch, W., Mischak, H., Pitt, A.R., (2005) The molecular make-up of a tumour, proteomics in cancer research, Clinical Science, **108**, 369–83. *Excellent insight in the role that sytems biology and proteomics can play in cancer research.*

Koo, B.S., Lee, D.Y., Ha, H.S., Kim, J.C. and Kim, C.W. (2005) Comparative analysis of the tear protein expression in Blepharitis patients using two dimensional electrophoresis, *J Proteome Res*, **4**, 719–724.

von Kriegsheim, A., Baiocchi, D., Birtwistel, M., Sumpton, D., Bienvenut, W., Morrice, N., Yamada, K., Lamond, A., Kalna, G., Orton, R., Gilbert, D. and Kolch, W. (2009) Cell fate decisions are specified by the dynamic ERK interactome, *Nature Cel Biol*, **11** (12), 1458–1466.

Kubota, K., Sato, Y., Suzuki, Y. *et al.* (2008) Analysis of glycopeptides using lectin affinity chromatography with MALDI-ToF mass spectrometry. *Anal Chem*, **80**, 3693–3698.

Kurczyk, M.E., Piehowski, P.D., Van Bell, C.T. *et al.* (2010) Mass spectrometry imaging of mating *Tetrahymena* show that changes in cell morphology regulate lipid domain formation. *Proc Natl Acad Sci*, **107** (7), 2751–2756.

Lafaye, A., Junot, C., Pereira, Y. *et al.* (2005) Combined proteome and metabolite-profiling analyses reveal surprising insight into yeast sulfur metabolism. *J Biol Chem*, **280** (26), 24723–24730.

Lau, T.Y., Power, K., Dijon, S., deGardelle, I., McDonnell, S., Duffy, J., Pennington, S.R. and Gallagher, W.M. (2010) Prioritization of candidate protein biomarkers from in vitro model system of breast tumor progression towards clinical verification. *J Proteome Res*, **9** (3), 1450–1459.

Lee, K.B. and Pi, K.B. (2010) Proteomic profiling combining solution-phase isoelectric fractionation with two-dimensional gel electrophoresis using narrow-pH-range immobilized pH gradient gels with slightly overlapping pH ranges. *Anal Bioanal Chem*, **396**, 535–539.

Leigh Anderson, N., Polanski, M., Pieper, R. *et al.* (2004) The human plasma proteome: A nonredundant list developed by combination of four separate sources. *Mol Cell Proteomics*, **3** (4), 311–326.

Li, N., Wang, N., Zheng, J. *et al.* (2005) Characterization of human tear proteome using multiple proteomics analysis techniques. *J Proteome Res*, **4**, 2052–2061.

Li, S.J., Peng, M.P., Liu, B.S. *et al.* (2008) Sys-BodyFluid: a systematical databases for human body fluid proteome research. *NAR*, **37** (Database issue), D907–D912.

Lohaus, C., Nolte, A., Blüggel, M. *et al.* (2007) Multidimensional chromatography: A powerful tool for the analysis of membrane proteins in mouse brain. *J Proteome Res*, **6**, 105–113.

Looso, M., Borchardt, T., Kruger, M. and Braun, T. (2010) Advanced identification of proteins in uncharacterized proteomes by pulsed in vivo SILAC. *Mol Cell Proteomics*, **9**, 1157–1166.

Luber, C.A., Cox, J., Lauterbach, H. *et al.* (2010) Quantitative proteomics reveals subset-specific viral recognition in dendritic cells. *Immunity*, **32**, 279–289.

Malmstrom, J., Beck, M., Schmidt, A., Lange, V., Deutsch, E.W. and Aebersold, R. (2009) Proteome wide cellular protein concentrations of the human pathogen *Leptospira interrogans*. *Nature*, **460**, 762–766.

McDonnell, L.A., van Remoortere, A., van Zeijl, R.J.M. *et al.* (2009) Automated imaging MS: toward high throughput imaging mass spectrometry. *J Proteomics*, **73** (6), 1279–1282.

Millioni, R., Miuzzo, M., Sbrignadello, S. *et al.* (2010) Delta2D and Proteomeweaver; performance evaluation of two different approaches for 2DE analysis. *Electrophoresis*, **31**, 1–7.

Minden, J.S., Dowd, S.R., Meyer, H.E. and Stuhle, K. (2009) Difference gel electrophoresis. *Electrophoresis*, **30**, S156–S161.

Mischak, H., Coon, J.J., Novak, J. *et al.* (2009) Capillary electrophoresis-mass spectrometry as a powerful tool in biomarker discovery and clinical diagnosis; an update of recent developments. *Mass Spec Rev*, **28** (5), 703–724.

Nesvizhskii, A.I., Vitek, O. and Aebersold, R. (2007). Analysis and validation of proteomic data generated by tandem mass spectrometry. *Nature Meth*, **4** (10), 787–797.

Nesvizhskii, A.I. and Aebersold, R. (2005) Interpretation of shotgun proteomic data. *Mol Cell Proteomics*, **4** (10), 1419–1440.

Olsen, J.V. and Mann, M. (2004) Improved peptide identification in proteomics by two consecutive stages of mass spectrometric fragmentation. *Proc Natl Acad Sci*, **101**, 13417–13422.

Olsen, J.V., de Godoy, L.M.F., Li, G. *et al.* (2005) Parts per million mass accuracy on an obitrap mass spectrometer via lock mass injection into a c-trap. *Mol Cell Proteomics*, **4** (12), 2010–2021.

Olsen, J.V., Blagoev, B., Gnad, F. *et al.* (2006) Global, in vivo and site-specific phosphorylation dynamics in signaling networks. *Cell*, **127**, 635–648.

Ow, S.Y., Salim, M., Noirel, J., Evans, C., Rehman, I. and Wright, P.C. (2009) iTRAQ underestimation in simple and complex mixtures; "the good the bad and the ugly". *J Proteome Res*, **8**, 5347–5555.

Paradela, A., Marcilla, M., Navajas, R. *et al.* (2010) Evaluation of isotope-coded protein labeling (ICPL) in the quantitative analysis of complex proteomes. *Talanta*, **80**, 1496–1502.

Parrish, J.R., Gulyas, K.D., Finley, R.L. Jr (2006) Yeast two-hybrid contributions to interactome mapping. *Curr Opin Biotech*, **17**, 387–393.

Pastorino, B., Boucomont-Chapeaublanc, E., Peyrefitte, C.N. *et al.* (2009) Identification of cellular proteome modifications in response to West Nile virus infection. *Mol Cell Proteomics*, **8** (7), 1623–1637.

Petrak, J., Ivanek, R., Toman, O. *et al.* (2008) Déjà vu in proteomics. A hit parade of repeatedly identified differentially expressed proteins. *Proteomics*, **8**, 1744–1749.

Picotti, P., Bodenmiller, B., Mueller, L.N., Domon, B. and Aebersold, R. (2009) Full dynamic range proteome analysis of S.cerevisiae by targeted proteomics. *Cell*, **138**, 795–806.

Pukala, T.L., Ruotolo, B.T., Zhou, M. *et al.* (2009) Subunit architecture of multiprotein assemblies determined using restraints from gas phase measurements. *Structure*, **17**, 235–1243.

Richardson, M.R., Liu, S., Ringham, H.N. *et al.* (2008) Sample complexity reduction for two-dimensional electrophoresis using solution isoelectric focusing prefractionation. *Electrophoresis*, **29**, 2637–2644.

Rohner, T.C., Staab, D. and Stoeckli, M. (2005) MALDI mass spectrometric imaging of biological tissue sections. *Mechanisms Ageing Development*, **126**, 177–185.

Schmidt, A., Donahoe, S., Hagens, K. *et al.* (2004) Complementary analysis of the Mycobacterium tuberculosis proteome by two-dimensional electrophoresis and isotope coded affinity tag technology. *Mol Cell Proteomics*, **3** (1), 25–42.

Schmidt, A., Kellermann, J. and Lottspeich, F. (2005) A novel strategy for quantitative proteomics using isotope coded protein labels. *Proteomics*, **5**, 4–15.

Speicher, D.W. (Ed) (2004) *Proteome Analysis, Interpreting the Genome*, Elsevier.

de Souza, G.A., de Godoy, L.M.F. and Mann, M. (2006) Identification of 491 proteins in the tear fluid proteome reveals a large number of proteases and protease inhibitors. *Genome Biol*, **7**, R72.

Stensballe, A., Jensen, O.N., Olsen, J.V., Haselmann, K.F. and Zubarev, R.A. (2000) Electron capture dissociation of singly and multiply phosphorylated peptides. *Rapid Commun Mass Spectrom*, **14**, 1793–1800.

Stoeckli, M., Staab, D., Schweitzer, A., Gardiner, J. and Seebach, D. (2007) Imaging of a beta-peptide distribution in whole body mice sections by MALDI mass spectrometry. *J Am Soc Mass Spectrometry*, **18**, 1921–1924.

Tabb, D.L. *et al.* (2009) Repeatability and reproducibility in proteomics identifications by LC-tandem MS. *J Proteome Res*, **9** (2), 761–776.

Thomas, C.E., Kelleher, N.L. and Mizzen, C.A. (2006) Mass spectrometric characterization of human histone H3: A bird's eye view. *J Proteomics Res*, **5** (2), 240–247.

Van Eyk, J.E. and Dunn, M.J. (Eds.) (2002). *Proteomic and genomic analyses of cardiovascular disease*, Wiley-VCH.

Vasudev, N.S., Ferguson, R.E., Cairns, D.A. *et al.* (2008) Serum biomarker discovery in renal cancer using 2DE and prefractionation by immunodepletion and isoelectric focusing; increasing coverage or more of the same? *Proteomics*, **8**, 5074–5085.

Veenstra, T.D. and Yates, III J.R. (2006) *Proteomics for biological discovery*, Wiley.

Vellaichamy, A., Sreekumar, A., Strahler, J.R. *et al.* (2009) Proteomic interrogation of androgen action in prostate cancer cells reveals roles of aminoacyl tRNA synthetases. *PLoS One*, **4** (9), e7075.

Washburn, M.P., Wolters, D. and Yates, J.R. III (2001) Large-scale analysis of the yeast proteome by multidimensional protein identification technology. *Nat Biotech*, **19**, 242–247.

Webb-Robertson, B.J.M. *et al.* (2009) A Bayesian integration model of high throughput proteomics and metabolomics data for improved early detection of microbial infections. *Pacific Symp Biocomputing*, **14**, 451–463.

Weckwerth, W. (2008) Integration of metabolomics and proteomics in molecular plant physiology – coping with the complexity by data-dimensionality reduction. *Phys Plantarum*, **132**, 176–89.

Weiss, M., Schrimpf, S., Hengartner, M.O., Lercher, M.J., von Mering, C. *et al.* (2010) Shotgun proteomics data from multiple organisms reveals remarkable quantitative conservation of the eukaryotic core proteome. *Proteomics*, **10**, 1–10.

Wiśniewski, J.R., Zougman, A., Nagaraj, N. and Mann, M. (2009) Universal sample preparation method for proteome analysis. *Nat Meth*, **6** (5), 359–362.

Wolf-Yadlin, A., Hautanlemi, S., Lauffenburger, D.A. and White, F.M. (2007) Multiple reaction monitoring for robust quantitative proteomic analysis of cellular signaling networks. *Proc Natl Acad Sci*, **104** (14), 5860–5865.

Wu, W.W., Wang, G., Baek, S.J. and Shen, R.-F. (2005) Comparative study of three proteomic quantitative methods, DIGE, cICAT, and iTRAQ, using 2D gel- or LC–MALDI TOF/TOF. *J Proteome Res*, **5**, 651–658.

Wu, C., Ifa, D.R., Manicke, N.E. and Cooks, R.G. (2010) Molecular imaging of adrenal gland by desorption electrospray ionization mass spectrometry, analyst, *Analyst*, **135**, 28–32.

Zabrouskov, V., Ge, Y., Schwartz, J. and Walker, J.W. (2008) Unraveling molecular complexity of phosphorylated human cardiac troponin I by top down electron capture dissociation/electron transfer dissociation. *Mass Spectr Mol Cell Proteomics*, **7**, 1838–1849.

Zhu, W., Smith, J.W. and Huang, C.M. (2010) Mass spectrometry-based label-free quantitative proteomics. *J Biomed Biotech*, article ID 2010:840518.

Zinn, N., Hahn, B., Pipkorn, R., Schwarzer, D. and Lehmann, W.D. (2009) Phosphorus based absolutely quantified standard (PASTA) peptides for quantitative proteomics. *J Proteome Res*, **8**, 4870–4875.

Index

Note: page numbers in *italics* refer to figures, those in **bold** refer to tables

Printed and bound by CPI Group (UK) Ltd, Croydon, CR0 4YY

27/10/2024

14580152-0004